Liquid and Vapor Flows in Porous Bodies

Topics in Chemical Engineering

A series edited by R. Hughes, University of Salford, UK

Volume 1 HEAT AND MASS TRANSFER IN PACKED BEDS
by N. Wakao and S. Kaguei

Volume 2 THREE-PHASE CATALYTIC REACTORS
by P.A. Ramachandran and R.V. Chaudhari

Volume 3 DRYING: Principles, Applications and Design
by C. Strumillo and T. Kudra

Volume 4 THE ANALYSIS OF CHEMICALLY REACTING SYSTEMS: A Stochastic Approach
by L.K. Doraiswamy and B.K. Kulkarni

Volume 5 CONTROL OF LIQUID-LIQUID EXTRACTION COLUMNS
by K. Najim

Volume 6 CHEMICAL ENGINEERING DESIGN PROJECT: A Case Study Approach
by M.S. Ray and D.W. Johnston

Volume 7 MODELLING, SIMULATION AND OPTIMIZATION OF INDUSTRIAL FIXED BED CATALYTIC REACTORS
by S.S.E.H. Elnashaie and S.S. Elshishini

Volume 8 THREE-PHASE SPARGED REACTORS
edited by K.D.P. Nigam and A. Schumpe

Volume 9 DYNAMIC MODELLING, BIFURCATION AND CHAOTIC BEHAVIOUR OF GAS-SOLID CATALYTIC REACTORS
by S.S.E.H. Elnashaie and S.S. Elshishini

Volume 10 THERMAL PROCESSING OF BIO-MATERIALS
edited by T. Kudra and C. Strumillo

Volume 11 COMPOSITION MODULATION OF CATALYTIC REACTORS
by P.L. Silveston

Please see the back of this book for other titles in the Topics in Chemical Engineering series.

Liquid and Vapor Flows in Porous Bodies
Surface Phenomena

N.V. Churaev
Institute of Physical Chemistry, Moscow, Russia

Foreword by **Professor J. Lyklema**
University of Wageningen, The Netherlands

English Edition Edited by **Dr. A. Galwey**
formerly of the Queen's University, Belfast, Northern Ireland

Translated from the Russian by **M.N. Churaeva**

CRC Press
Taylor & Francis Group
Boca Raton London New York

CRC Press is an imprint of the
Taylor & Francis Group, an informa business

Originally published in Russian in 1990 as ФИЗИКОХИМИЯ ПРОЦЕССОВ МАССОПЕРЕНОСА В ПОРИСТЫХ ТЕЛАХ by Khimiya, Moscow

Amsteldijk 166
1st Floor
1079 LH Amsterdam
The Netherlands

British Library Cataloguing in Publication Data

Churaev, N. V. (Nickolai Vladimirovich)
Liquid and vapor flows in porous bodies : surface phenomena. – (Topics in chemical engineering ; v. 13)
1. Porous materials – Permeability 2. Surface chemistry
I. Title II. Galwey, A.
620.1′16

ISBN 90-5699-149-3

CONTENTS

INTRODUCTION TO THE SERIES		vii
FOREWORD TO THE ENGLISH EDITION		ix
LIST OF SYMBOLS		xi
INTRODUCTION		1
Chapter I	**POROUS BODIES AS DISPERSE SYSTEMS**	3
§1	Forces acting among particles in porous bodies	4
§2	Models of porous structure	18
§3	Methods of determination of the structural characteristics of porous bodies	20
Chapter II	**EQUILIBRIUM OF LIQUIDS IN POROUS BODIES**	27
§1	Adsorption layers	29
§2	Wetting films	31
§3	Isotherms of disjoining pressure of aqueous wetting films	35
§4	Equilibrium between menisci and films	37
§5	Isotherms $W(\Delta\mu)$ for porous bodies	41
Chapter III	**PROPERTIES OF LIQUIDS IN NARROW PORES**	51
§1	Microstructure of liquids in thin layers	51
§2	Viscosity of liquids in thin interlayers	56
§3	Solvent action of liquids in fine pores. Viscosity of solutions	60
§4	Density of liquids in fine pores	66
Chapter IV	**FLOW IN PORES FILLED WITH LIQUID**	69
§1	Filtration	71
§2	Electro-osmotic flow	80
§3	Capillary osmosis	92
§4	Reverse osmosis	98
§5	Thermo-osmosis	104
§6	Flow of non-freezing water in porous bodies	112
Chapter V	**FILM FLOW**	123
§1	Hydrodynamics of wetting films	123
§2	Thermocapillary film flow	132
§3	Thermocrystallization film flow	137
§4	Wetting film flow driven by a concentration gradient	141
§5	Electro-osmotic flow of wetting films	142

Chapter VI VAPOR FLOW IN POROUS BODIES 147
§1 Free-molecular, viscous flow and a transient regime 147
§2 Vapor diffusion through gas 156
§3 Vapor diffusion in a non-uniform electric field 161
§4 Flow of vapor–gas mixture 164

Chapter VII COMBINED TRANSFER OF LIQUID AND VAPOR 171
§1 Film flow and vapor diffusion 172
§2 Film flow during evaporation into vapor atmosphere 182
§3 Combined flow of vapor, films and condensate 187
§4 Evaporation from communicating capillaries 188

Chapter VIII NON-STATIONARY MASS TRANSFER. EVAPORATION, DRYING 191
§1 Evaporation of electrolyte solutions 192
§2 Evaporation of polymer solutions 196
§3 Influence of surfactants on water evaporation from porous bodies 200
§4 Evaporation in the field of corona discharge 205
§5 Study of the mechanism of internal mass transfer by radiotracers 207
§6 Evacuation of non-volatile liquids from porous bodies 210
§7 Rate of drying 214
§8 Coefficients of internal mass transfer 220
§9 Deep drying 226
§10 Thermo-transfer of moisture 231

Chapter IX SHRINKAGE OF NON-RIGID POROUS BODIES DURING DRYING 237
§1 Shrinkage of a periodic colloid structure 238
§2 Experimental investigations of shrinkage mechanism during drying 245

Chapter X WETTING AND CAPILLARY IMPREGNATION 249
§1 Contact angles 249
§2 Hysteresis of contact angles 260
§3 Dynamic contact angles 263
§4 Capillary rise and imbibition 271
§5 Acceleration of capillary impregnation of hydrophobic porous systems by the addition of surfactants 287

CONCLUSION 301

REFERENCES 305

INDEX 321

Introduction to the Series

The subject matter of chemical engineering covers a very wide spectrum of learning and the number of subject areas encompassed in both undergraduate and graduate courses is inevitably increasing each year. This wide variety of subjects makes it difficult to cover the whole subject matter of chemical engineering in a single book. The present series is therefore planned as a number of books covering areas of chemical engineering which, although important, are not treated at any length in graduate and postgraduate standard texts. Additionally, the series will incorporate recent research material which has reached the stage where an overall survey is appropriate, and where sufficient information is available to merit publication in book form for the benefit of the profession as a whole.

Inevitably, with a series such as this, constant revision is necessary if the value of the texts for both teaching and research purposes is to be maintained. I would be grateful to individuals for criticisms and for suggestions for future editions.

R. HUGHES

Foreword to the English Edition

This book is the English translation of the Russian monograph *Fiziko-khimiya protsessov massoperenosa v poristykh telakh* (Physicochemical mass transfer processes in porous bodies), written by N.V. Churaev, and published in 1990. Presenting an English version has much to do with fostering access to a wealth of information that is still largely hidden from those who do not understand Russian.

Churaev's book is an exponent of the Russian habit of condensing the collected experience of the author(s) into monographs. Such books are usually printed on cheap paper and available at very low prices, so that they are accessible to everybody interested. As such they serve as a mode of communication between Russian scientists. In fact, the present monograph refers frequently to other ones. I myself possess about 50 such monographs and for me they are a valuable source of information. Because of the usually specialized and very personal nature of such books, there is a wide disparity between them with respect to contents and presentation.

The present book is written by N.V. Churaev, who after the demise of Derjaguin may be considered the Grand Old Man of Russian–Ukrainian Colloid and Interface Science. His scientific output is very broad, covering a range of topics and including both theoretical and experimental contributions. The present monograph is not the only one to which he has contributed. For political reasons his qualities have until recently remained virtually unknown in the West, except to those few who were allowed to travel behind the Iron Curtain. One could therefore say that the English edition of his most recent book makes much accessible that has remained hidden hitherto.

The book is presented systematically. It starts with an introduction to the basic surface forces, then expands by applying these to topics of increasing complexity: thin films on flat surfaces, on curved surfaces, statics first, then dynamics. The author does not eschew discussions of very complicated systems and situations, such as those exhibiting mixtures of hydrodynamic, electrokinetic, capillary and thermal flows in two-phase systems. A variety of ideas and theoretical elaborations are presented with experimental illustration, many of which have not yet been published in the West. The emphasis is on the models, some of them elaborated in great detail, rather than on the experiments. As the book progresses, the contents become more practice-oriented.

One should not consider a book like this as a counterpart of contemporary western text books. Rather it is an anthology of hitherto inaccessible, but in places very useful and valuable Russian scientific literature, of which the bibliography at the end gives an impression. At the same time it acknowledges the achievements of an eminent scientist who has served the international scientific community under adverse conditions for many years.

J. LYKLEMA
Wageningen, The Netherlands

List of Symbols

A	Hamaker constant
a	adsorption, g/cm^3
a_m	coefficient of liquid diffusion
C	solution concentration
c	vapor concentration, g/cm^3
D	diffusion coefficient
D_s	coefficient of surface diffusion
d	diameter of pores
E	voltage
e	electron charge
F	force
f	mass force
G	Gibbs free energy
g	gravity acceleration
H	distance between particles surface, thickness of a slit pore
h	thickness of a flat film or interlayer
$\hbar$	Planck's constant
j_i	frequency of jumps of surface molecules
K_i	permeability coefficient
K_s	parameter of structural forces
$K(x)$	local curvature
k	Boltzmann constant
L	capillary length
l	length of liquid column in a capillary
m	active porosity
n	porosity
n_i	number of molecules per unit volume
P	pressure in the liquid film
P_0	pressure in the bulk liquid
P_c	capillary pressure
p	vapor or gas pressure
p_s	saturated vapor pressure
Q	heat of adsorption

Q_i molar flux
q mass flux
R gas constant
R_i curvature radius
r capillary radius
T absolute temperature
T_0 temperature of ice melting
t° temperature, °C
t time
U potential energy of molecules
V volume
v flow rate
v_m molar volume
W mass content
w relative mass content
x coordinate
z ion valency

Greek

α capillary conicity
α_{ij} phenomenological coefficient
α_0 coefficient of condensation
β shrinkage coefficient
Γ adsorption, g/cm^2
γ distribution coefficient
γ_0 density of dry porous body
δ molecular size
δ_T thermogradient coefficient
Δ symbol of difference
ε_i dielectric permittivity
ε_0 dielectric constant
ε^* depth of potential well
ζ electrokinetic potential
η viscosity
θ contact angle
θ_A advancing contact angle
θ_d dynamic contact angle
θ_R receding contact angle
κ inverse Debye length
λ molecular free path
λ_i heat of ice melting
λ_v heat of sublimation
μ chemical potential
μ_i mass of mole
ξ frequency

Π	disjoining pressure
π	osmotic pressure
ρ	density
σ	surface tension
σ_i	surface charge
τ	shear stress
τ_0	yield stress
φ	rejection coefficient
φ_0	relative air humidity
χ	thermo-osmotic coefficient
ψ	electrical potential
ϑ	line tension

Introduction

Both natural and artificial porous bodies of different structure and composition are widely used in modern engineering and chemical technology. Therefore, the problem arises to describe more satisfactorily the properties of porous bodies and associated processes like evaporation, drying, wetting and capillary impregnation, filtration and capillary condensation.

Contrary to general approaches that apply the theory of heat and mass transfer, developed for continuous media, to porous bodies, in this book porous bodies will be considered as heterogeneous systems, the disperse state of which is responsible for features and the mechanism of processes of mass transfer. This approach is based on physicochemical and microhydrodynamic analysis of transfer processes of gases, liquids and their vapors in pores. The field of surface forces acting here not only changes the fluid properties but influences the kinetics of mass transfer processes. In turn, forces acting among particles in a porous body depend on the fluid state, which results in changes of porous structure in the course of transfer processes. The typical examples are the known phenomena of swelling and shrinkage, loss of firmness and destruction of porous bodies. Processes of mass transfer and structure formation are interrelated since both are controlled by the particle size and the field of surface forces forming a particular structure of a porous body according to external conditions. The features of this structure (pore size and shape, kinds of their interconnection) govern fluid distribution in a porous network and the kinetics of mass transfer processes.

Structure peculiarities, equilibrium and properties of liquids in porous bodies are considered step-by-step in this monograph. Then various mechanisms of mass transfer (liquid flow in pores and films, gas diffusion, combined transfer of liquid and vapor, convective diffusion in solutions, structure formation, capillary phenomena and wetting) are discussed. Examples are presented demonstrating how the results obtained can be applied in the analysis of various technological processes.

The monograph systematizes and summarizes the results of research undertaken with the author's participation and in cooperation with a number of scientific groups: Kalinin Polytechnical Institute (at present, Twer Technical University), Moscow Institute of Chemical Engineering, Leningrad Agrophysical Institute and the Moscow Technology Institute of Food Industry. Most of the model experiments and calculations were carried out in the laboratory of thin liquid layers, Moscow Institute of Physical Chemistry, Russian Academy of Sciences.

The author wishes to express his deepest gratitude to Academician B. V. Derjaguin for his constant attention to this area of surface phenomena, and also to his colleagues and postgraduate students K. Aitkuliev, S. S. Barer, S. A. Bardasov, V. V. Berezkin, I. G. Ershova, A. P. Ershov, O. A. Kiseleva, I. N. Kornil'ev, A. V. Novikova, Ya. I. Rabinovich, I. P. Sergeeva, V. D. Sobolev, V. M. Starov and Z. M. Zorin for their active and interested participation without which these works would hardly have been accomplished.

Chapter I

Porous Bodies as Disperse Systems

Porous bodies represent one of the disperse states of matter studied by physical chemistry of colloid and surface phenomena. Such disperse systems, where close contacts between the particles of the solid phase provide a constancy of shape, are referred to as rigid porous bodies. Among these there exist also easily deforming non-rigid porous bodies, for example pastes and powders. Here solid particles represent a disperse phase, and a liquid or gas represents the continuous phase (dispersion medium). More rigid structures develop in the case of closer contacts between particles, up to their concretion and formation of a continuous solid phase. Spaces between solid particles (pores) are filled completely or partially with a liquid. The strength of a porous body, peculiarities of its structure and pore sizes depend on the forces acting between particles. The modern theory of surface forces and experimental research in this field are presented in the monograph [1].

In this book, only some concepts and conclusions from the theories contributing to understanding the processes of mass transfer in porous bodies will be discussed.

We start with the theory of surface forces developed for a simple system of two interacting flat particles. In this case the so called disjoining pressure Π (defined as the force per unit area) may be calculated as a function of the thickness h of a flat interlayer. A thermodynamic definition of the disjoining pressure on the basis of the excess free surface energy of a thin interlayer is given in the next section. Respective equations and methods of calculation of molecular, electrostatic and structural components of the disjoining pressure are also described.

It will be shown also how to pass from the interactions of plane surfaces to curved ones. Expressions for forces of pair interaction of spherical particles are used for calculation of the potential curves which determine the equilibrium states of interacting particles forming a porous body.

Consideration of liquid and gas fluxes through porous bodies requires some information about the geometry of the porous network, and the shapes and sizes of pores. In the last two sections of this chapter some models of porous structures and methods for their characterization are discussed.

§1 Forces Acting Among Particles in Porous Bodies

Interparticle interaction giving rise to the formation of the structure of a porous body is governed by surface forces. These forces are long-range in comparison with the forces of chemical bonds whose radii of action are comparable with atom sizes. The long-range action of surface forces is associated with electrostatic and electromagnetic fields that are generated by surface charges and oscillating atomic or molecular dipoles. The action of these fields is detectable at distances up to 10^{-5} cm beyond the surface, exceeding by many orders the radius of action of chemical bond forces.

Disjoining pressure

Disjoining pressure $\Pi(h)$ may be defined as the first derivative of the Gibbs free energy of a thin flat interlayer with respect to its thickness h:

$$\Pi(h) = -(\mathrm{d}G/\mathrm{d}h)_{P_0,\, T,\, \mu_i} \tag{I.1}$$

at constant values of the pressure P_0 in the bulk phase, temperature T and chemical potentials μ_i of dissolved components. The concept of disjoining pressure as a basic thermodynamic characteristic of thin interlayers was introduced by Derjaguin.

The free energy $G(h)$, for the above mentioned conditions, equals the reversible work of the interlayer thinning in the course of an isothermal–isobaric process from infinity to some finite thickness h:

$$G(h) = -\int_{\infty}^{h} \Pi(h)\mathrm{d}h\,. \tag{I.2}$$

The stable state of the interlayer corresponds to a minimum value of the $G(h)$ function.

Let us consider first the interaction between two particles in the form of plates placed in a liquid dispersion medium (Fig. 1.1), assuming the area of

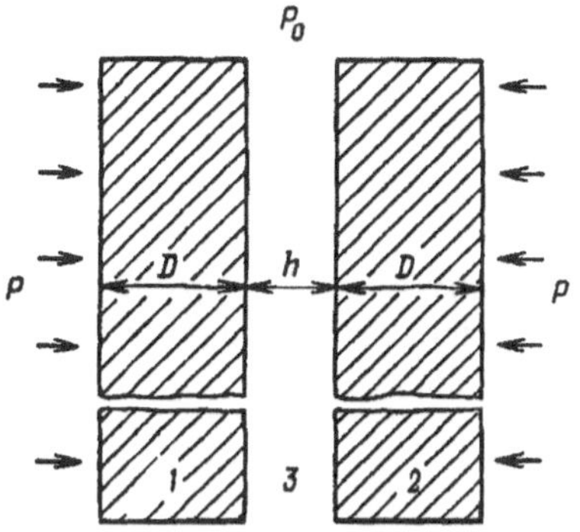

Figure 1.1 Interaction between two particles *1* and *2* through the plane interlayer *3*.

the plates to be so large that edge effects may be neglected. If the thickness of the plates D is considerably larger than that of the liquid layer h, we can reduce the interaction to that of two semi-infinite bodies *1* and *2* separated by the interlayer of a medium *3*.

To maintain the plates at rest when they are brought together at a distance smaller than the radius of action of surface forces, a force should be applied producing on the outer surfaces the pressure, P, higher (in the case of repulsion) or lower (in the case of attraction) than the pressure, P_0, in a bulk ambient liquid. The pressure difference, $P - P_0$, defines the disjoining pressure Π of the interlayer which depends on its thickness h. The function $\Pi(h)$ is called the isotherm of the disjoining pressure. It contains all the information about the field of surface forces acting in a given system. Disjoining pressure is positive ($\Pi > 0$) when the plates are repelled and negative ($\Pi < 0$) when they are attracted to one another.

Let us consider the various kinds of surface forces and associated components of disjoining pressure in thin interlayers of a gas or a liquid. The van der Waals forces acting between individual atoms and molecules are related to the dipole moments (inductive and orientational interaction) including those of oscillating quantum dipoles (dispersion interaction). In the case of a condensed medium, only the latter effect manifests itself summarizing interactions of a great number of molecules. Forces of molecular attraction of macroscopic bodies represent the result of action of all interatomic dispersion forces. Inductive and orientational interactions are of less importance because of the random dipole distribution in bulk phases.

Molecular attraction appears when the plates are so close to each other that the fluctuating electromagnetic fields, existing near to the interface and exponentially decaying away from it, overlap. These alternating electromagnetic fields are generated by electron motion in atoms representing the quantum oscillators.

Dispersion forces

The macroscopic theory of dispersion forces pioneered by E. M. Lifshits [2] leads to the following asymptotic expression for the molecular component of disjoining pressure at $h < 10$ nm [3]:

$$\Pi_m(h) = -\frac{\hbar}{8\pi^2 h^3}\int_0^\infty \frac{(\varepsilon_1-\varepsilon_3)(\varepsilon_2-\varepsilon_3)}{(\varepsilon_1+\varepsilon_3)(\varepsilon_2+\varepsilon_3)}d\xi = -\frac{A_{132}}{6\pi h^3}. \tag{I.3}$$

Here $\hbar$ is the Planck constant divided by 2π; $\varepsilon(i\xi)$ is the dielectric permittivity as a function of imaginary frequency $i\xi$. Indices 1, 2 and 3 refer to the plates 1 and 2 and to the interlayer 3, respectively (Fig. 1.1).

Forces of molecular attraction of the plates decrease with distance less abruptly (as $1/h^3$) than forces of dispersion attraction of individual molecules (proportional to $1/r^7$) which can be explained by the collective nature of the interactions of the many molecules that constitute condensed bodies.

Equation (I.3) links molecular forces with spectral characteristics of interacting bodies expressed by functions $\varepsilon(i\xi)$. The reason for this dependence is that the fluctuating electromagnetic fields are generated by virtual photons with energies close to the differences between the energy levels of atoms and molecules that are responsible for the spectral properties of the interacting bodies. Frequencies within the visible and UV range, that is, the frequencies of electron transitions in atoms, provide a major contribution to the constant of molecular forces A_{132} known as the Hamaker constant. Table 1 shows calculated values of A_{132} for a number of practically important cases.

Table 1 A_{132} values calculated by eq. (I.3).

Interacting bodies	$A_{132}\ 10^{20}$, J
M–air–M	40
M–water–M	26
M–air–Q	18
M–P–M	17
M–water–P	3.2
M–water–Q	2.7
Q–air–Q	7.9
Q–water–Q	1.4
P–air–P	6.4

Table 1 (Continued)

Interacting bodies	$A_{132}\,10^{20}$, J
P–water–air	1.0
P–water–P	0.9
Q–water–air	1.1
M–water–air	11

Index M denotes metals, Q denotes quartz, P denotes polymer materials and nonpolar liquids. Values of A_{132} do not differ much for various metals and polymers (if the interacting plates are similar). Presented values refer to gold and polystyrene; for iron and copper A_{132} is approximately two times lower. For Teflon which is less polar than polystyrene the constant is lower by 35%, and higher by 60% for the more polar nylon.

As seen from Table 1, in a symmetrical case (for similar bodies 1 and 2) the attraction forces are stronger when the bodies are separated by a gas interlayer, because in this case ε_3 has a minimal value equal to 1. A liquid interlayer sufficiently depresses the molecular forces. In a symmetrical case A_{131} is always positive, which gives rise to forces of attraction. In a non-symmetrical case, when bodies 1 and 2 are different and one of them is less polar than the interlayer 3, A_{132} becomes negative, as seen from eq. (I.3). The necessary condition is expressed by the inequality $\varepsilon_1 < \varepsilon_3$ (or $\varepsilon_2 < \varepsilon_3$) (in the range of frequencies which gives the major contribution to the integral in eq. (I.3)). In this case dispersion forces are repulsive. As a first approximation, the index of refraction $n \cong \sqrt{\varepsilon}$ can be used to characterize the polarity of the medium in the visible range of frequencies.

For interlayers with thickness $h > 10$ nm, the effect of electromagnetic retardation of molecular interaction, caused by the finite rate of propagation of information about the state of oscillating quantum dipoles, should be taken into account. In the limiting case of thick interlayers (h > 30–40 nm), dispersion forces decrease more sharply with rise in h:

$$\Pi_m(h) = -\frac{B}{h^4}, \tag{I.4}$$

where B is the constant of totally retarded dispersion forces.

For dielectric materials, B values are determined by the static values of dielectric permittivity ε_0:

$$B = \frac{3\hbar c}{16\pi^2\sqrt{\varepsilon_{03}}}\frac{(\varepsilon_{01}-\varepsilon_{03})(\varepsilon_{02}-\varepsilon_{03})}{(\varepsilon_{01}+\varepsilon_{03})(\varepsilon_{02}+\varepsilon_{03})}, \tag{I.5}$$

where c is the speed of light.

In the transition range of thicknesses (from 10 to 30–40 nm), an approximate solution may be used. In the case of identical bodies 1 and 2 and aqueous interlayers [1]:

$$\Pi_m = -\frac{A_{131}}{6\pi h^3}\left(1 - \frac{1 + bl}{[1 + (bc/2h)]^3}\right), \tag{I.6}$$

where $b = 6.5 \times 10^{-17}$ s/rad and $l = 3.3 \times 10^{15}$ rad/s. Here A_{131} is the value from eq. (I.3).

For identical plates of thickness D (commensurate with h), forces of molecular attraction decrease because of a reduced number of interacting molecules. This may take place in porous bodies containing thin plate particles (for instance clays). The following expression for the molecular component of disjoining pressure may be used, in this case [4]:

$$\Pi_m(h) = -\frac{A_{131}}{6\pi h^3}\left(1 - \frac{2}{(1 + a)^3} + \frac{1}{(1 + 2a)^3}\right), \tag{I.7}$$

where $a = D/h$.

As seen from this equation, the attractive force per unit area decreases by 20% at $a = 1$, and by half at $a = 0.5$.

Forces of molecular attraction also change when complete adsorption layers of surfactants, oxide films or film coatings of thickness δ exist on the particle surfaces. Surface layers less polar than the particle substance reduce molecular attractive forces, and, in contrast, metallization of dielectric particles increases their molecular attraction. When $h \leq \delta$, the plates interact as if they were made from the coating material or an adsorption layer. For example, adsorption layers of surfactants (δ = 1–3 nm) drastically decrease molecular attraction of metals at short distances. In the case of dielectric materials the screening effect is less pronounced [5]. Surface layers that are loose and impregnated with liquid also reduce forces of molecular attraction.

Electrostatic forces

When the surfaces of the bodies 1 and 2 immersed in a liquid medium are charged, their interaction includes electrostatic forces [1, 6]. Electrostatic forces depend on the values and signs of surface potentials ψ_1 and ψ_2. Surfaces become charged due to preferential adsorption of ions of the same sign or due to dissociation of surface ionogenic groups: –OH, –COOH, etc. Calculations of the surface charge σ are based on the Langmuir equation of

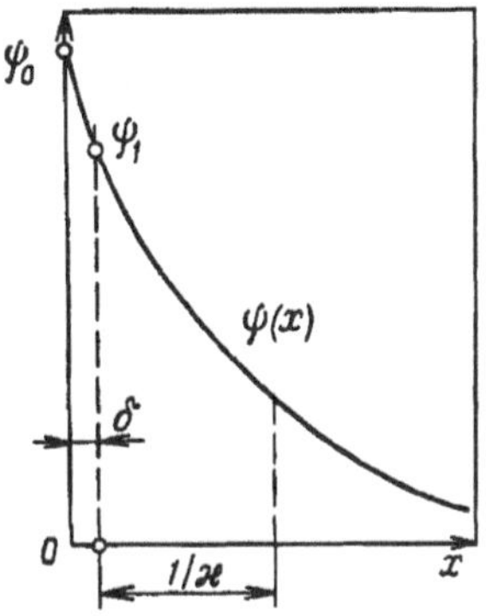

Figure 1.2 Potential distribution $\psi(x)$ in the vicinity of a plane charged surface.

adsorption applied to ionic systems. However, theoretical methods fail to take into account the complex interactions between ions and real surfaces. Therefore, to determine surface charge and potentials, experimental methods are often used: measurement of particle electrophoretic mobility, potentiometric titration of suspensions and sols. In the case of porous bodies, rates of electro-osmotic flow and streaming potentials are usually measured.

Distribution of the potential $\psi(x)$ in an electrical double layer (EDL) in the vicinity of a single charged surface is shown in Fig. 1.2. Here ψ_0 characterizes the surface potential and ψ_1 is the potential of the Stern layer consisting of adsorbed ions. The thickness of the Stern layer is close to the size of hydrated ions. The potential decreases exponentially beyond the adsorption layer:

$$\psi(x) = \psi_1 \exp(-\kappa x). \quad \text{(I.8)}$$

The potential ψ_1 is generated by the surface charge density:

$$\sigma = \frac{\varepsilon_0 \kappa \psi_1}{4\pi}. \quad \text{(I.9)}$$

Here $1/\kappa$ is the Debye radius, i.e. the distance at which the potential decreases e times. The κ-values depend on the static dielectric permittivity of the liquid medium ε_0 and ionic concentration C of the solution. In the case of a symmetrical electrolyte:

$$\frac{1}{\kappa} = \sqrt{\frac{\varepsilon_0 RT}{8\pi z^2 F^2 C}}, \quad \text{(I.10)}$$

where F is Faraday, R is the gas constant, z is the ionic valency, T is the temperature, and C is the electrolyte concentration measured in mol/cm^3.

It follows from eqs. (I.10) and (I.8) that the radius of action of a charged surface increases with reducing electrolyte concentration and rise in the value of ε_0. Thus, in aqueous 1 – 1 electrolyte solutions (for example NaCl) the $1/\kappa$-value achieves 0.1 μm at 10^{-5} mol/l. In concentrated electrolyte solutions the surface electric field penetrates a distance far less into the dispersion medium. At 10^{-1} mol/l the Debye radius does not exceed several molecular diameters.

The overlap of EDL when solid surfaces approach each other leads in the case of equal surface potentials $\psi = \psi_1 = \psi_2$ to the appearance of repulsive forces and a positive electrostatic component of the disjoining pressure $\Pi_e > 0$. For a binary electrolyte when $z_+ = z_- = z$ and with a weak EDL overlap, i.e. at $\kappa h > 1$, the analytical expression obtained for Π_e [6] was:

$$\Pi_e(h) = 64CRT\gamma^2 \exp(-\kappa h), \tag{I.11}$$

where $\gamma = \tanh(zF\psi/4RT)$.

For high surface potentials (>100 mV), $\gamma \to 1$ and Π_e becomes independent of the value of the surface potential. At low surface potentials, when the tangent function is approximately equal to its argument, eq. (I.11) transforms into the form:

$$\Pi_e(h) = \frac{\varepsilon_0 \kappa^2 \psi^2}{2\pi} \exp(-\kappa h). \tag{I.12}$$

In this case Π_e is a quadratic function of the surface potential ψ.

Increasing electrolyte concentration reduces electrostatic repulsion because changes of κ, being in the exponent, influence Π_e more strongly. For highly overlapped EDL, $\kappa h < 1$, and when surface potentials are high, the electrostatic repulsive forces follow the power law:

$$\Pi_e(h) = \frac{\pi \varepsilon_0}{2h^2}\left(\frac{RT}{zF}\right)^2. \tag{I.13}$$

More complex cases, when surface potentials of particles are different, have been studied in the Derjaguin theory of heterocoagulation [7].

When surface potentials have different signs (e.g. $\psi_1 > 0$ and $\psi_2 < 0$), which may be the case for a porous body formed from a mixture of different particles, electrostatic forces cause attraction of the surfaces at any distance h and $\Pi_e < 0$. Thus, in the case of equal and low potentials ($\psi_1 = -\psi_2$), for

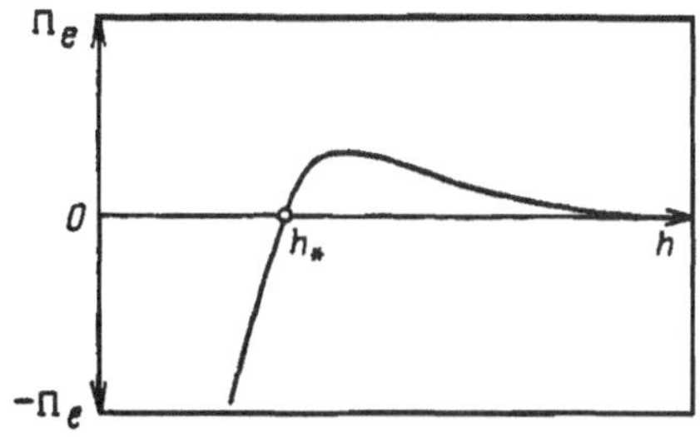

Figure 1.3 The change of sign of the electrostatic component of disjoining pressure at surface potentials of the same sign but of different values.

$\kappa h > 1$ we have:

$$\Pi_e(h) = -4RTC\left(\frac{zF\psi}{RT}\right)^2 \exp(-\kappa h) . \qquad (I.14)$$

When surface potentials of interacting particles are of the same sign but of different values, the Π_e dependence on h is more complex (Fig. 1.3). The particles repel each other in thick interlayers and attract in thin ones. The disjoining pressure changes sign at the thickness:

$$h_* = \frac{1}{\kappa}\ln\left(\frac{\psi_1}{\psi_2}\right) . \qquad (I.15)$$

As seen from this equation, the more surface potentials differ in values and the lower is the electrolyte concentration responsible for κ, the greater is h_*. For example at $z = 1$, $\psi_1 = 100$ mV and $\psi_2 = 50$ mV: $h_* = 0.7/\kappa$, i.e. it is about 70 nm at electrolyte concentration of 10^{-5} mol/l and only 2 nm at 10^{-2} mol/l.

Because the adsorption energy of ions is usually high ($U >> kT$, where k is Boltzmann constant), electrostatic forces weakly depend on temperature. Molecular forces are also insensitive to temperature changes, for they depend on electron motion in atoms taking place even at $T = 0$ because of quantum effects. Structural forces are more temperature-sensitive [8–10].

Structural forces

These forces are associated with structural changes of liquids in thin interlayers. The study of liquid structure by methods of Monte Carlo and molecular dynamics showed that boundary liquid layers, i.e. the layers adjacent to a solid surface, have structures that differ in density, mobility, molecule orientation, energy and number of intermolecular bonds from that of the bulk liquid. For simple liquids these changes occur within a small number of molecular layers. In polar liquids the effect can be more extended due to dipole–dipole interaction. Liquids with hydrogen bonds, for example water, form structurally changed boundary layers of much greater thickness. A network of oriented hydrogen bonds increases the thickness of ordered layers of water in contact with hydrophilic solid surfaces.

Structural forces significantly depend on the lyophilicity of the substrate and its affinity to the liquid phase. Thus, the effective thickness of structurally-changed water layers increases with hydrophilicity of the substrate rising to several nanometers. The extent to which the structure changes depends on the number and topography of surface centers capable of forming hydrogen bonds with water molecules.

Overlapping of single boundary layers, associated with approaching surfaces, gives rise to repulsive structural forces in the case of lyophilic surfaces and to attractive ones in the case of lyophobic ones. This effect is associated with the different liquid structures in the vicinity of lyophilic and lyophobic substrates. In the former case the density of the liquid is increased and in the latter it is decreased compared to the bulk. Different structures of water boundary layers in the vicinity of surfaces with different hydrophilicity is the reason for different signs of structural forces. In the intermediate range of hydrophilicity the structure of the boundary layers may not differ markedly from the bulk phase. There are no structural forces here, and there are only molecular and electrostatic interactions between particles. Structural forces are strong in the case of surfaces characterized by water contact angles $\theta_0 < 15°$ (hydrophilic repulsion) and $\theta_0 > 60°$ (hydrophobic attraction). Molecular and electrostatic forces prevail at moderate hydrophilicity of the surfaces ($15° < \theta_0 < 60°$). In this case interaction of particles is well described by the DLVO (Derjaguin–Landau–Verwey–Overbeek) theory developed for colloidal systems [1, 6].

The dependence of the structural component of disjoining pressure on the interlayer thickness can be expressed as [1, 8–10]:

$$\Pi_e(h) = K_s \exp(-h/l), \tag{I.16}$$

where K_s and l are constants. Representative literature values are shown in Table 2 for a number of systems. The value of K_s characterizes the force per unit area at $h \to 0$, and l is the characteristic length of structural ordering.

The degree of hydrophobicity of a surface is characterized here by the advancing contact angle θ_A. As seen from the table, K_s is positive for hydrophilic surfaces and negative where these are hydrophobic. The K_s-values seem to depend on electrolyte concentration decreasing with an increase of the latter. This dependence may arise as a result of action of the electric field of ions on water structure that brings the structure of the boundary layer closer to that of the bulk electrolyte solution. A similar effect is caused by elevated temperature loosening of the network of hydrogen bonds and depressing the long-range action of the hydrophilic surface. Structural forces probably do not reveal themselves in water at temperatures higher than 65–70°C.

Table 2 Experimental values of K_s and l.

System	K_s, N/m^2	l, nm
Quartz–water solution of KCl, 10^{-4} mol/l	10^6	1.0
Quartz–water solution of KCl, 10^{-3} mol/l	8.0×10^5	1.0
Mica–water solution of KNO_3, 10^{-2}–10^{-4} mol/l	10^7	1.0
Mica–water solution of NaCl, 5×10^{-4} mol/l	1.4×10^7	0.9
Mica–water solution of NaCl, 5×10^{-3} mol/l	3.0×10^6	0.9
Montmorillonite, water solution of NaCl, 10^{-4} mol/l	2.0×10^6	2.2
Hydrophobized mica ($\theta_A = 65°$)–water	-2.0×10^7	1.0
Hydrophobized mica ($\theta_A = 94°$)–water	-4.0×10^7	1.3
Hydrophobized mica ($\theta_A = 113°$)–water	-5.0×10^7	2.5

Recently, experimental evidence has been found for hydrophobic attraction forces with much larger radius of action [10]. They follow the same exponential law (I.16) but with l-values from 10 to 100 nm. The nature of these forces is not yet determined. Their role may be of great importance in colloids consisting of hydrophobic particles, in emulsions and flotation phenomena.

Steric forces

When the surfaces of particles are covered with adsorption layers of surfactants or polymers (their thickness may achieve tens of nm) and the particles approach each other, the overlapping of the adsorption layers gives rise to a steric component of disjoining pressure. For complete surfactant monolayers a simple model of mutual impenetrability, a steric restriction of the ability to approach each other is often used. In the case of more loose and less uniform polymer adsorption layers, the dependence of attraction forces on the interlayer thickness is close to exponential. The forces caused by overlapping and layer deformation may exceed in value other components of surface forces. At a low degree of overlapping, the interaction is elastic. The modulus of elasticity is about 10^6–10^7 N/m^2 at $h = 50$–100 nm. Further deformation at a high degree of overlapping occurs slowly and is partially irreversible. To reach an equilibrium state between overlapped adsorption layers may take a

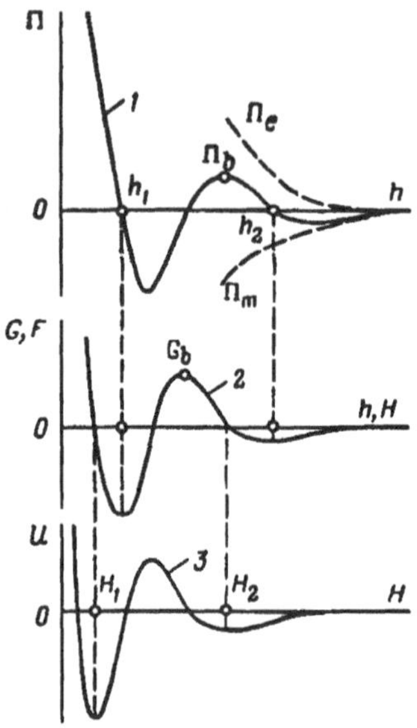

Figure 1.4 Qualitative form of isotherms of disjoining pressure, $\Pi(h)$, and free energy, $G(h)$, of plane interlayers (curves *1* and *2*). Isotherms of forces acting between spherical particles $F(H)$ and the energy of their interaction $U(H)$ are shown by curves *2* and *3*, respectively.

long time. The time necessary to complete the deformation depends on the rheological properties of the polymer in adsorption layers.

Isotherms of disjoining pressure and potential curves

The resulting action of all components of the surface forces is called the total isotherm of disjoining pressure of the interlayer, $\Pi(h)$. As a first approximation, this action may be considered additive: $\Pi(h) = \Sigma\Pi_i(h)$. Curve *1* in Fig. 1.4 provides a typical example of such a total isotherm. Dashed curves, Π_m and Π_e, represent the qualitative form of two components of disjoining pressure: molecular and electrostatic. Molecular attractive forces generally prevail when particles are well apart: at such distances $\Pi < 0$ and particles tend to come closer. With further thinning of the interlayer, repulsive electrostatic forces increase and exceed the molecular attraction forces. When $h = h_2$, this causes Π to change the sign that gives rise to a force barrier Π_b. In far thinner interlayers, the molecular forces prevail again due to their sharp increase at short distances. As a result, the repulsion is replaced by an attraction. However, short-range repulsive forces appear again at $h < h_1$. In the case of hydrophilic surfaces they are associated with structural repulsive forces. For lower surface hydrophilicity or in the case of nonpolar liquids, the repulsion arises because of the necessity to overcome the barrier caused by the removal of several ordered monolayers strongly bound to the surface. In the case of lyophobic surfaces, the contact may be "dry". Here the repulsion is caused by the Born forces, the reversible elastic repulsion by the electron shells of surface atoms. When the overlapping of electron shells results in binding, irreversible chemical bonds may be formed between particles which corresponds to so-called crystallization contacts. In many cases, the equilibrium thickness h_1 corresponding to force curves in Fig. 1.4 may be achieved only after a long time because of high viscous resistance of the interlayers to thinning.

The equilibrium interparticle distances in a porous body correspond to the minimum free energy of the system. The specific free energy (per unit area) of two plane particles $G(h)$ is expressed by eq. (I.2). The potential curve $G(h)$ related with the isotherm $\Pi(h)$ is also shown in Fig 1.4 (curve *2*). Generally it has two minimum values: the outer minimum (at $h = h_2$) corresponding to relatively large interparticle distances about several nm, and the closer and the deeper minimum corresponding to the thickness $h = h_1$, of about some monolayers. The first state is metastable because of higher values of G. The particles can spontaneously pass to a stable state corresponding to the deeper potential well. The rate of transition depends on the height of the potential barrier G_b separating regions of minimum and on the thermal energy of particles, kT. Under the action of an external force the transition may go faster.

Due to fluctuations of the particle energy, the reverse transition to the outer, secondary well is also possible, however they have to overcome a much higher barrier. As a result, a dynamic equilibrium between occupied states in primary and secondary potential wells is established. Such processes occur in colloids when particles can move freely (Brownian motion). Since in rigid porous bodies the thermal mobility of particles is restricted, the kinetics of the transitions between these two states is impeded. Porous bodies formed, for example, by the drying of dispersions, usually inherit a structure acquired in the course of evaporation of a liquid. This structure depends on interparticle forces specified by the isotherm $\Pi(h)$, on the rate of drying and the processes of relative particle motion. The sediments forming in a liquid medium acquire a non-rigid porous structure while intense forces of attraction are acting. In this case the residue is formed by precipitating aggregates. More dense sediments are formed when there are weak interactions which do not prevent gravitational or other inertial forces packing particles closely. Therefore, the potential curves $G(h)$ determine the forming structure of a porous body, the extent of defects and deviations from the ideal close-packed structure corresponding to the minimum free energy of the whole system, i. e. the realization of the maximum number of contacts in the primary potential well.

When a rigid porous body is formed by dissolution of inclusions or defects in a continuous solid body (for example, porous glass or nuclepore filters) or by sintering of particles, surface forces are responsible only for internal tensions in the hard skeleton of the solid body. Disjoining pressure in narrow pores is balanced by the elastic reaction of the skeleton. In pores filled with air, forces of molecular attraction play a major role giving rise to internal stresses in walls separating neighbouring pores. If pores are filled with a liquid and a surface charge is generated, forces of electrostatic repulsion arise. Then a change in the sign of the stresses may cause local destruction of regions of defective structure. In narrower pores, the repulsion of lyophilic surfaces may be caused by structural forces. Here the pressure applied to walls is greater for cylindrical pores than for slits of the same linear size. When there is incomplete filling of a porous body by a liquid, the stresses in the skeleton may be also caused by capillary forces. The strength of porous bodies is determined by the simultaneous action of both surface and capillary forces (see Chapter IX).

Interaction of spherical particles

The model of plane particles and flat interlayers (Fig. 1.1) discussed above is a simplification not always realized in practice. Non-planar contact of two hard spheres is a more realistic model. In this case the Derjaguin approximation [11] is applicable when the radius of curvature is much larger than that of the radius of action of surface forces. For hard spheres of radii R_1 and R_2, the interaction force F is expressed through free energy of interaction of plane surfaces:

$$F(H) = 2\pi R_1 R_2 G(h)/(R_1 + R_2), \tag{I.17}$$

where H is the shortest distance between particle surfaces (Fig. 1.5). For similar particles of radius R_0:

$$F(H) = \pi R_0 G(h). \tag{I.18}$$

Calculations have shown that in the case of molecular forces eq. (I.18) is valid when $H < R_0/100$, and in the case of electrostatic forces when $\kappa R_0 > 10$. Thus, at $\kappa = 10^6$ cm^{-1} corresponding to the concentration of $1-1$ electrolyte of 10^{-5} mol/l, eq. (I.18) is valid for particles of $R_0 > 0.1$ μm.

Because $F(H)$ differs from $G(h)$ only by a constant factor, these functions are qualitatively the same (Fig. 1.4, curve 2). The larger the particle radius the stronger is the force of interaction F. To calculate the free energy of pair interaction of two spherical particles the forces $F(H)$ over distance H between particles are integrated:

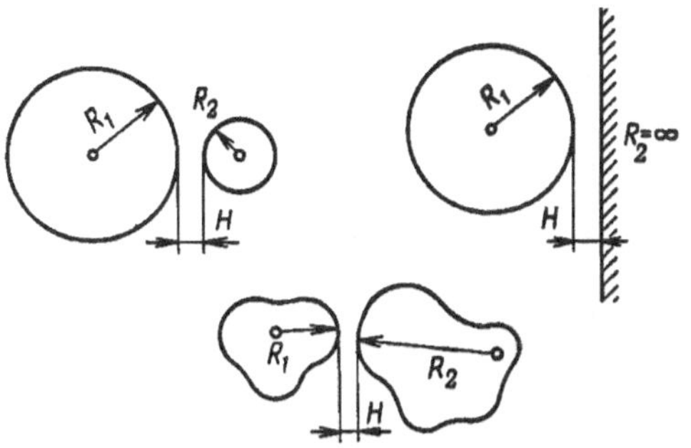

Figure 1.5 The calculation of forces acting between non-planar particles.

$$U(H) = \int_{H}^{\infty} F(H)\mathrm{d}H\,. \tag{I.19}$$

The curve $U(H)$ (Fig. 1.4, curve *3*) differs from the similar potential curve for plane interlayers (Fig. 1.4, curve *2*) in that the minima of the free energy at H_1 and H_2 are shifted to shorter interparticle distances and their depth depends on the radius of particles. Therefore, in the case of larger particles they may be fixed in the secondary potential well which becomes sufficiently deep.

Equilibrium of small particles in the secondary potential well corresponds to a relatively easily deformable non-rigid porous body. The strength of the structure sharply increases when the particles are fixed in the primary potential well. The binding energy of a particle in a porous body is defined by the number of contacts with neighbors. According to Rehbinder *et al.*, the structure strength is equal to [12]

$$P_s = \chi p_1, \tag{I.20}$$

where χ is the number of contacts per unit area in the shear plane and p_1 is the strength of an individual contact. The p_1-value for spherical particles is determined by the difference of the F values in maximum and in the primary minimum of the force curve $F(H)$.

For particles of irregular shape, the force F may be calculated from eq. (I.17), where R_1 and R_2 are local curvature radii at the point of contact (Fig. 1.5). When one of the radii is equal to infinity (contact: sphere–plane), the coefficient at $G(h)$ in eq. (I.17) is equal to $2\pi R_1$. Thus, it follows that in forming porous bodies the favorable contacts are peak–plane or plane–plane.

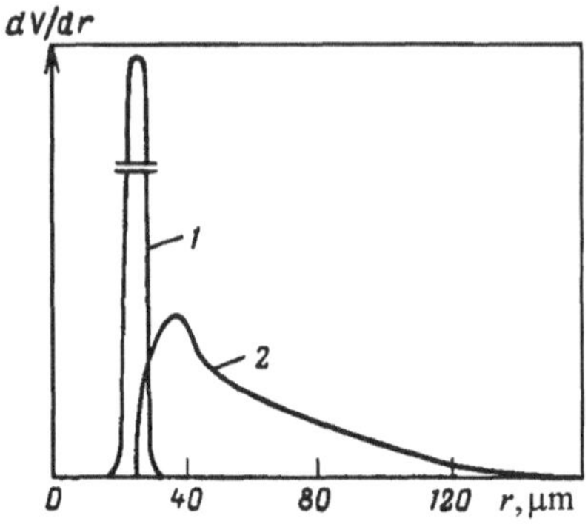

Figure 1.6 Pore-size distribution in quartz sand obtained in filtration experiments (curve *1*) and from the curves of capillary retention (curve *2*).

The form of the curves $G(h)$ and $U(H)$ depends on particular contributions of the different components of disjoining pressure to the total isotherm $\Pi(h)$. For example for uncharged particles (when $\Pi_e = 0$) the potential curve at $h > h_1$ lies in the region $G < 0$ and has only the closer minimum. When particles are highly charged, the secondary minimum may disappear and the height of the potential barrier, preventing close contacts, may be larger.

Various additives to the dispersion medium which change the charge or hydrophilicity of particle surface are used to control the binding forces and porous body structure. Such additives may be salts, acids and bases, surfactants and polyelectrolytes or soluble polymers.

§2 Models of Porous Structure

A detailed description of the structural peculiarities of a porous body is very complex and tedious. Moreover, full information, if it were available, would be excessive. To describe the kinetics of mass transfer processes, it is sufficient to know some averaged structure parameters. The point is that the overall linear dimensions of a porous body generally are much greater than the mean pore size. Therefore, liquids and gases flowing in a porous body pass many times through repeated regions of similar microstructure: in particular the widening or narrowing of pore bores. Thus, to characterize the integral effect, we can use the effective or averaged pore size.

In descriptions of mass transfer processes in real systems, the latter are usually modelled by a network of identical capillaries or packed spherical particles. Various properties of a porous body may be described by different characteristic features of pore space geometry. For example a pathway-averaged size of the pores is responsible for filtration properties of specimens,

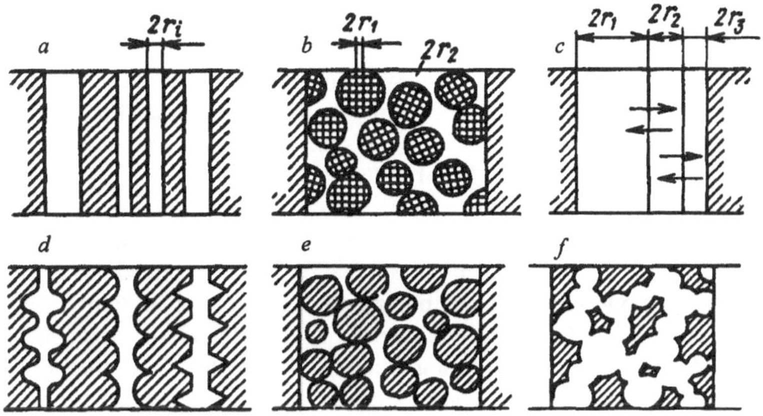

Figure 1.7 Various models of porous body structure.

and pore space configuration in points of contacts and the manner of widening and narrowing communication is responsible for liquid capillary retention.

Figure 1.6 shows the pore-size distribution obtained from filtration experiments (curve *1*) and from the data on capillary retention (curve *2*) for the same sample of sand (fraction 0.1–0.25 mm). Calculations were made for the same models of porous medium: an array of cylindrical capillaries of various radii r. The distribution of percolation pores is narrow, reflecting the effect of path-averaging (sample length is 25 cm), as seen from Fig. 1.6. The distribution of curvature radii of capillary menisci is significantly wider. It reflects the peculiarities of contacts between particles. The maximum pore size approaches the size of sand grains (possible size of structure defects); the minimum size corresponds to percolation pores.

It is convenient to consider filtration phenomena in porous bodies with reasonably uniform pores on the basis of capillary models (Fig. 1.7). One of the most widely used models (Fig. 1.7, *a*) is formed from straight parallel capillaries (in the shape of a slit or a cylinder) defined by a distribution function $F(r)$ or some average value of radius r_c. If a porous body is extremely non-uniform, several characteristics of capillary sizes are used. Biporous structures represent an example when a porous body is composed of porous particles (Fig. 1.7, *b*). For this structure two characteristic average pores diameters are defined: the first set inside the particles, and the second set between them. A set of interconnected capillaries that differ in size constitute another model of non-uniform pores. This accounts for mass exchange between the capillaries within the specimen (Fig. 1.7, *c*).

Liquid retention in a porous body and liquid equilibrium in the case of incomplete saturation with liquid are conveniently described by special models taking into account the existence of widening and narrowing. The simplest are models of bead pores (Fig. 1.7, *d*) and the model of solid spheres randomly or orderly distributed in porous media (Fig. 1.7, *e*). In many cases "inverse" models are also used in which randomly distributed empty overlapped spheres form a system of pores, and the spaces between them are occupied by the solid phase (Fig. 1.7, *f*).

Modern computer methods enable consideration of much more complicated models of random packing of spheres, rods and particles of various shapes. Another line of computer simulation is the application of lattice models and models of branching pores using the theory of random processes and percolation [13]. In many cases the application of such complicated models has proven itself: for example solving the problem of displacement of one fluid by another from porous media (mercury porosimetry, porous electrodes, extraction of residual oil). Studying physicochemical mechanisms of mass transfer processes, we may restrict our consideration to simpler models of porous bodies.

§3 Methods of Determination of the Structural Characteristics of Porous Bodies

Methods for the determination of the parameters of porous structure are based on various approaches: simple capillary models are generally used. The most popular methods are based on obtaining the adsorption isotherm and that of capillary condensation [14, 15]. When inert gases (nitrogen, argon) interacting with solid bodies only by dispersion forces are used as an adsorbate, the isotherms mainly depend on the peculiarities of the porous structure of specimens, which enables comparative experiments to be performed. Absolute, quantitative values are less reliable, because the values obtained for pores size are related to an assumed shape of pores.

Capillary pressure

The model of cylindrical pores completely wetted with adsorbate is most frequently used. The pore radius r is related to the relative pressure of adsorbate vapor p/p_s by the Kelvin equation:

$$r = \frac{2\sigma v_m \cos\theta}{RT} \ln\frac{p_s}{p}, \tag{I.21}$$

where σ is surface tension of a condensate, v_m is its molar volume and θ is the contact angle.

The pore-size distribution is obtained from the desorption isotherms of the capillary condensate $V(p/p_s)$. Equation (I.21) expresses the variations of radii

Δr corresponding to the change of condensate volume ΔV with vapor pressure decreasing by $\Delta(p/p_s)$. The problems arising in application of eq. (I.21) are connected to the uncertainty of contact angle values in narrow pores, where fields of surface forces overlap, and the finite thickness of adsorption films should be taken into account. The use of the standard adsorption equipment has promoted the wide application of this method for pore size analysis.

Another variation of this method is the suction of a capillary retained liquid by putting a porous body in contact with a reference for which the relation between mass content W and capillary pressure P_c is known. At an equilibrium between the subject body and the reference, the pressure P_c inside the bodies is equal. A measured change of volume ΔV of the liquid retained in capillaries of the subject body, related to a range of pore radii Δr, corresponds to the variation of pressure ΔP_c caused by drying of the reference body. Values of r are determined from the Laplace equation:

$$r = \frac{2\sigma \cos\theta}{P_c}. \tag{I.22}$$

This differs from the desorption method in that the pressure differences and therefore changes of the chemical potential are set up for the liquid phase rather than for the vapor. A convenient series of devices was developed to carry out routine measurements by this method [16].

To investigate the porous structure of film materials (for example polymer membranes), one can apply the method of displacement of a liquid from a completely saturated specimen by a gas or another immiscible liquid. Gradually increasing pressure P of the displacing fluid, one must register that at some pressure P_1 the fluid starts to pass through a network of wider pores, where the capillary pressure $P_c = P_1$. Having assumed some pore model, r_{max} values are obtained from the Laplace equation $r_{max} = 2\sigma/P_1$, where σ is surface or interface tension. As the pressure P grows, the displacement of a fluid from narrower pores occurs until it completes at some pressure P_2. This pressure corresponds to the minimum size of open pores $r_{min} = 2\sigma/P_2$. The pore-size distribution can be obtained by analyzing the relationship $V(P)$ between the volume V of passing fluid and the applied pressure P [17]. In the case of easily deformable specimens, this method should use displacing liquids with low interface tension. The well known method of mercury porosimetry applied for rigid porous bodies is a variant of this method.

Adsorption and gas flow

To determine the amount of penetrable open pores in fine porous bodies, a combination of adsorption and filtration methods is applicable [18]. The

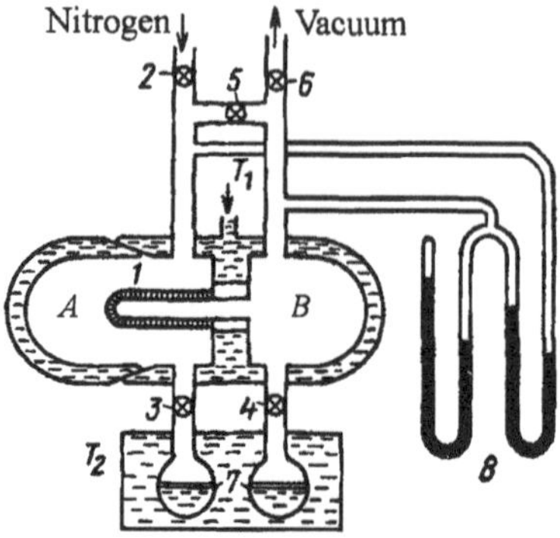

Figure 1.8 The experimental apparatus for determining the pore-size distribution by the adsorption-filtration method.

liquid completely filling the pores is desorbed into surrounding medium where a gradually changing vapor pressure p/p_s is maintained. Empty open pores are detected by the change in the rate of flow of an inert gas through the specimen.

The experimental apparatus is shown in Fig. 1.8. The subject porous body *1* separates compartments *A* and *B* of a detachable glass chamber thermostatted at temperature T_1. The chamber is pre-evacuated by a vacuum pump with valves *2*, *3* and *4* shut and valves *5* and *6* open. Valves *5* and *6* are then closed and the adsorbate vapor passes through valves *3* and *4*. Equal relative pressure in compartments *A* and *B* is created by holding the liquid in ampules *7* at temperature $T_2 < T_1$. The degree of filling of pores with adsorbate is determined from the nitrogen flow passing through valve *2* into compartment *A*. The diminution of the pressure difference with time is registered by manometer *8*. Measurements are carried out from the value of $p/p_s = 0$, when all pores are empty, to p_0/p_s values corresponding to the termination of nitrogen flow, when all pores are filled with the adsorbate. For example the Kelvin radius $r = 50$ nm corresponds to $p_0/p_s = 0.98$ for water vapor. Using this method, it was demonstrated that in porous glass membranes the radii of the main portion of open pores were $r < 3$ nm.

Methods based on adsorption measurements of the specific surface S (cm^2/cm^3) of dry rigid porous bodies [14, 15, 19] are widely used together with methods of rarefied gas flow [20, 21]. The average radius of open pores $r_c = n/S$ can be determined using the S values and those of the porosity n.

When a gas flows in the Knudsen regime [21], the specific surface area measured using this method is equal to the outer surface of powder particles. This enables one, by comparison with adsorption data, to determine separately inner and outer particle surface areas when inner pores are present.

Obviously, the minimum pore size is defined by the discreet molecular structure of the substance. Cavities of molecular size cannot be considered as pores, they are a constituent part of real solids. However, a transition from such defects of molecular structure to pores is not sharp: in many cases it is difficult to distinguish between them. It is known that several monolayers may exhibit properties of the condensed phase [22]. On this basis, cavities with linear size larger than 0.5–0.7 nm may be considered as pores. Adsorption measurements set nearly the same lower limit. The notion of the specific surface area and its filling layer-by-layer with an adsorbate fails for pores with radii considerably less than 1.5 nm where the fields of surface forces overlap. Here sorption processes are considered rather as the space filling of micropores [15], that is for instance characteristic of zeolites.

Liquid flow

For non-rigid porous bodies irreversibly changing their structure during liquid evacuation, the data on porous structure can be obtained by measuring the rate of the liquid flow. Using the model of cylindrical capillaries of equal radii r, we can relate r to the filtration coefficient K (mol/cm) in the Darcy equation (see eq. (IV. 3)):

$$r = (8Kv_{\mathrm{m}}/n)^{1/2}, \tag{I.23}$$

where v_{m} is the molar volume of the liquid and n is the porosity of the specimen. This equation determines the equivalent pore radius of the model giving the same values of the liquid flow rate for the capillary and for pores of the specimen. However, it is difficult to identify the part of porous space participating in the flow. Values of the active porosity m may significantly differ from the total porosity including stagnant zones. To bring measured values of r closer to real ones, one should use values of m instead of n in eq. (I.23). Calculations based on the total porosity give underestimated values of the average pore size.

Radiotracer technique

The average velocity of the liquid flow in pores $v_{\mathrm{p}} = v/m$ can be measured in another experiment including not only the measurements of the flow rate v but the observation of tracing ions or molecules passing through the porous system. The flow rate is equal to $v = V/t$, where V is the volume of permeate and t is time. The tracer introduced into the flow should move at the same rate as the liquid and should not be adsorbed by pore surface. For bodies with wider pores ($\kappa h >> 1$), when EDL do not overlap, the application of ions as

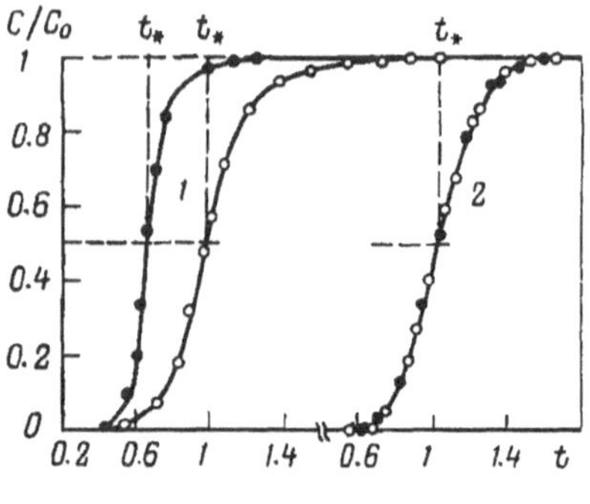

Figure 1.9 Change in concentration, C, of the outflowing radiotracer solution ($S^{35}O_4^{2-}$ – black points; HTO – light points) with time $t = V/V_0$ for the samples formed from silica gel particles (0.2–0.4 mm) (curves *1*) and samples of glass threads $d = 0.25$ mm (curve *2*).

tracers, having the same sign as the pore surface that excludes absorption, gives good results. Radioactive tracers are more often used [23] introduced at very low concentration that does not change the properties of a liquid or a porous body.

The tracer is introduced into the input solution where its concentration is kept at a constant level C_0. The outflowing solution is analyzed for the tracer content. As a result, the relation between the outflowing concentration C and the time t passed from the beginning of filtration may be obtained. Results for two different porous bodies are shown in Fig. 1.9. Experiments were carried out for two kinds of tracers: ionic ($S^{35}O_4^{2-}$) and molecular (HTO), where T is tritium, a radioactive isotope of hydrogen. The differences in the tracer flow are observed only in the case of biporous structure packing of granules of silica gel (curves *1*). Due to the charge effects, $S^{35}O_4^{2-}$ anions do not penetrate into the narrow pores of silica gel granules and pass only through the system of larger interparticle channels. On the contrary, HTO molecules are able to penetrate also into narrow pores. As a result, the time necessary to completely replace the liquid in the pores by the traced liquid here is much longer. The time it takes to achieve the concentration C_0 in the outflowing solution or relative concentration $C/C_0 = 1$ corresponds to the complete displacement. The dimensionless time $t = V/V_0$ is plotted on the abscissa axis (Fig. 1.9), where V is the volume of permeate and V_0 is the volume of the liquid in the specimen.

The average time t_* for the tracer to pass through the specimen corresponds to the concentration $C = 0.5C_0$. For a packing of glass threads (curve *2*), the time t_* corresponds to the rate of flow $v_p = L/t_*$ equal to $v_p = v/n$, where L is the length of the specimen, v is the measured flow rate and $n = 0.912$ is the total porosity of the specimen. This means that here the active porosity is equal to the total porosity $n = m$, and therefore there are no stagnant zones. The corresponding filtration radii can be found from eq. (I.23).

For porous bodies with blind or closed pores, one can use the values of t_* to determine the active porosity from the equation $m = t_*v/L$. So, for silica gel granules (curve *1*, black points) the t_* values for the ionic tracer not penetrating narrow pores are less than for the molecular species. The active porosity m corresponding to the intergrain space is (in this case) less than the total porosity n which includes narrow pores. Calculations give $m = 0.482$ and $n = 0.715$. The part of the volume occupied by narrow pores $f = 1 - (m/n) = 0.326$ obtained by the tracer method agrees with the independently measured porosity of the silica gel granules.

For the model of equal cylindrical pores, the displacement front must be step-like, because the blurring of the front caused by molecular diffusion is negligible at high flow rates. Therefore, the experimentally observed blurring of the displacement front (Fig. 1.9) reflects the distribution of tracer rates caused by the non-uniformity of the porous space. Non-uniform structure could be characterized by linear size (in the direction of the flow) of the unit volume λ that possesses all the structure peculiarities of the porous body. Parameter λ is determined by the derivative $(dc/dt)_*$ at $t = t_*$:

$$\lambda = L/4\pi(dc/dt)_*^2\,, \tag{I.24}$$

where the concentration c is expressed in the dimensionless form: $c = C/C_0$.

For granulated materials, the λ value is commensurate with the average size of particles. So, for uniform sand $\lambda \cong (1.5-2)d$, where d is the average diameter of sand grains. When the size distribution of particles is wider or particles have a non-isometric shape, the ratio λ/d increases reflecting a higher degree of non-uniformity of porous space. For example, packing of glass threads of a diameter $d = 0.25$ mm and length of 5 mm have the ratio λ/d of about 7.

If the specimen thickness L is much higher than λ, the filtration pathways are averaged and the average filtration radius of pores is a statistically acceptable characteristic of the porous body. On the contrary, the model of averaged filtration channels is not applicable at $L \sim \lambda$. Here the fractal characteristics of the porous body should be taken into account [24, 25].

In conclusion we would emphasize that the method chosen to characterize the structure of a porous body under consideration should comply with the investigation purposes. The method must determine those structural peculiarities that exert the major role under the given conditions.

Chapter II

Equilibrium of Liquids in Porous Bodies

The distribution of both liquid and vapor within a porous body in equilibrium with its environment depends on the sizes and geometry of the porous space and the external conditions. The general condition for equilibrium is the balance of changes of chemical potential $\Delta\mu$ (J/mol) of molecules in liquid and vapor phases compared to the standard state of the bulk liquid at the same temperature T:

$$\Delta\mu = -v_m \Delta P = RT \ln (p_s - \Delta p)/p_s = RT \ln p/p_s, \qquad \text{(II.1)}$$

where ΔP is the pressure decrease in the liquid placed in a porous body and $\Delta p = p_s - p$ is the corresponding decrease in vapor pressure, where p_s is the saturated vapor pressure over the plane surface of the bulk liquid. The decrease in the pressure may be caused by various reasons: curvature of the liquid–gas boundary generating a capillary pressure ($\Delta P = P_c$); action of a surface force field in thin liquid layers which in the case of wetting films can be reduced to the disjoining pressure ($\Delta P = \Pi$). The difference in pressure for solutions and pure solvents equals the osmotic pressure $\Delta P = RT\gamma C$, where C is concentration of the solute and γ is the activity coefficient. In all these cases, a decreased vapor pressure $p < p_s$ is established, according to eq. (II.1), over the liquid surface with pressure decreased by ΔP. We do not consider here the cases of $\Delta P > 0$ and $p > p_s$, because the states of supersaturation are unstable and rarely realized in porous bodies. For solutions of volatile components, the condition of equality of chemical potentials of all components of the liquid should be added to eq. (II.1).

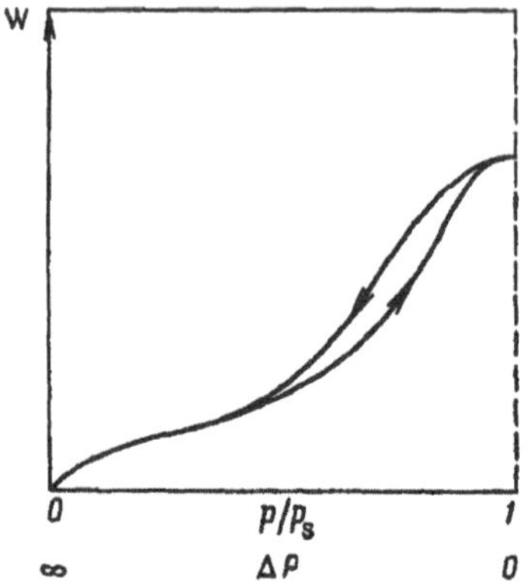

Figure 2.1 Qualitative example of the isotherm $W(\Delta\mu)$ for porous bodies.

The following states of a fluid in a porous body can be recognized: (a) adsorbed molecular layers on the pore surface with thickness from a fraction of a monolayer to several monolayers; (b) wetting films $h > 0.8-1$ nm thick, to which the theory of the liquid state and disjoining pressure are applicable; (c) pores filled with liquid; (d) vapor phase. Each of these states will require a particular approach to formulate equilibrium conditions because of the differences in surface force fields. In the first case, these are mainly short-range forces of intermolecular bonds. In the second case, the theory of long-range surface forces becomes applicable; wetting films may be regarded as a portion of a thinned liquid phase bounded on one side by solid substrate and on another by vapor. The state of the liquid filling the pores depends on the capillary pressure of confining menisci and the surface force field generated by pore walls. The general gas laws are applicable to the vapor phase except the case of narrow pores that are commensurate with molecule sizes.

Some characteristics of a liquid state in a porous body may be represented by the isotherms $W(\Delta\mu)$, where W is mass content of the liquid per unit mass of the solid phase. A typical example of such an isotherm is presented in Fig. 2.1, where p/p_s values or corresponding ΔP-values, in accordance with eq. (II.1), are plotted on the abscissa axis.

For porous bodies with average pore size $r \leq 0.1$ μm (with relative vapor pressure $p/p_s \leq 0.99$), the whole isotherm can be obtained by experimentally measuring the mass content W at various values of p/p_s. In the case of wider pores, measurements of capillary pressure ΔP are required to obtain the isotherm region relating to pore filling, because it is difficult to achieve and maintain relative vapor pressure near to the saturation point with a precision up to a thousandth part or less of p/p_s. A combination of adsorption and capillary methods enables the complete isotherm $W(\Delta\mu)$ to be obtained.

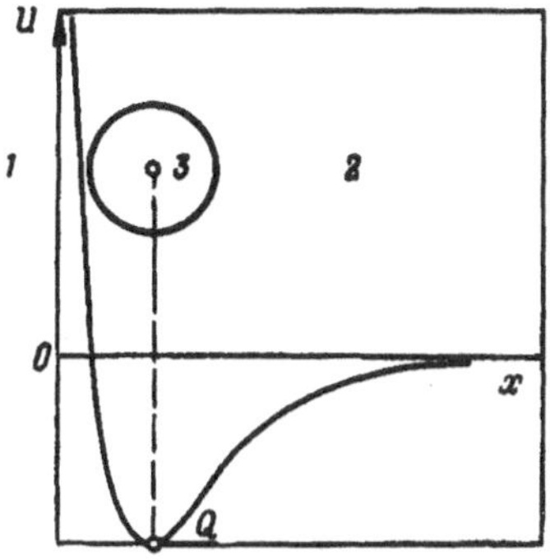

Figure 2.2 Potential energy $U(x)$ of interaction between the molecule and a flat wall.

Various states of liquids in porous bodies, depending on relative vapor pressures, are considered in the following sections.

§1 Adsorption Layers

In the region of relatively low vapor pressures, individual molecules adsorb on a pore surface. There is still no strict universal theory capable of describing quantitatively the adsorption on real surfaces. Let us consider the simplest case of adsorption of nonpolar molecules on a homogeneous molecular smooth substrate. The position of a molecule 3 in the region of a potential minimum corresponds to its adsorbed state (Fig. 2.2). The depth of the minimum Q is determined by the combined action of dispersion forces of attraction and repulsive forces caused by the overlap of electron shells of the molecule and molecules of the wall. Dispersion force theory gives the following equation for the energy of attraction between the molecule and the uniform wall *1* in the media *2*:

$$U_m = (\hbar/4\pi x^3)\int_0^\infty [(\varepsilon_2-\varepsilon_1)\alpha/(\varepsilon_2+\varepsilon_1)\varepsilon_2]d\xi, \qquad \text{(II.2)}$$

where x is the distance from the wall; $\varepsilon_1(i\xi)$ and $\varepsilon_2(i\xi)$ are frequency dependences of dielectric permittivities of the wall and the media; $\alpha(i\xi)$ is the frequency characteristic of polarizability of the molecule *3* in the medium *2*. In vacuum and in air, where $\varepsilon_2 = 1$, the equation simplifies. However, strictly speaking the theory of dispersion forces is applicable only to distances x much greater than molecule size. It is also difficult to describe quantitatively the forces of Born repulsion increasing approximately as $1/x^9$. Consequently, the curve $U(x)$ can be expressed only approximately, and various approaches have been used [19, 22].

When the adsorption of a particular molecule can be considered as a process that is independent of all other molecules, the Boltzmann equation is applicable for the calculation of molecular concentration in the field of the adsorbent:

$$C(x) = C_0 \exp[-U(x)/kT], \tag{II.3}$$

where $C(x)$ is molecule concentration (mol/cm^3) at a distance x from the wall and C_0 is the concentration in the bulk vapor, at $x \to \infty$, where $U(\infty) = 0$. The extent of adsorption Γ (mol/cm^2) can be determined as follows:

$$\Gamma = \int_0^\infty (C - C_0)\mathrm{d}x = C_0 \int_0^\infty \{\exp[-U(x)/kT] - 1\}\mathrm{d}x. \tag{II.4}$$

Expressing C_0 through vapor pressure ($C_0 = p/RT$), we find a linear dependence of adsorption on vapor pressure: $\Gamma = Kp$. This isotherm is called the Henry isotherm. Quantitative theoretical estimation of the Henry constant K requires, as seen from eq. (II.4), a knowledge of the full potential curve $U(x)$.

The Langmuir equation describes localized adsorption on energetically equal centers where sites already occupied are inaccessible for adsorption. This equilibrium is expressed by:

$$\Gamma(p) = \Gamma_\infty K_0 p/(1 + K_0 p). \tag{II.5}$$

Here Γ_∞ is the total number of sites where adsorption may occur. When $p \to \infty$, this equation transforms into eq. (II.4) with the Henry constant equal to $K = K_0\Gamma_\infty$.

The Langmuir equation allows the isotherm to be expanded up to the region where the adsorption becomes saturated and all sites on the surface are occupied at $pK_0 >> 1$. K_0 values in the Langmuir equation are related to the adsorption energy Q (Fig. 2.2) by $K_0 = b_0\exp(-Q/kT)$, where b_0 is a constant that depends on the molecule site area and temperature.

Brunauer, Emmett and Teller (BET) expanded the Langmuir approach to multilayer molecular adsorption introducing the assumption that the adsorption energy is equal to Q only for the first monolayer and for the next layers it is equal to the heat of condensation of the bulk liquid Q_0. The BET equation has the following form:

$$\Gamma(p) = -B\Gamma_{\infty}(p/p_s)/\{(1 - p/p_s)\,[1 + (B - 1)\,p/p_s]\}, \tag{II.6}$$

where $B = \exp[(Q_0 - Q)/RT]$. Using this equation, one can extend the adsorption isotherm up to much higher p/p_s.

Calculations considering the adsorption of multiatomic molecules require the description of their conformation in the adsorbed state taking into account the atom–atom potentials [19]. Microheterogeneity of a solid substrate adds significant difficulties to this problem. As a result, experimental rather than theoretical methods are generally used to investigate the isotherm $\Gamma(p/p_s)$. Attempts have been made to solve the inverse problem, that of determining the parameters of atom–atom potentials characterizing the energy of adsorption, comparing the isotherms of different molecules adsorbed on the same, well characterized smooth surface [19].

§2 Wetting Films

When the vapor pressure approaches saturation, the film thickness grows. At $h \geq 0.8-1.0$ nm, the adsorbed material should be considered to be wetting films, because they can be regarded as a part of a thinned liquid phase [1, 26]. To calculate isotherms $h(p)$ or $h(\Delta P)$ for wetting films, equations of the long-range force theory are used (Chapter I, §1). For wetting films of nonpolar liquids interacting with a substrate only through dispersion forces, the isotherm $\Pi(h)$ takes the form:

$$\Pi(h) = -A/6\pi h^3 = -(RT/v_m)\ln p/p_s. \tag{II.7}$$

The constant $A = A_{130}$ is equal to -2×10^{-20} J for hydrocarbon films on quartz and one order of magnitude higher on metal surfaces. For water on quartz, this constant equals -7×10^{-21} J, but in water films one needs to take into account electrostatic and structural forces in addition to dispersion ones.

Wetting films of hydrocarbons with molar volume $v_m = 150$ cm^3/mol on quartz surface are approximately 1 nm thick which corresponds, in accordance with eqs. (II.8) and (II.1), to the relative vapor pressure $p/p_s = 0.948$. The thickness $h = 0.8$ nm corresponds to $p/p_s = 0.88$. Therefore wetting films are formed when the relative vapor pressure in the surrounding media is $p/p_s > 0.9$. At lower vapor pressures, the surface bonded material should be considered rather as adsorption films. The analysis of isotherms of water vapor adsorption on quartz [1, 26] shows that the thickness of water films at $p/p_s \leq 0.9$ does not exceed a few molecular layers.

It follows from eq. (II.1) that the condition for equilibrium between a wetting film and a meniscus is the equality of capillary (under the meniscus) and disjoining pressure (in the film): $P_c = \Pi(h)$. In a wide plane slit (Fig. 2.3)

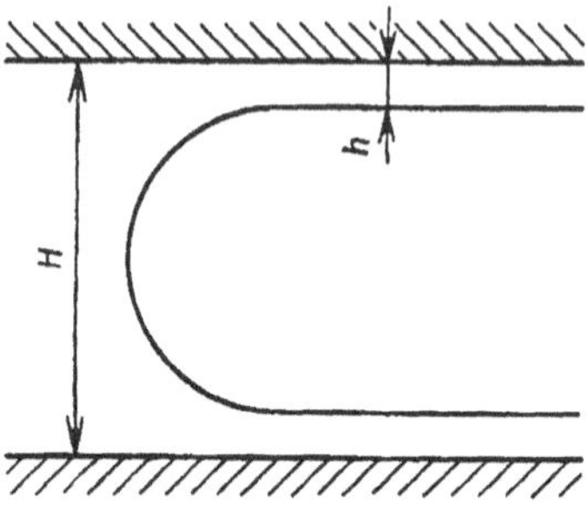

Fig. 2.3

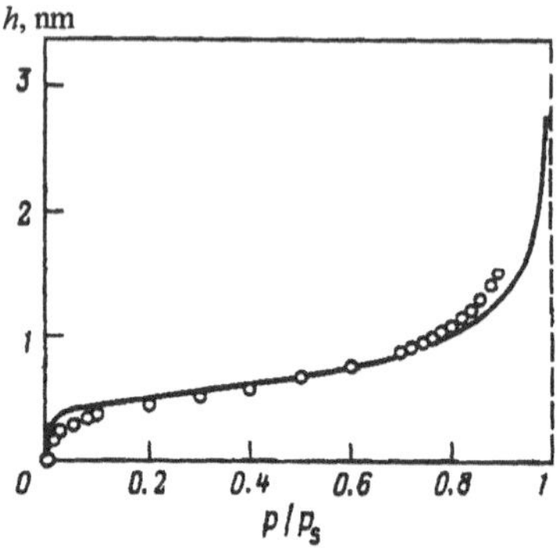

Fig. 2.4

Figure 2.3 Equilibrium of the capillary meniscus with the wetting films in a plane slit.

Figure 2.4 Isotherm for low temperature adsorption of nitrogen on nonporous Al_2O_3 (points). Solid curve is plotted using eq. (II.7).

this condition for a nonpolar completely wetting liquid can be written in the form:

$$2\sigma/H \cong -A/6\pi h^3. \tag{II.8}$$

This equation gives the dependence of the film thickness h in a slit pore on thickness $H >> h$ [26]. For nonpolar liquids, at $H = 10$ μm the thickness h is about 2 nm, increasing to 10 nm in slits 100 μm thick partly filled with condensate. Therefore, even in bodies with wide pores, the thickness of non-polar liquid films does not exceed tens of nm.

It has been shown experimentally that the lower limit of applicability of the isotherm (II. 8) is limited by $h \approx 0.5\text{–}0.6$ nm. For example, this takes place for argon films on graphite, helium films on quartz, polymers and fluorite, decane on Al_2O_3 [26]. However, experimental isotherms rather follow a power law $\Pi(h) \sim 1/h^n$ with the exponent ($n = 2.4\text{–}2.6$) less than 3 (see eq. (II.8)). There are several reasons for these deviations. The analysis of microstructure of polymolecular adsorption films shows that the film density is not uniform. Near to a solid substrate the density is higher and near to the boundary with the gas phase it gradually decreases. Considering the film to be a bilayer structure consisting of a dense inner monolayer and a less dense outer part leads to a decrease of the isotherm exponent to 2.2–2.6 [26]. The theory of molecular forces, on the basis of which eq. (II.8) was obtained, assumes that the film density is uniform throughout the film and that the liquid is incompressible.

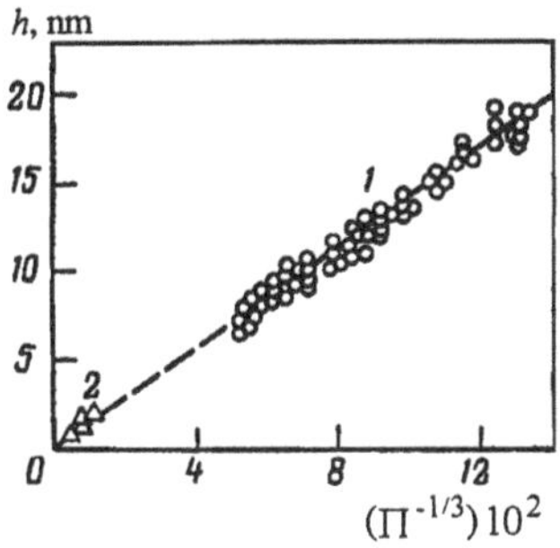

Figure 2.5 Isotherm of disjoining pressure of tetradecane wetting film on a molecular smooth mica obtained by thinning of the liquid interlayer (*1*) and by vapor adsorption (*2*). The values of Π are expressed in dyn/cm^2.

The second reason for possible deviations is that eq. (II.8) is valid only for molecularly smooth substrates. Surface roughness may cause deviations from the equation to the region of thicker films, resulting in a decrease in n. The influence of the roughness is significant if the film thickness is commensurate with the height of the roughness Δ. When $h >> \Delta$ and $h << \Delta$, the deviations are small because in the first case the film–gas interface remains practically planar and in the second case the film follows irregularities and its thickness remains uniform.

In Fig. 2.4, the experimental isotherm (points) of polymolecular adsorption of nitrogen (at 78 K) on nonporous alumina [15] is compared with theoretical values calculated according to eq. (II.8) (solid line). The constant A, obtained by fitting the experimental points, is equal to -2.8×10^{-19} J which is close to the theoretical value for liquid nitrogen on quartz. Equation (II.9) describes the isotherm sufficiently well up to the film thickness of 0.5–0.6 nm. Discrepancies at $p/p_s > 0.7$ relate to the onset of the influence of capillary condensation in pores.

The experimentally measured isotherm $h(\Pi)$ for nonpolar tetradecane films on a molecular smooth mica surface is shown in Fig. 2.5. Experimental points fit the isotherm (II.8) with $A = -5\times10^{-21}$ J coinciding with the constant calculated from spectral data for mica and tetradecane [26]. As seen from the graph, the isotherm is linearized in coordinates $h(\Pi^{-1/3})$ and its slope is proportional to $(-A/6\pi)^{1/3}$.

For wetting films covering spherical particles or inside cylindrical capillaries, the influence of film curvature should be taken into account by incorporating the contributions from capillary forces in addition to surface ones. The pressure P_f in a wetting film on a spherical surface with the radius r is equal to:

$$P_f = \Pi(h) - (2\sigma/r). \qquad \text{(II.9)}$$

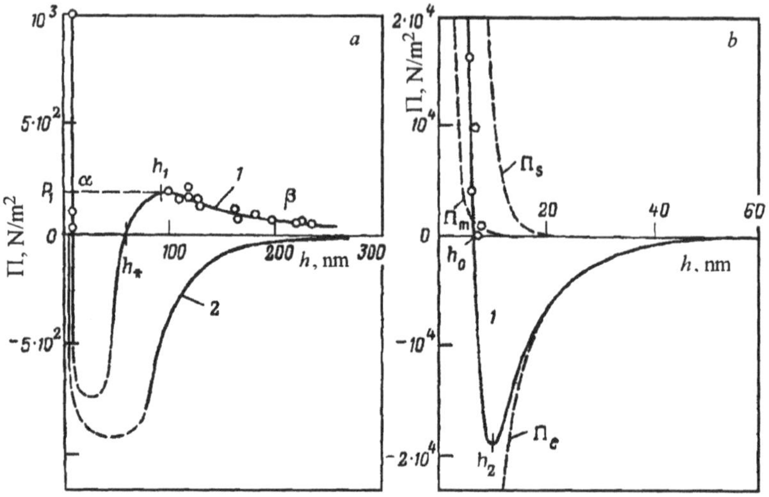

Figure 2.6 Isotherm of disjoining pressure $\Pi(h)$ of water films on quartz surface: *a*) region of high thickness (β-films), *b*) region of low thickness (α-films).

It is seen from the equation that the film thickness remains limited on a convex surface even at saturation, when $p/p_s = 1$ and $\Delta P = 0$. From the latter condition and eq. (II.9), one can determine the thickness of a nonpolar liquid film at saturated vapor pressure:

$$h_s = (Ar/12\pi\sigma)^{1/2}. \tag{II.10}$$

The film thickness does not exceed 5 nm for particle radii of about 0.1 μm.

Thickness of films covering internal surfaces of cylindrical capillaries of the radii r, when the surface is concave, is higher than that of the plane film under the same external conditions. Here we have:

$$P_f = \Pi(h) + (\sigma/r). \tag{II.11}$$

In the general case, the local thicknesses $h(x)$ of a film on a rough surface, in equilibrium with the vapor, may be found from the equation stipulating the constancy of pressure in all parts of the film:

$$P_f = \Pi(h) + \sigma K(x) = \text{const.} \tag{II.12}$$

where $K(x)$ is the local curvature of the film free surface, $\Pi(h)$ is the local

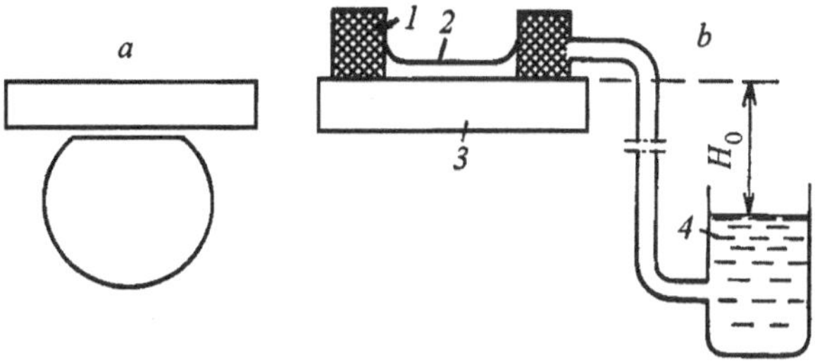

Figure 2.7 The experimental methods used to obtain isotherms of disjoining pressure of wetting films.

disjoining pressure. The film thickness is lower over convex and higher over concave regions as compared to flat areas. Nevertheless, such rough film surface corresponds to its equilibrium with the vapor phase because, according to eq. (II.1), constant vapor pressure p/p_s = const corresponds to constant pressure in the liquid phase $\Delta P = P_f$ = const.

Isotherms of disjoining pressure are more complex for wetting films of polar liquids where, besides molecular forces, electrostatic and structural components can act as well. Different signs of these components make the films unstable in some range of thicknesses that leads to the film rupture. The criterion of stability, according to Derjaguin, is the condition $\partial\Pi/\partial h < 0$ [1, 26].

§3 Isotherms of Disjoining Pressure of Aqueous Wetting Films

The experimental isotherm $\Pi(h)$ of water films on silica surface is presented in Fig. 2.6. The region of greater thicknesses $h > h_1$ was studied by the method of a floating bubble (Fig. 2.7, *a*) or by the suction of water from the film *2* (surrounded by meniscus) through the filter *1* (Fig. 2.7, *b*). A plane film was formed on the polished substrate *3*. The values of disjoining pressure can be determined from the capillary pressure for an equilibrium meniscus, P_c, that equals the hydrostatic pressure:

$$\Pi = P_c = \rho g H_0, \qquad \text{(II.13)}$$

where g is the gravity acceleration, ρ is the density of the liquid and H_0 is the distance from the level of liquid in the vessel *4*, movable in a vertical direction. The film thickness was measured by microinterference methods.

At the capillary pressure of the meniscus $P_c > P_1$ (Fig. 2.6, *a*), the films lost the stability (at $h < h_1$) and spontaneously transformed into the more stable state of thin films ($h < h_2$, Fig. 2.6, *b*). Because the thick film state is metastable, they are called β-films, and the thermodynamically stable thin films are called α-films. The region of the isotherm corresponding to thin α-films is presented on a larger scale in Fig. 2.6, *b*. Experimental points for α-films were obtained by ellipsometric measurements of their thickness at various relative vapor pressures.

There are no experimental points in the region of unstable film thickness $h_2 < h < h_1$ (where $\partial\Pi/\partial h > 0$). Furthermore, the region of supersaturation at $p/p_s > 1$, from h_0 to $h = h_2$, was not quantitatively investigated. It follows from qualitative observations that the thickness of an α-film, being in equilibrium with the drops condensed over it, changes negligibly and, therefore, it is confirmed that the isotherm $\Pi(h)$ crosses the thickness axis.

The isotherm $\Pi(h)$ was extrapolated into these experimentally unstudied regions using theoretical data [1, 9]. Equation (II.8) with $A = -7.2\times10^{-20}$ J calculated from spectral data for water was used to determine molecular forces and the results of numerical calculations with $\psi_1 = -150$ mV and $\psi_2 = -55$ mV were used to calculate electrostatic forces. Also the structural component was taken into account for which in eq. (I.16) $K_s = 2\times10^6$ N/m^2 and $l = 2$ nm. Consideration of the three components of surface forces resulted in the theoretical isotherm presented in Fig. 2.6 (solid lines *1*). As seen from the graphs, this satisfactorily fits the experimental points obtained for β- and α-regions of the isotherm. The contribution of the individual components is shown in Fig. 2.6, *b* by dashed lines.

The emergence of unstable thicknesses is related to a sign change of electrostatic forces at $h = h_* = 60$ nm (see eq. (I.15)). The return of the isotherm to the region of positive Π-values occurs because molecular forces Π_m, and mainly the forces of structural repulsion Π_s, prevail over those of electrostatic attraction Π_e of film surfaces. Because in the region $h > 20$ nm mainly the electrostatic forces are acting, one can control β-films stability varying surface potentials of the film. In the case of water or quartz, OH^- and H^+ ions determine these potentials, therefore a decrease of pH leading to the decrease of ψ_1 and ψ_2 potentials will make β-films unstable. The fact that at pH < 4 β-films do not form at all was confirmed experimentally. A rise in electrolyte concentration acts likewise, reducing the long-range electrostatic forces due to the growth of κ. At $C > 10^{-2}$ mol/l of KCl, α-films form immediately and the β-branch of the isotherm does not arise. In these cases the isotherm $\Pi(h)$ takes the form presented by curve 2 in Fig. 2.6, *a*.

The α-films are stable due to the contribution from the structural forces. Therefore, to control their thickness, one should first of all change the hydrophilic character of the substrate (for example, using methylation, surfactant adsorption, additives influencing water structure, temperature changes). A temperature rise up to 65–70°C diminishes the thickness of α-films to a monolayer. Hydrophilization of quartz surface, on the contrary, leads to a significant growth of the thickness of the α-film.

Because the critical pressure P_1 does not exceed 10^4-10^5 N/m^2, metastable β-films can exist only in bodies with relatively wide pores ($r > 0.1-1$ μm) and under the condition that a porous body originally saturated with water is gradually dehydrated. In the course of vapor adsorption only the thermodynamically stable α-films can form. Therefore in the majority of cases, capillary condensed water in porous bodies maintains direct contact with thin α-films.

The methods of obtaining isotherms of disjoining pressure $\Pi(h)$ for wetting films and the results of these investigations are described in more detail in the monographs [1, 26].

§4 Equilibrium between Menisci and Films

The content of liquid in a porous body W at rather high p/p_s is determined by the condition of equilibrium between adsorption or wetting films and the capillary condensate. These conditions are more complicated than was presented in Fig. 2.3. First, there is a transition zone between the film and the meniscus where surface and capillary forces act together; this changes the meniscus shape and makes the radius of curvature less than $(1/2H - h)$. Secondly, in narrow pores ($H < 0.1$ μm), where the fields of surface forces overlap, the meniscus is no longer a surface with a constant curvature. This does not allow the Laplace equation to be applied in the forms (I.22) and (II.9).

General equilibrium conditions for menisci and films were obtained from the variations of Gibbs free energy in the system meniscus–film in a plane parallel slit of thickness H taking into account the overlapping of surface force fields inside the slit [27]. In narrow pores, there is mutual attraction of films covering opposite surfaces of the slit due to molecular forces. Consequently, the film thickness depends not only on the relative vapor pressure in the outer media but also on the slit thickness H:

$$\Pi(h, H) = \Pi(h) + 2B_1/(H - 2h)^3 + B_2/(H - h)^3 = P_c = -(RT/v_m)\ln p/p_s. \quad \text{(II.14)}$$

Here $\Pi(h)$ is the isotherm of disjoining pressure of the film on a single surface (at $H \to \infty$), B_1 and B_2 are parameters depending on spectral characteristics of the slit material and the liquid: $\varepsilon_1(i\xi)$ and $\varepsilon_2(i\xi)$. In the case of complete wetting both parameters are positive. Equation (II.14) represents the expression for the isotherm for wetting films $h(p/p_s, H)$ and simultaneously defines the equilibrium capillary pressure of the meniscus without any assumption concerning its shape.

An analogy to the Laplace equation defining the relation between the slit thickness and the capillary pressure of the meniscus is the expression:

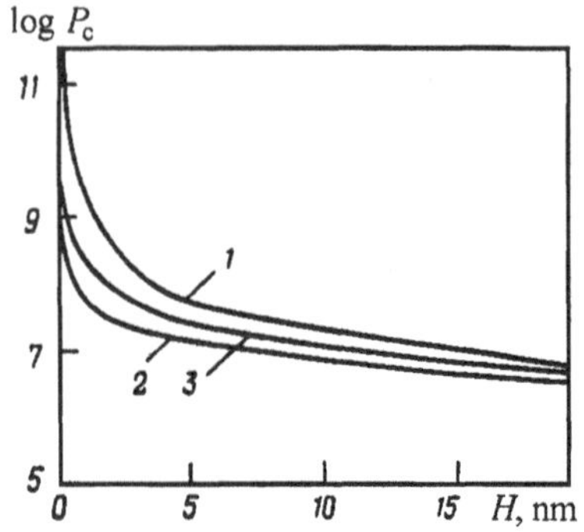

Figure 2.8 Dependences of the meniscus capillary pressure P_c (dyn/cm^2) on the width H of the plane slit calculated using various equations.

$$P_c = [2\sigma + g(h, H) - G(H)]/(H - 2h), \qquad \text{(II.15)}$$

where $g(h, H)$ and $G(H)$ are the excess specific free energies of interaction between different regions of the slit coated with films and filled with the condensate, respectively. Values of $g(h, H)$ can be determined from eq. (II.14). By definition (see eq. (I.15)):

$$g(h, H) = 2\int_h^\infty \Pi(h, H)\mathrm{d}h\,,$$

where the factor 2 reflects the existence of two films on opposite surfaces of the slit. Correspondingly,

$$G(H) = \int_H^\infty \Pi(H)\mathrm{d}H\,,$$

where $\Pi(H)$ is the isotherm of disjoining pressure in the liquid interlayer between slit surfaces.

The results of calculations of the meniscus capillary pressure P_c, as a function of the slit thickness H according to the exact equilibrium equation (II.15) (curve *1*) and to commonly used equations ignoring the overlapping of surface force fields: $P_c = 2\sigma/H$ (curve *2*) and $P_c = 2\sigma/(H - 2h)$ (curve *3*), are presented in Fig. 2.8. Comparison of these curves shows that the overlapping becomes significant when $H < 20$ nm.

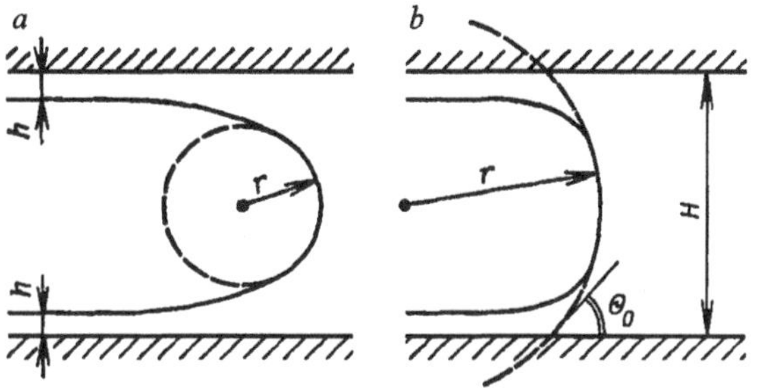

Figure 2.9 Complete (*a*) and partial (*b*) wetting of the slit surfaces with liquid.

Correction for the finite film thickness ($H - 2h$) brings P_c values closer to the curve *1*, but the disagreement with the exact theory still remains considerable. In pores of width H = 3–5 nm, the real capillary pressure is greater by one order of magnitude than values calculated using $2\sigma/H$ and 2–3 times higher than values calculated using $2\sigma/(H - 2h)$.

In wider pores ($H > 20$ nm), the interaction of films separated by the gap and the interaction of slit surfaces separated by a liquid can be neglected, i.e. $G(H) = 0$. In this case, $g(h, H)$ is not already a function of H and $g(h, H) = 2g(h) = 2\int_h^\infty \Pi(h)dh$. As a result, instead of (II.15) we obtain:

$$P_c = [2\sigma + 2\int_h^\infty \Pi(h)dh\,]/(H - 2h). \quad \text{(II.16)}$$

For films of nonpolar liquids, when the isotherm $\Pi(h)$ can be represented by eq. (II.8) of the theory of molecular forces, we have approximately:

$$P_c = 2\sigma/(H - 3h). \quad \text{(II.17)}$$

The difference between the radius r of meniscus curvature and $(H - 2h)/2$ is related to the transition region between the meniscus and the film (Fig. 2.9, *a*). This case corresponds to complete wetting: the meniscus radius is less than the slit half width and the contact angle fails to form. As seen from (II.16), the reason is a positive value of the integral in eq. (II.16), since the whole isotherm $\Pi(h)$ lies in the region $\Pi > 0$.

Partial wetting (Fig. 2.9, *b*), when the continuation of the meniscus forms the contact angle θ_0 with slit surfaces, occurs when the integral in eq. (II.16) is negative. This is possible when the isotherm $\Pi(h)$ (for example, in Fig. 2.6) goes to the region of negative disjoining pressures. The radius of the meniscus curvature r becomes greater than the slit half width H and its capillary pressure is equal to $P_c = \sigma/r$.

To find the contact angle θ_0, let us rewrite eq. (II.16) as

$$(P_cH/2) - P_c h = \sigma + \int_{h_0}^{\infty} \Pi(h)\mathrm{d}h\,.$$

It follows from geometrical considerations that $P_cH/2 = \sigma H/2r = \sigma\cos\theta_0$. Substituting this expression into the first term of the left side and bearing in mind that $P_c = \Pi$, we obtain:

$$\sigma\cos\theta_0 = \sigma + \int_{h_0}^{\infty} \Pi(h)\mathrm{d}h + \Pi_0 h_0. \qquad \text{(II.18)}$$

This equation relates the contact angle θ_0 to the equilibrium film thickness h_0, capillary pressure of the meniscus $P_c = \Pi_0$ and the isotherm of disjoining pressure of wetting films $\Pi(h)$. This is known in the theory of wetting as the Frumkin–Derjaguin equation, and these authors first gave its thermodynamic justification.

This equation may be extended to adsorption films. In this case it takes the form:

$$\sigma\cos\theta_0 = \sigma + RT \int_{p_0}^{p_s} \Gamma(p)\mathrm{d}\ln p, \qquad \text{(II.19)}$$

where $\Gamma(p)$ is the adsorption isotherm, R is the gas constant, T is the temperature. Equation (II.19) is derived by application of the Gibbs adsorption equation to the free surface energy of a solid covered with an adsorption film. Equation (II.19) transforms into (II.18) when the film thickness is substituted for the adsorption term $\Gamma = hv_m$ and the expression for vapor pressure p is taken from eq. (II.1).

To determine the contact angle, one should use the theoretically calculated isotherms $\Pi(h)$, because it is difficult to get experimental points in the region $\Pi < 0$ (Fig. 2.6). For example, calculations according to eq. (II.18) using the isotherm $\Pi(h)$ presented in Fig. 2.6 for water on silica surfaces give the contact angle $\theta_0 = 5°$ in agreement with experimental results. The calculation of contact angles using eq. (II.18) will be considered in more detail in Chapter X.

§5 Isotherms $W(\Delta\mu)$ for Porous Bodies

Having formulated the equilibrium conditions for the capillary liquid, adsorption and wetting films, we can consider the isotherms $W(\Delta\mu)$ for model and real porous bodies which reflect the dependence of porous body mass content W on the variations of the chemical potential $\Delta\mu$. These isotherms characterize the equilibrium state of a liquid in a porous body. They are also necessary for calculations of the coefficients determining the kinetics of mass exchange processes.

A typical dependence of W on $\Delta\mu$ was presented in Fig. 2.1. In the region of low p/p_s, the form of the graph is governed by vapor adsorption on the particle surfaces of the porous body. When p/p_s values increase, capillary condensation becomes possible within narrow pores and at grain contacts. Assuming that the interlayers with thickness $H \geq 0.8-1$ nm may possess the properties of a liquid phase, we can consider the phenomenon of capillary condensation. For example, water capillary condensation starts at $p/p_s \geq 0.3$. For nonpolar liquids ($\sigma = 35$ mN/m, $v_m = 100$ cm^3/mol), these values decrease to $p/p_s \geq 0.1$. The filling of micropores ($H < 0.8-1$ nm) should be treated by methods of statistical mechanics for molecular systems through computer modelling or in the framework of the theory of micropore space filling [15]. As distinct from adsorption, adsorbed molecules are disposed in a potential well in the middle of pores, and the process becomes similar to dissociation in a continuous condensed media.

Fisher and Israelachvili [28] showed experimentally that in the case of non-polar liquids with meniscus curvature radius $r \geq 0.6$ nm and $p/p_s \geq 0.1$, the equations for capillary condensation properly describe the process of filling space between two crossed cylinders, of radii about 1 cm, covered with molecular smooth sheets of mica. This is close to the estimations presented above. For water, deviations from the Kelvin equation took place even at $r \leq 5$ nm that were a consequence of changes of properties in thin layers.

It has been shown by methods of molecular dynamics that the discrete molecular structure of the liquid manifests itself even in droplets of radii 1–2 nm [29]. This leads to inconsistency of Kelvin and Tolman equations relating surface tension to the curvature of liquid surface. According to Tolman:

$$\sigma = \sigma_0[1 \pm (2\delta/r)], \tag{II.20}$$

where δ is the molecule size, σ_0 is the surface tension of the plane interface.

For concave capillary menisci, the surface tension increases with decreasing r, and vice versa for droplets. However, these effects may be compensated by appropriate changes in density of the liquid (its decrease in the case of the capillary condensate) because of changes of the capillary pressure. In this connection, one can hardly use the Tolman equation for correction of the surface tension values of the capillary condensate.

If in narrow pores ($H \cong 1$ nm) the adsorption monolayer is incomplete and the effect of the transition layer may be neglected, then the overlapping of surface force fields will cause, in the case of nonpolar liquids, only an increase of the equilibrium contact angle [27]:

$$\cos\theta = \cos\theta_0 - [(A_{101} - A_{131})/4\sigma H^2], \qquad \text{(II.21)}$$

where θ_0 is the contact angle at $H \to \infty$, determined by eqs. (II.18) and (II.19), A_{101} and A_{131} are molecular forces constants for slit surfaces interacting through the gas and the liquid, respectively. The second term of eq. (II.21) is always positive, because $A_{101} > A_{131}$ as it follows from eq. (I.3). For example, the second term equals 2% of the value of the first in the case of a decane thin layer between two quartz surfaces (at $H \cong 1$ nm). Although the bulk liquid decane wets quartz completely ($\cos\theta_0 \cong 1$), in a narrow slit its meniscus can form a finite contact angle $\theta \cong 18°$, and the vapor pressure over it correspondingly increases.

When the influence of the adsorption layer in equilibrium with the condensate cannot be neglected, the general equation of the theory (II.15) should be used. For rather wide pores (H or r is greater than 20–30 nm), the usual Kelvin equation is applicable with the bulk values of molar volume and liquid surface tension.

Capillary hysteresis

The reason for the hysteresis of capillary condensation, manifesting itself by the appearance of two branches of the isotherm $W(\Delta\mu)$ (Fig. 2.1), is the possibility of metastable states for the condensate and films. Condensate metastable states are caused by geometrical peculiarities of pore space such as the alternating widening and narrowing or blind pores of a special form (Fig. 2.10). States a and b correspond to the same value of $\Delta\mu$ and to the same capillary pressure $P_c = -\Delta\mu/v_m$, for a liquid bound in a pore by the menisci of equal curvature. However, the states b corresponding to more matter in the system are thermodynamically less favorable and, therefore, are metastable. The state a with a smaller amount of adsorbate relates to lower free energy of the system. The transition $b \to a$ may occur spontaneously, but the potential barrier must be overcome. The time of transition depends on the height and

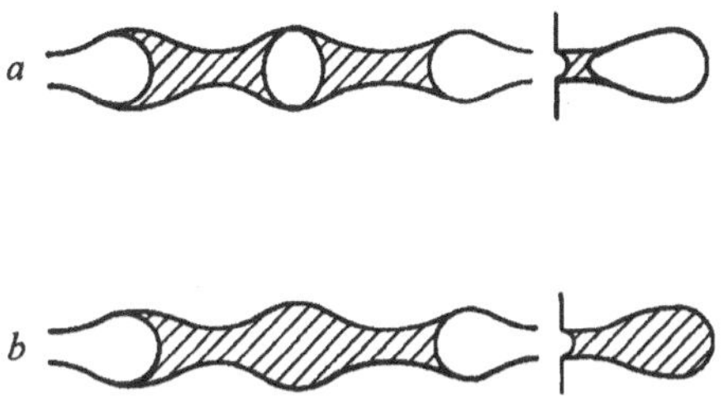

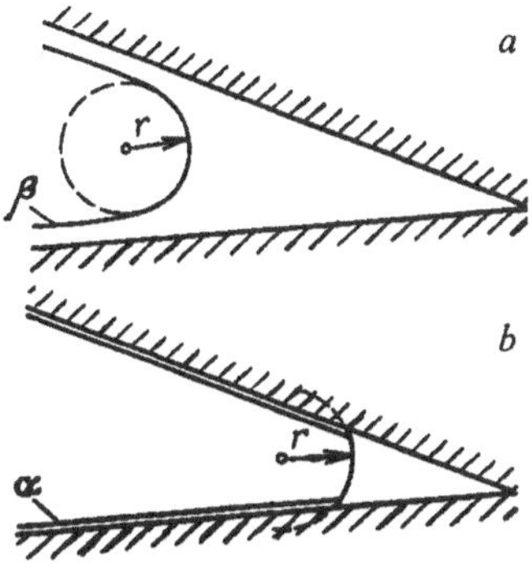

Fig. 2.10 Fig. 2.11

Figure 2.10 Peculiarities of porous structure leading to the hysteresis of capillary condensation.

Figure 2.11 Hysteresis of capillary condensation connected with the metastability of wetting β-films.

the width of the barrier. If this barrier is high, the metastable states (Fig. 2.1, top curve) can exist for a long time and become a physical reality. It has been proven experimentally that the transition can be accelerated by X-rays which create gas bubbles inside the metastable phase.

Another reason for hysteresis of the capillary condensation may be the metastable state of wetting films. As shown in Fig. 2.6, metastable β-films and thermodynamically stable α-films may exist at the same p/p_s-value when $\Pi < P_1$. The influence of β-film metastability appears in rather wide pores. Metastable β-films providing complete wetting can emerge behind the receding meniscus (Fig. 2.11, *a*), when the liquid was evaporated from pores. At the same vapor and capillary pressure, the equilibrium state with the α-films (Fig. 2.11, *b*) relates to less mass content *W* and, therefore, to the low branch of the isotherm $W(\Delta\mu)$. The film transition $\beta \rightarrow \alpha$ accompanied by the contact angle growth has been observed experimentally [26]. This has considerable impact on the rate of water evaporation from capillaries (see Chapter V).

The barrier between metastable β-state and α-state of the films lowers as P_c approaches P_1. Also the probability of nucleation of a more stable α-phase grows with the increasing area of β-films. This critical area depends on P_c and for capillaries of $r = 10$ μm is about 3×10^{-3} cm^2 growing to 1.5×10^{-2} cm^2 at $r = 30$ μm when the capillary pressure is lower. Films with greater area are unstable and may transform quickly into more stable α-films by breakage or thinning.

In cylindrical pores, hysteresis may be related to the metastability of the

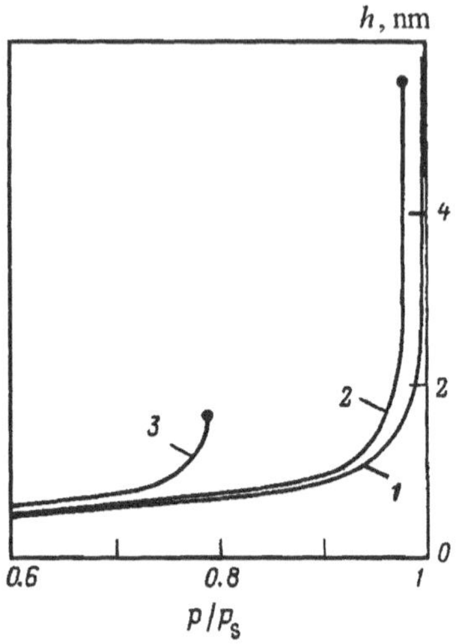

Figure 2.12 Changes in film thickness of a nonpolar liquid on plane surface (*1*) and in capillaries of radius $r = 100$ nm (*2*) and 10 nm (*3*) with increasing relative vapor pressure.

condensate in respect of the state of the wetting films. Dependence of film thickness on vapor pressure is expressed by eq. (II.11) which can be rewritten in the form:

$$\Pi(h) + [\sigma/(r-h)] = \Phi(h) = -(RT/v_m)\ln p/p_s, \qquad \text{(II.22)}$$

where $\Pi(h)$ is the isotherm of disjoining pressure for plane films.

The films lose the stability at $\partial\Phi/\partial h < 0$. This condition shows that the loss of stability, leading to film coalescence and to the filling of the pore by the condensate, occurs long before the thickness h approaches the capillary radius.

Dependences of the thickness h of films on a surface of the pore with $r = 10$ nm (curve *3*) and 100 nm (curve *2*), calculated by eq. (II.22), are presented in Fig. 2.12. For comparison, the isotherm for plane films is presented on the same graph (curve *1*). Calculations were performed for films of nonpolar decane ($v_m = 195$ cm^3/mol, $\sigma = 23$ mN/m) in quartz capillaries at 293 K, taking into account only molecular forces. The constant A in eq. (II.8) is taken equal to -3×10^{-19} J. The end point of the isotherms $h(p/p_s)$ in Fig. 2.12 corresponds to the loss of film stability. The filling of a capillary of $r = 10$ nm occurs at $h = 1.8$ nm and $p/p_s = 0.79$, as a result of a sharp growth of thickness fluctuations with loss of the stability of the films. Evacuation of the condensate begins at lower vapor pressure $p/p_s = 0.66$ corresponding to $P_c \cong$

$2\sigma/(r-h)$. When $p/p_s < 0.66$, the condensate becomes unstable in respect to the state corresponding to less mass content of the porous body (Fig. 2.12, *a*). Therefore, in the region of p/p_s between 0.79 and 0.66 (for the example presented) on the desorption branch the pores remain filled with condensate. Only the curvature of the restricting menisci changes leading to a small variation of mass content.

When a porous body contains cylindrical pores of various radii, a wide hysteresis loop of capillary condensation will arise. Because the capillary filling with condensate due to film coalescence or condensate breakage proceeds through vapor diffusion, these transitions may be retarded. As a result, equilibrium states of the condensate and the films cannot be quickly achieved, especially in the case of rapid evaporation or condensation. For this reason, equilibrium isotherms $W(p/p_s)$ are often useless for description of fast mass exchange processes.

Let us continue to consider the results presented in Fig. 2.12. In a pore of a larger radius (curve 2), the evacuation of the capillary will occur at $p/p_s = 0.96$, and its filling at $p/p_s = 0.98$. Consequently, it follows that the effect under consideration is more obvious in narrow cylindrical capillaries and pores.

Another reason for the hysteresis may be the metastability of adsorption film in respect to the condensate state in very narrow pores [27]. The situation is similar to that presented in Fig. 2.11. The only difference is that the reason for the loss of stability is a strong mutual attraction of the films through the space of a narrow pore. Exact calculations performed for pores having the shape of slits have shown that the effect appears when the slit width is below 2.5 nm. Both low and high boundaries of the capillary condensation hysteresis also arise. For cylindrical pores, a series of approximate solutions are known.

Thus, the hysteresis loop of isotherms $W(\Delta\mu)$ may be related both to geometrical effects and to metastability of states of wetting and adsorption films.

We can now distinguish some specific regions on the isotherm $W(\Delta\mu)$. At $p/p_s < 0.1$–0.2, the form of the isotherm is determined primarily by a vapor adsorption process on a pore surface. This region can be described by the appropriate respective adsorption isotherms $\Gamma(p/p_s)$ (see §1). The relation between the values of adsorption and mass content is given by the equation: $W(p/p_s) = S\mu_a\Gamma(p/p_s)$, where S (cm^2/g) is the specific surface area of the porous body related to the solid phase mass, and μ_a is the molecular mass of the adsorbate.

At relative vapor pressures, $0.1 < p/p_s < 0.9$, there is equilibrium between the capillary condensate and the adsorption films (§2). The form of the isotherm $W(\Delta\mu)$ in this region depends on peculiarities of porous structure; two specific examples will be considered further below. At $p/p_s > 0.9$, the sharp rise of film thickness (Fig. 2.4) leads to a change of equilibrium

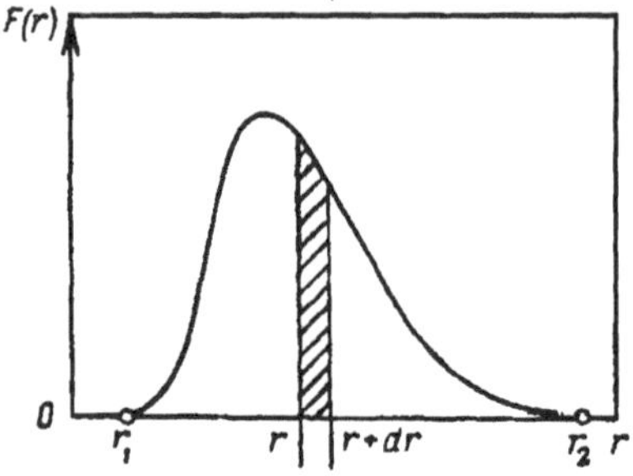

Fig. 2.13

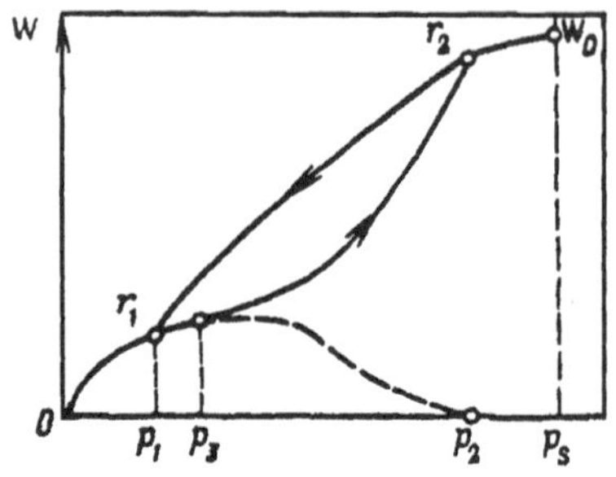

Fig. 2.14

Figure 2.13 Distribution function of cylindrical pores by radii.

Figure 2.14 Qualitative representation of the isotherm $W(p/p_s)$ for a porous body formed from cylindrical capillaries of different radii.

conditions between films and capillary menisci. In this region wetting films emerge; their behavior and properties are characterized by the isotherms of disjoining pressure.

Let us consider the desorption isotherm of a porous body, all pores of which are filled with a liquid restricted at $p/p_s = 1$ by menisci with plane surface. As p/p_s values decrease, at first there is only distortion of menisci at the outer surface of the body. Significant changes of W take place at those p/p_s values at which the evacuation of wider pores becomes feasible. Consider two models of porous bodies contrasting different features of pore space geometry. The first one, the model of straight cylindrical capillaries (Fig. 1.7, *a*), can illustrate the hysteresis phenomenon related to metastable states of wetting films (Fig. 2.11). Let us specify a distribution function $F(r)$ of pores by radii (Fig. 2.13) normalized so that:

$$\int_0^{\infty} F(r)\mathrm{d}r = 1 . \quad \text{(II.23)}$$

The product $F(r)\mathrm{d}r$, equal to the shaded region in Fig. 2.13, characterizes the portion of pore volume in the interval from r to $r + \mathrm{d}r$. As p/p_s decreases below 1, when all the capillaries are filled with a liquid and $W = W_0$, a change of mass content will occur at p/p_s determined from the Kelvin equation:

$$p_2/p_s = \exp\left[-2\sigma v_m/RT(r_2 - h)\right]. \quad \text{(II.24)}$$

Here r_2 is the largest capillary radius and h is the equilibrium film thickness determined by eq. (II.11) on the basis of the isotherm of disjoining pressure $\Pi(h)$ at $\Pi_0 = \sigma/r_2$. These calculations are correct only for completely wetting liquids and not too narrow pores.

At $p/p_s < p_2/p_s$, there is progressive removal of the liquid from capillaries of smaller radii being completed at a p_1/p_s value corresponding, according to eq. (II.24), to the smallest radius r_1 of cylindrical capillaries (Fig. 2.14). When $p/p_s < p_1/p_s$, wetting or adsorption films remain on the capillary surface. The isotherm region at $p < p_1$ is governed by reversible physical adsorption; and $W(p)$ only depends on the value of specific surface area and on the shape of the isotherm $\Gamma(p)$.

The filling of the capillaries with the condensate will occur with a delay, therefore the adsorption and desorption branches will not coincide. The region of the isotherm between p_1 and p_3 determines the changes in the mass content W which are caused by the growth of film thickness (Fig. 2.14). Feasible changes in the content W_f of liquid in the films are shown schematically by the dashed line extending the above-mentioned region. The mass content W_f decreases as pores are being filled with the condensate, despite continuing film thickness growth. Competition between these trends, thickness growth and the decrease of area occupied by films, produces a maximum on the dashed curve $W_f(p/p_s)$. The difference $W - W_f$ determines the amount of the capillary condensate in the porous body. The filling of the widest capillary will occur accordingly at vapor pressure $p > p_2$.

Therefore, to determine pore radii from the isotherm $W(p/p_s)$, one should use the desorption branch which describes the metastable states of both films and the condensate. All the information on pore radii distribution is contained in the region between p_2 and p_1 of the desorption isotherm. Certainly, this only applies to the model of straight capillaries. The presence of the conic pores will make such a hysteresis impossible or will cause significant narrowing of the hysteresis loop. For example, the capillary hysteresis discussed does not take place for completely wetting liquids in porous structures represented by a set of conic capillaries modelling the regions where particles contact.

Regular packings of spherical particles with equal radii represent a second model of a porous body (Fig. 2.15). Here the amount of the capillary condensate may be determined by summation of volumes v_i of the liquid bridges in the points of contact. Then the mass content of the condensate can be expressed by the equation:

$$W_c = v_i \rho N/2M, \qquad \text{(II.25)}$$

where N is the number of contacts per particle, ρ is the density of the liquid and M is the mass of a particle. For spheres with equal radii r and solid phase density ρ_s, taking the average number of contacts equal to $N = 6$, we have:

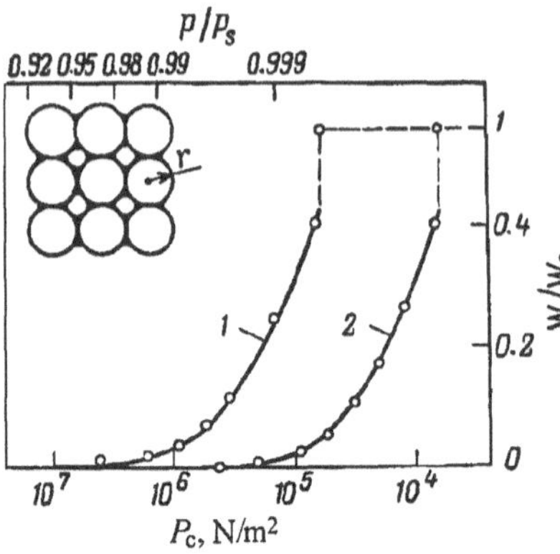

Fig. 2.15

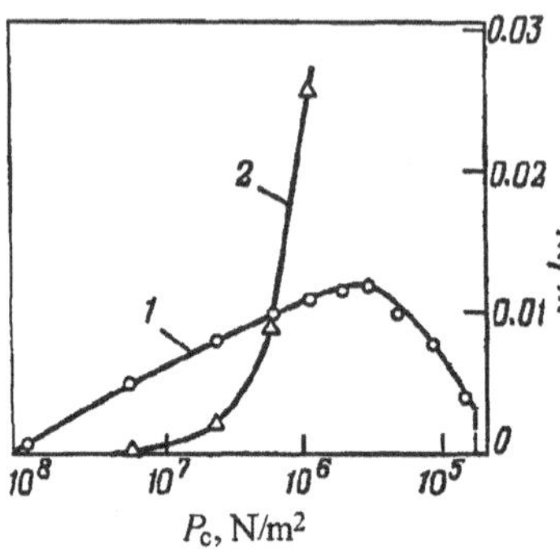

Fig. 2.16

Figure 2.15 Isotherms of relative mass content $W/W_0(P_c)$ for the model of hard spheres of radius $r = 1$ μm (*1*) and 10 μm (2).

Figure 2.16 Dependence of the relative volume of film liquid (*1*) and capillary condensate (2) on the capillary pressure P_c for the model of hard spheres of radius $r = 1$ μm.

$$W_c = [3\rho v_i/(4/3)\pi r^3 \rho_s] = 9 v_i \rho / 4\pi r^3 \rho_s. \qquad \text{(II.26)}$$

Calculations of v_i as a function of capillary pressure P_c may be performed on the basis of well-known numerical solutions described in [30, 31].

The results of calculations of $W/W_0(P_c)$ for the case of complete wetting are presented in Fig. 2.15, where W_0 is the mass content when all the pores are filled with the liquid. The values of W represent the sum of mass contents of capillary (W_c) and film (W_f) liquids. The latter is determined as:

$$W_f = 3h\rho(1 - s)/r\rho_s, \qquad \text{(II.27)}$$

where h is the film thickness and $(1 - s)$ is the part of particle surface coated with films. The values of s were found knowing the sphere area covered by the condensate. The film thickness was calculated according to eq. (II.8) at $\Pi = P_c$ taking $A = -10^{-19}$ J. The calculations were made for the liquid with $\rho = 1$ g/cm^3, $\sigma = 72$ mN/m and for two radii of particles $r = 1$ μm and 10 μm; and $\rho_s = 2.5$ g/cm^3.

A decrease in capillary pressure P_c and corresponding approach of p/p_s to 1 results in an increase in W values. At $W/W_0 = 0.4$, a coalescence of the capillary menisci takes place leading to spontaneous filling of the empty space with the liquid. This process is shown in Fig. 2.15 by the vertical dashed line. An increase of particle diameter results in the decrease of the mass content W at the same P_c or p/p_s values.

Part of the isotherm W/W_0 in the region of low mass contents is shown in Fig. 2.16 in a larger scale for the case $r = 1$ μm. The curve *1* corresponds to the mass of the film liquid W_f and the curve *2* to that of the capillary condensate W_c. At first, the amount of the film liquid increases, because of the film thickness growth with increasing p/p_s, then it declines because the area of the particle surface occupied by films reduces. When the liquid fills all the pores, $W_f = 0$. At high capillary pressure ($p/p_s < 0.985$) the amount of the film liquid exceeds the amount of the condensate. Although at $p/p_s >$ 0.985 the film mass is relatively small, the films generate (up to $W/W_0 \leq 0.4$) the main hydraulic resistance to the mass transfer, for the capillary bridges are separated by film regions. The peculiarities of condensation in the model of solid spheres, with account of transition zone between films and menisci, are described in more detail in [32].

It follows from this analysis that the shape of isotherms $W(\Delta\mu)$ differs for different models of porous structure. This enables one to choose the appropriate model of the porous body on the basis of experimentally obtained isotherms.

In conclusion, we note that the phenomenon of capillary condensation can be observed not only in the liquid–vapor systems, but also in the liquid–liquid systems, for example, in liquid binary mixtures near to the line of phase separation.

Chapter III

Properties of Liquids in Narrow Pores

In Chapter II the action of surface forces controlling the distribution of liquid in a porous body was discussed. However, the kinetics of mass transfer processes is also influenced by some changes of physical properties of liquids in thin layers such as viscosity, dielectric permittivities, density, surface tension and solvent power.

§1 Microstructure of Liquids in Thin Layers

The application of computers has made it possible to gain new information about the molecular structure of liquids by the methods of molecular dynamics and Monte-Carlo simulation.

The sense of the numerical experiments consists in setting definite intermolecular potentials to determine the time-averaged distribution of molecules in the interlayer that corresponds to the minimum free energy of the system and, therefore, its equilibrium state under the given conditions. First the potentials of molecular forces were tested on models of bulk liquids in the absence of surface forces, and those potentials were chosen which most closely represented the main physical properties of the bulk liquids. A good approximation for simple liquids, where molecules can be approximated by hard spheres of radii δ, is the Lennard–Jones potential of intermolecular interaction $U_1(r)$:

$$U_1 = -4\varepsilon^*[(\delta/r)^6 - (\delta/r)^{12}], \tag{III.1}$$

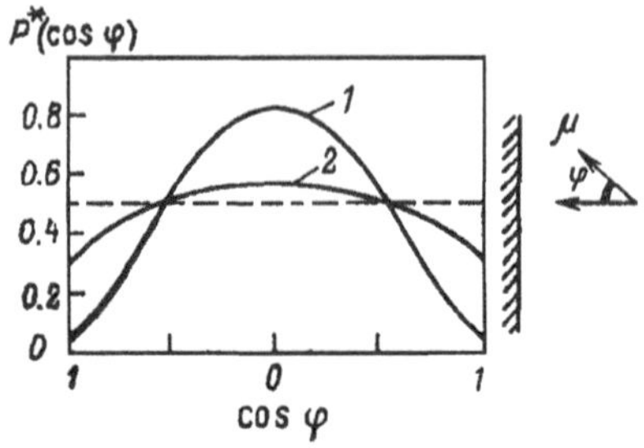

Figure 3.1 Distribution of probabilities P^* of realization of different angles φ between the dipole direction and the normal to the wall for the molecular layer near to the hydrophobic wall (*1*) and in the center of the interlayer (2) [37].

where r is the distance between molecular centers and ε^* characterizes the depth of the potential well.

The energy of molecule interaction with a wall, U, is expressed by eq. (II.2) or by various modifications of eq. (III.1). In the latter case, the integration of pair interactions between an individual molecule and all the molecules of the plane wall should be carried out, leading to the following equation:

$$U = -(4\pi\varepsilon^* n\delta^3/15)\,[(\delta/x)^3 - 1/3(\delta/x)^9], \qquad \text{(III.2)}$$

where n is the number of atoms per unit volume of the wall and x is the distance along the normal to the surface. The first term of the equation refers, as in eq. (II.2), to forces of molecular attraction. The second term characterizes the repulsion caused by overlapping of atom electron shells.

The case of polar associated liquids forming intermolecular hydrogen bonds is more complicated. For water, several expressions for the potential U were found which more or less successfully reproduce the physical properties and anomalies of the bulk water. Hydrogen bonds may be modelled, for example by a rigid system of four charges (two $q = +0.33e$ and two $q = -0.33e$) with addition of the Lennard–Jones potential for atom–atom interactions (O–O, O–H, H–H) [33]. The orientation of hydrogen bonds depends on the given positioning of charges.

The results of calculations [33–38] show that the microstructures of water boundary layers near to hydrophilic and hydrophobic surfaces are different. Near to a hydrophilic surface, water density is higher and the tangential mobility of molecules is lower. Near to a hydrophobic surface the opposite applies: the density is lower compared to the bulk and the tangential mobility is higher. The pressure P acting on the walls of a thin interlayer of thickness h may be considered, at a first approximation, proportional to the number of molecules, n_s, in the unit volume of the near-to-wall liquid layer: $P = n_s kT$, where k is the Boltzmann constant. For hydrophilic surfaces, $P > P_0 = n_0 kT$, where n_0 is the number of molecules in the near-to-wall liquid layer at $h \to \infty$. In water interlayers between two hydrophobic surfaces $P < P_0$, for in this case $n_s < n_0$. This gives a molecular-kinetic explanation for the different signs of a structural component of disjoining pressure $\Pi_s(h) = P - P_0$ and different signs of K_s in eq. (I.16) for water interlayers between hydrophilic and hydrophobic surfaces.

Structural changes in the water boundary layers consist not only (and not mainly) of density changes, but rather of the changes of their relative positions, the orientation of dipole molecules. Near to a flat solid hydrophobic wall, water molecules group by layers, with planes and dipoles for the most part oriented parallel to the wall. The distribution of probabilities P^* of dipole orientations characterized by $\cos\varphi$, where φ is the angle between the direction of the dipole μ and the normal to the wall, is presented in Fig. 3.1. The curve *1* shows the distribution of probabilities $P^*(\cos\varphi)$ for a molecular layer near to the wall and the curve *2* shows that for the center of the water interlayer of thickness h = 25 nm at 283 K [37]. In the bulk water phase all the dipole directions are equally probable, i.e. $P^* = 0.5$ (dashed line). As seen from the graph, near to hydrophobic surface the probability that $\cos\varphi = 0$ is higher corresponding to $\varphi = 90°$, i.e. to the dipole orientation parallel to the surface. In this case the dipoles are oriented in the opposite directions giving the resultant dipole moment equal to zero. The anisotropic, orientationally ordered, water state is retained in the center of the interlayer (curve *2*), although the extent of the orientational order is lower here.

Another orientational structure is generated in the vicinity of hydrophilic surfaces. Here the dipoles orient mainly perpendicular to the surface. The values $\cos\varphi = -1$ correspond to the molecule orientation with oxygen atoms towards the substrate. The reason for this orientation is the hydrogen bonding with surface OH-groups. The dipole orientation leads to the formation of the dipole potential.

Density oscillations caused by formation of a layered molecular structure are common for all, including nonpolar, liquids. The oscillations reflect the discrete liquid structure, the finite size of its molecules. The oscillations reduce near to deformable surfaces (liquid–gas) or in the case of rough solid walls. This reduction also takes place in the case of nonspherical molecules, for example those of ellipsoidal shape.

Even in the absence of surface forces, a flat solid boundary is a strong structuring factor acting like a defect disturbing the structure of liquid layers adjacent to the surface.

Surface curvature also influences the distribution of molecules near to a solid. In the case of a plane interlayer (dashed curve, Fig. 3.2), the density oscillates with a period close to the molecular diameter showing both

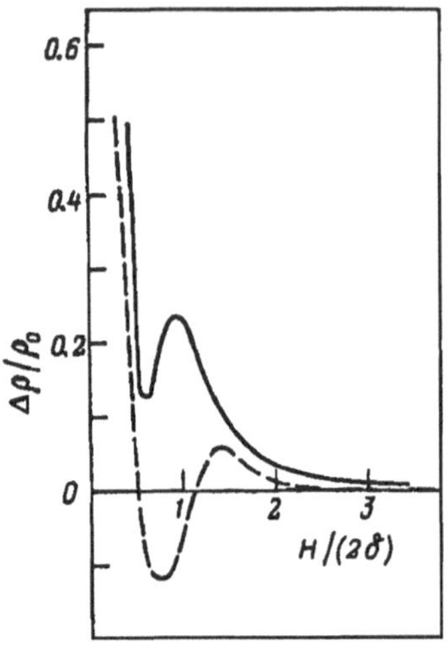

Figure 3.2 Dependence of relative changes $\Delta\rho/\rho_0$ of the average density of liquid (ρ_0 is the bulk density) on the characteristic linear size H of the system.

increased and decreased values compared to the bulk. The density of the liquid placed in a cylindrical pore (solid curve) is only increased and the period of oscillations is lower. These data refer to the model liquid consisting of rigid spherical molecules in the absence of intermolecular forces [39]. The values of H in Fig. 3.2 correspond to the specific linear size: halfwidth of the plane interlayer or the pore radius.

The existence of density oscillations and corresponding oscillations of disjoining pressure in thin interlayers is confirmed by direct experiments. The experimental results [40] obtained for the forces acting between two cylinders, of radius $R_0 = 1$ cm, coated with molecular smooth mica sheets and crossed at the angle of 90° are presented in Fig. 3.3. The period of oscillations of the force F is close to the molecule diameter of liquid. The F/R_0-values plotted on the y-axis are proportional, in accordance with eq. (I.18), to the specific energy of interaction of plane surfaces. The regions of the graph $F/R_0(H)$ corresponding to directly measured values of the force are shown by the solid lines. The points corresponding to maximum and minimum force values are connected by dashed lines related to the regions of instability. The points of intersection of solid lines with the axis $F = 0$ correspond to the stable state of the system for the crossed cylinders. The repulsion forces ($F > 0$) increase up to the peak values as the cylinders approach one another. When a local maximum is overcome, the transfer to the next solid line occurs as a result of extrusion of the molecular layer from the gap H between the surfaces.

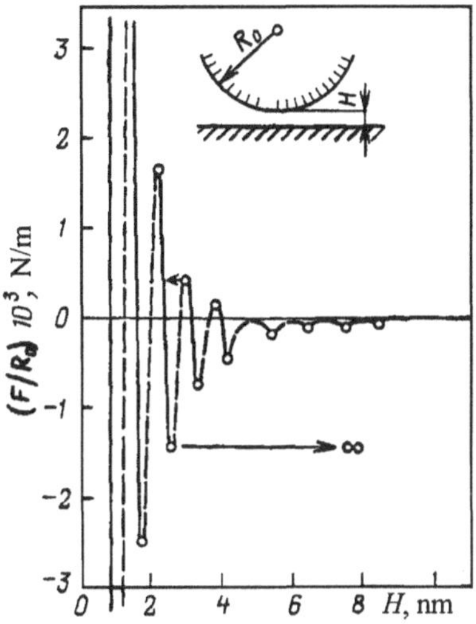

Figure 3.3 Results of measurements of the forces $F(H)$ acting between crossed cylinders of radius R_0, when they approach one another in nonpolar liquid with near-spherical molecules ($2\delta = 1$ nm) [40].

To separate the cylinders, it is necessary to overcome attractive forces ($F < 0$) acting between them. Having approached a force minimum, the cylinders separate spontaneously to the next equilibrium state with $F = 0$. To bring the cylinders together and to approach direct contact, it is necessary to overcome a series of force barriers. Thus, the extrusion of the liquid from the thinning interlayer proceeds in a stepwise manner.

The number of oscillations observed in liquids with quite rigid spherical molecules ranges from 6 to 10. The interlayer thickness from which the oscillations begin to become detectable lies in the range from 5 to 10 nm depending on the molecular diameter. For the more flexible, more easily deformable *n*-octane molecules, the oscillations manifest themselves only in the two closest surface layers.

Microstructural forces inside the interlayer bounded by hydrophilic surfaces give rise to potential barriers which must be overcome to establish a close contact between particles. These barriers may be, at least partly, kinetic in nature and associated with the viscous resistance to extrusion of several residual layers of molecules.

Layering effects are less distinct in the case of chemically heterogeneous or rough surfaces. In this case structural parameters change practically monotonically leading to exponential changes in structural forces with interlayer thickness according to eq. (I.16). The effect of oscillations of microstructural forces also follows from the results of computer simulations.

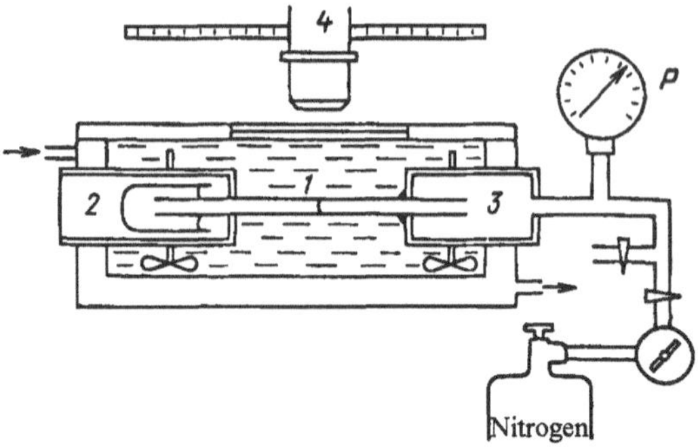

Figure 3.4 The experimental apparatus used for measuring the viscosity of liquids in microcapillaries.

§2 Viscosity of Liquids in Thin Interlayers

It is difficult to study experimentally the distribution of viscosity, density and other properties of a liquid near to a solid surface. In experiments with single capillaries or model porous bodies, only the values averaged over the total cross-section are measured. Some information about the distribution pattern may be gained from experiments with capillaries of various radii or porous bodies differing only by the mean pore size. Layer-by-layer measurements of boundary viscosity may only be carried out for nonvolatile liquids using a blow-off method [41]. This method was successfully used to study relatively thick boundary layers of heavy oils and polymethylsiloxanes on metal substrates.

The most reliable information on the viscosities of such volatile liquids as water, benzene and CCl_4 in thin layers was obtained by a method based on the measurement of flow rates in fine strictly cylindrical quartz capillaries of constant radius [42] which is a variant of capillary viscosimetry. The experimental scheme is presented in Fig. 3.4. A fine capillary *1* is placed into a thermostatted bath. One end of this capillary is dipped into a Teflon ampoule *2* with the liquid under investigation, another end is put into a chamber *3* where various gas pressures can be created and maintained. Nitrogen with controlled pressure was usually used for this purpose. Thermostatting was maintained by semiconductor thermoelements with a precision of 0.05°C.

The rate of liquid flow through the capillary v was determined by the time necessary to displace the meniscus to a distance Δx measured by a horizontal comparator. The apparatus was mounted under a reference microscope *4* on a movable table of the comparator. The results of measurements of water flow rates v as a function of applied pressure difference in left and right chambers,

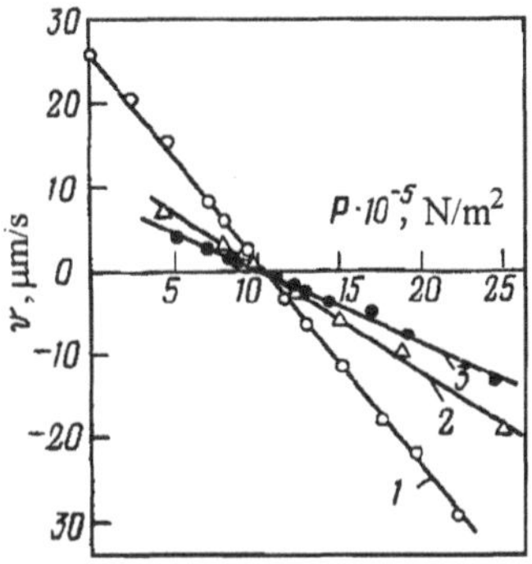

Fig. 3.5

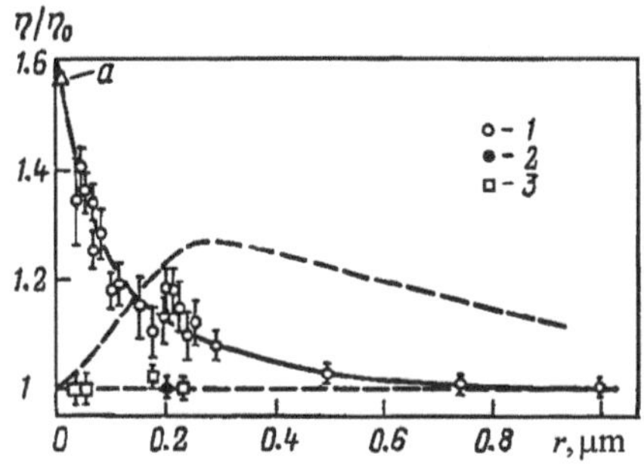

Fig. 3.6

Figure 3.5 Dependences of the rate of meniscus displacement $v(P)$ in the capillary ($r = 0.051$ μm) filled to various lengths l: 1.16 cm (*1*), 2.51 cm (*2*) and 3.75 cm (*3*).

Figure 3.6 Average viscosity η of water (*1*), CCl_4 (*2*) and benzene (*3*) vs. capillary radius; η_0 is the viscosity of bulk liquids.

P, are presented in Fig. 3.5. The measurements were carried out while filling the capillary at various length $l = 1.16$ cm (curve *1*), 2.51 cm (curve *2*) and 3.75 cm (curve *3*). Because the meniscus displacement Δx in the experiments did not exceed 0.01–0.02 cm, the corresponding changes of l were neglected. The dependences $v(P)$ obtained were linear and crossed the pressure axis at $P = P_c$, where $P_c = 2\sigma/r$ is the capillary pressure of the meniscus. When the meniscus advanced, complete wetting was achieved by the preliminary displacement of the meniscus at Δx backward to form a wetting film. The meniscus advanced ($v > 0$) at $P < P_c$ and receded ($v < 0$) at $P > P_c$. The linearity of the plots $v(P)$ was evidence of the absence of hysteresis and allowed the Poiseuille equation to be used to calculate the mean viscosity of liquid in the capillary:

$$\eta = r^2 \Delta P / 8 v l, \tag{III.3}$$

where $\Delta P = P - P_c$. The capillary radius was simultaneously determined from $P_c = 2\sigma/r$. Measurements of r by capillary pressure of nonpolar liquids completely wetting quartz (CCl_4, C_6H_6) and directly using a microscope (for $r > 5$ μm) gave results which essentially coincided with those obtained from P_c. The intersection of the plots $v(P)$ at one point for the various values of l confirms that the capillary had a strictly cylindrical bore.

The results obtained using this method for water (*1*) and two nonpolar liquids, carbon tetrachloride (*2*) and benzene (*3*), are shown in Fig. 3.6. As seen from the graph, the viscosity of water in fine capillaries is increased. The viscosity of nonpolar liquids in the same capillaries retain their bulk values. These measurements refer to capillary radii of $r \geq 30$ nm. As in the case of nonpolar liquids, the thickness of boundary layers does not exceed several molecular diameters, so that the changes in viscosity could be detected only in capillaries of much lower radius.

Recently, such measurements were performed by the method of picosecond technique in quartz glass with uniform pores of mean radius $r = 2.2$ and 5.4 nm [43]. The relaxation time t of the molecular rotational reorientation caused by a short light impulse ($\lambda = 590$ nm) was measured. The t-values are proportional to the viscosity: $\eta = tkT/v_0$, where v_0 is the molecule volume and k is the Boltzmann constant. For polar nitrobenzene completely wetting pore surface, the viscosity of surface layers 1.2 nm thick appears to be increased by a factor of three. The effect of the increased viscosity disappeared on lyophobization of the porous glass. Also no change in viscosity was found for nonpolar carbon disulfide partially wetting the pore surface.

Attempts to produce single quartz capillaries with radii less than 30 nm have failed. Glass capillaries cannot be used to measure water viscosity because of the strong influence of leaching and dissolution of the walls. One can reach the region of smaller radii r using porous bodies with an insoluble skeleton, for example silica gels and glass membranes. However, in this case the r values should be regarded as an estimation. The results of measurements show that water viscosity persists to grow (Fig. 3.6, point a) with radii decreasing down to 3–4 nm, and the rise in viscosity in the narrowest pores achieves some 60–65%. The increased water viscosity was found also in silica gel pores of radii 2–5 nm [44] and in nuclepore filters with cylindrical pore radii smaller than 15 nm [45]. It has been shown by methods of non-equilibrium molecular dynamics [46] that a marked increase in viscosity also takes place in a simple Lennard–Jones fluid in narrow slit pores. The reason for this viscosity rise is an increase in the fluid density near to the surface and the layered structure of the boundary layers. The viscosity approached the bulk values at slit thickness $H > 10\delta$, where δ is the molecular diameter.

The experimentally observed growth of viscosity agrees well with the results of computer simulations (§1) which predict decreased molecule mobility near to a hydrophilic surface, such as quartz. The good wetting properties arise through interaction of water molecules with surface hydroxyls by hydrogen bonds.

In hydrophobized capillaries the observed water slippage can be partly interpreted as a viscosity decrease in near-to-wall water layers [47]. This phenomenon will be considered in detail in Chapter IV. The enhanced mobility of boundary layers near to hydrophobic surfaces also appears in the results of computer simulations [37].

The observed increase of water viscosity in fine capillaries can be partly explained by the counter osmotic water flow in the capillary arising due to the

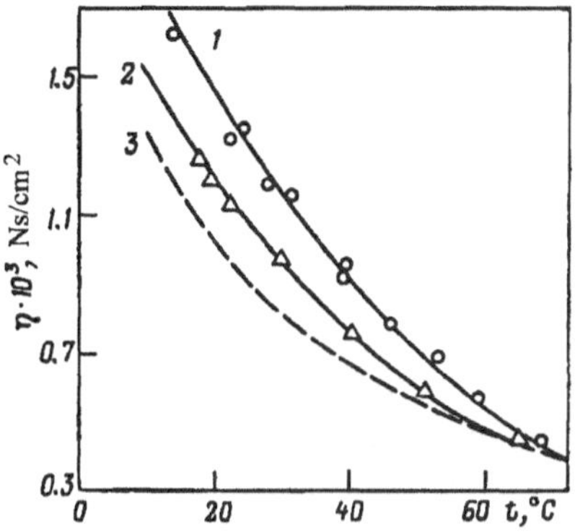

Figure 3.7 Temperature dependence of water viscosity η in quartz capillaries $r = 0.05$ μm (*1*), $r = 0.17$ μm (*2*) and in the bulk (*3*).

streaming potential, i.e. by the electroviscosity phenomenon. Theoretically, the effect is more pronounced at $\kappa r \sim 1-3$, where $1/\kappa$ is the Debye radius. A possible influence of electroviscosity on water flow ($\kappa = 10^5$ cm^{-1}) in quartz capillaries is shown in Fig. 3.6 by the dashed line. In calculations the value of surface potential $\psi = -150$ mV obtained for water in the same quartz capillaries was used. It follows from the theory [48] that the influence of electroviscosity should be most conspicuous in capillaries of radii 0.2–0.4 μm and become negligible at $r \leq 50$ nm. The disagreement of the dashed line and the experimental results of viscosity measurements shows that the electroviscosity effect was apparently absent under the conditions of the experiments. This might be connected with the fact that at a meniscus displacement by $\Delta x = 100-200$ μm the volume of flowing liquid is only about 5×10^{-12} cm^3 (at $r = 10^{-5}$ cm) which is quite insufficient for the generation of a considerable streaming potential. In capillaries of $r < 10^{-5}$ cm, the volume is still lower.

Therefore, the data on water viscosity in fine capillaries, shown in Fig. 3.6, demonstrate the existence of structural changes of boundary water layer and have probably no relation to electroviscosity. To be more certain, a theory of electroviscosity for the conditions of a single moving meniscus must be developed. In fine porous bodies, where $\kappa r \leq 0.035$, the electroviscosity plays a still smaller role. At these κr values the influence of electroviscosity does not exceed several percent even with the correction for nonlinearity of the Poisson–Boltzmann equation [48].

The structural mechanism of changes of water viscosity in fine pores is also supported by its temperature dependence. The dependence of measured values of the mean viscosity on the temperature, increasing from 20 to 70°C, is shown in Fig. 3.7. The curve *1* corresponds to a quartz capillary of radius $r = 0.05$ μm, and the curve *2* corresponds to $r = 0.17$ μm. The dashed curve *3* shows the change in the viscosity of bulk water with increasing temperature. Water viscosity in fine capillaries experiences sharper decrease with temperature, as seen from Fig. 3.7, due to the destruction of the modified structure of the boundary layer as a consequence of the enhanced thermal motion of the molecules. At temperatures higher than 65–70°C the destruction is completed, and water viscosity in fine capillaries practically does not differ from that of bulk water. The same effect of thermal destruction of modified water layer was found in narrow pores of silica gel [49], in finely porous glass [50], and in nuclepore filters [45]. The effect of temperature on water viscosity in thin capillaries and pores cannot be explained in the framework of the theory of electroviscosity.

Molecules of liquid crystals create many more extended ordered boundary layers near to surfaces. Their flow in fine capillaries was studied by the same method (Fig. 3.4). In wider capillaries, molecules of nematic liquid crystal MBBA are oriented normal to the surface creating a structure with the highest shear viscosity. At $v = 10^{-2}$ cm/s, the thickness of the oriented layers was $h_0 =$ 3.4 μm in the capillary with $r = 20$ μm and decreases in capillaries of smaller radii: $h_0 = 1.7$ μm at $r = 11.8$ μm and $h_0 = 1.2$ μm at $r = 8.3$ μm [51]. At higher flow rates the oriented structure begins to break down under the action of enhanced shear stress.

In much narrower capillaries with $r < 8$ μm, the nematic liquid formed a planar structure with the molecule axes oriented parallel to the capillary surface. Here the viscosity was the lowest and did not depend on the shear velocity, as for common Newtonian liquids.

Thick boundary layers near to lyophilic surfaces may be created not only by liquid crystals, but also by other polar liquids, including polymers. The thickness of normally oriented boundary layers of nitrobenzene and its derivatives makes up to 40–60 nm at room temperature [52].

§3 Solvent Action of Liquids in Fine Pores. Viscosity of Solutions

The problem of the influence of capillary radius (or mean pore size) on solution viscosity is very complicated. An essential feature is that a decrease of pore size causes changes of solute concentration compared to that of the bulk solution outside the pores. This is connected with a change in the solvent action of structurally modified water in boundary layers and fine pores. Therefore, the equilibrium composition and concentration of aqueous solutions in communicating narrow and wider pores may differ considerably.

The change in the solvent action of boundary water layers near to hydrophilic surfaces was, in particular, used for the determination of the content of so-called "bound water" [53]. In making the relevant quantitative estimations, it was assumed that the bound water completely loses its solvent power. However, proceeding from modern concepts of water microstructure

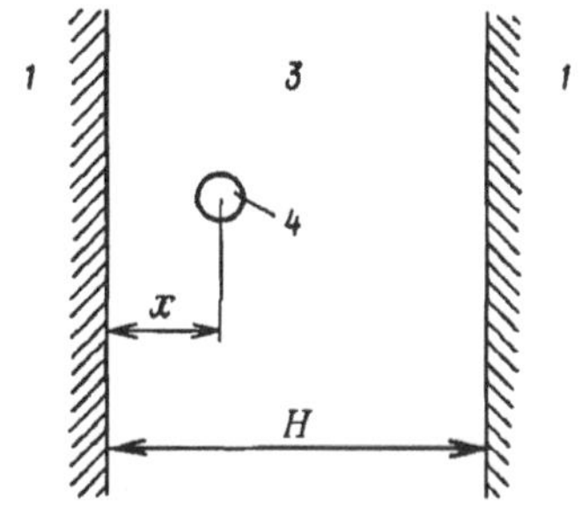

Figure 3.8 For calculations (in the text) of interactions between solute molecules and surfaces of a plane slit pore.

in thin layers, one can assume that water solvent power, like density and viscosity, changes gradually as the surface is approached. The increased density of water packing in boundary layers and the destruction of intermolecular hydrogen bonds must lower its solvent power due to the diminution of a number of voids, as for bulk water under high pressure.

However, water structuring is not the only reason for the change of solute concentration in fine pores. The second reason is the field of surface forces leading to redistribution of concentration of dissolved molecules across a narrow pore. This phenomenon takes place in the absence of solvent structural changes and may appear in all liquid mixtures.

Consider the interaction of a neutral molecule, *4*, of a solute with the surfaces of a planar thin slit (Fig. 3.8). The theory of molecular forces gives the following expression for the energy of a molecule in a narrow pore:

$$U(x) = -(\hbar/4\pi)[1/x^3 + 1/(H-x)^3]\int_0^\infty [(\varepsilon_1 - \varepsilon_3)\alpha_{43}/(\varepsilon_1 + \varepsilon_3)/\varepsilon_3]d\xi, \quad \text{(III.4)}$$

here x is the distance from the wall and H is the slit width; ε_1 and ε_3 are frequency dependences of dielectric permittivities of the walls and the solvent, respectively; α_{43} is the polarizability of the molecule 4 of the solute in the solvent environment.

If intermolecular interactions of the solute are neglected, i.e. at low concentration, the distribution of solute molecules across a slit pore may be expressed by the Boltzmann equation:

$$C(x) = C_0 \exp[-U(x)/kT], \quad \text{(III.5)}$$

where C_0 is solution concentration outside the field of surface forces (at $U(x) = 0$), for example at the center of a wide pore or in the bulk solution.

When $U(x) < 0$, solution concentration in pores will be increased. As seen from eq. (III.4), this occurs at $\varepsilon_1 > \varepsilon_3$, when the material of walls is more polar than the solvent. The reverse condition, $\varepsilon_1 < \varepsilon_3$, corresponds to a decreased solution concentration. To determine the average solution concentration in a pore, one should integrate eq. (III.5) over the range of x from $x = \delta$ to $x = H - \delta$, where δ is the molecule radius of the solute. Theoretical estimations show that for mixtures of nonpolar liquids the concentration of the less polar component can be approximately halved if the width of the interlayer between two silica surfaces is 1.5–2 nm [54]. Experiments with fine-porous membranes show that the concentration of water solution of sucrose in porous glass ($r = 1.8$ nm) is half that of the equilibrium bulk solution [55].

In the case of electrolyte solutions, the interaction between ions and the wall is mainly determined by electrostatic forces. When the pore surface is not charged, one of the effects is the action of image forces. The change of ion energy due to the image forces in a slit pore of width H is:

$$U(x) = [(ez)^2 (\varepsilon_{30} - \varepsilon_{10})/4\varepsilon_{30}(\varepsilon_{30} + \varepsilon_{10})][1/x + 1/(H - x)], \qquad \text{(III.6)}$$

where e is the electron charge, z is the ion valency; ε_{10} and ε_{30} are static values of dielectric permittivity of the slit material and the media surrounding the ion. Using again the Boltzmann equation (III.5), one can find the distribution of ion concentration $C(x)$ across the pore. In this case, a rise or decrease of ion concentration in a pore also depends on the sign of the difference between static dielectric permittivities.

The phenomenon of modified solvent structure, for example water, may lead to an additional change of ion energy related to the difference between the static dielectric permittivity in a fine pore ε_{p0} and its bulk value $\varepsilon_{30} = 80$. According to the Born equation:

$$U(x) = (ez)^2 [1/\varepsilon_{p0}(x) - 1/\varepsilon_{30}]/2r_i, \qquad \text{(III.7)}$$

where r_i is ion radius. To make calculations using this equation, one should know the dependence of ε_{p0} on the distance x from the walls. Taking as a first approximation some average value $\varepsilon_{p0}(H) < \varepsilon_{30}$ for given interlayer width, H, we see that in aqueous solutions the effect of Born energy also causes a decrease in electrolyte concentration in fine pores. Decreased dielectric permittivity of thin water interlayers was experimentally established for clays, silica gels, porous acetate cellulose membranes [1]. Although the estimations of ε_{p0} are not particularly reliable, for interlayers 2–3 nm thick different measurements gave values in a range from 20 to 30. A decrease in rotational mobility of water molecules in thin interlayers, pointing to a decreased static dielectric permittivity, is evident from microstructural computer simulations [36]. Experiments with reverse osmotic membranes show that electrolyte concentration in narrow uncharged pores ($r = 1-2$ nm) is reduced by a factor of 10–20 in agreement with calculations by eq. (III.6) and (III.7) [56].

If the surface of a pore is charged, the ion distribution across the interlayer changes. The direct action of the electric field should be added to the forces discussed above:

$$U(x) = ez\psi(x), \tag{III.8}$$

where $\psi(x)$ is a local value of the electric field potential. With this, the diffuse layers of counter ions compensating the surface charge of the walls are formed (Chapter I, §2). The theory of the electric double layer and corresponding electrosurface phenomena are described in detail in the monographs [48, 57].

A non-uniform distribution of solute concentration (in both molecular and ion forms) across a fine pore influences the viscous flow of solutions. A solution flowing into fine pores changes its concentration and consequently its viscosity. Particularly, the technology of reverse osmosis rejection [58] is based on the changes of solution concentration flowing through narrow pores.

If the structure of a pure solvent in fine pores is changed, the solute going into the pores may rebuild solvent microstructure and thereby influence the viscosity of the liquid in the pores. The same effects are known for bulk liquids. For example, the viscosity of aqueous solutions may be increased or decreased compared to the bulk [59] even due to the action of small amounts of additives. In the first case, the action of dissolved molecules and ions is said to be structure-stabilizing; in the second, structure-destroying. However, the same solutes may affect the viscosity of bulk water and its boundary layers differently, because water structure in fine pores differs from that of the bulk.

The results of measurements of the viscosity, η, of KCl solutions in porous glass with a mean pore radius about 10 nm [60] are shown in Fig. 3.9. The viscosity of solutions of low concentrations in pores considerably exceeds the viscosity of the bulk solution, η_0, because of the change of solvent structure. The viscosity of pure water was here 1.6 times higher than in the bulk (see Fig. 3.6, point *a*). The increase of concentration leads to a decrease in the solution viscosity in fine pores. At concentrations higher than 1 mol/l, the viscosity is practically not different from that for the bulk solution. Introduction of K^+ and Cl^- ions thus tended to destroy modified water structure in the fine pores.

In the case of bulk water, K^+ and Cl^- ions are also structure-destroying. However, the viscosity of the bulk solution, η_0, reduces only by 2% with increasing KCl concentration up to 1 mol/l. In fine pores, the destroying

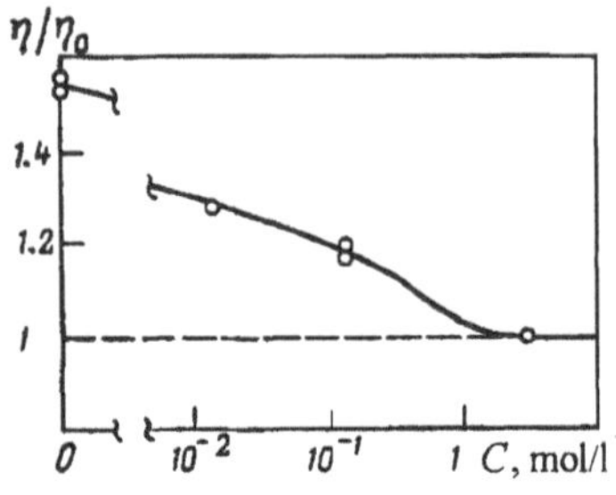

Fig. 3.9

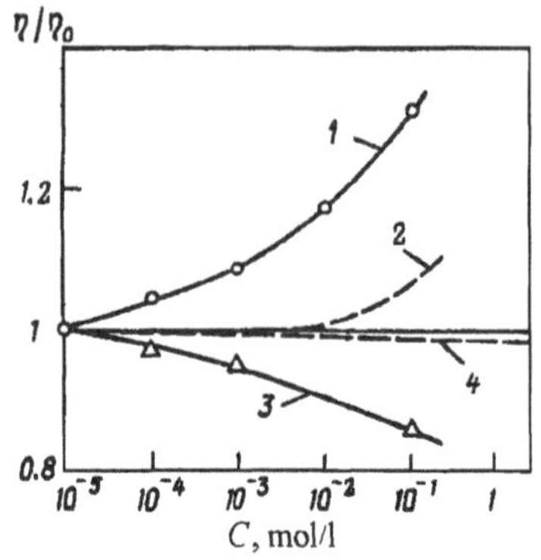

Fig. 3.10

Figure 3.9 Influence of KCl concentration on the relative viscosity of aqueous solutions in fine porous glass.

Figure 3.10 Change in viscosity η of sucrose aqueous solutions (*1*, *2*) and KCl (*3*, *4*) as compared with the viscosity of the bulk water η_0 for bulk solutions (*2*, *4*) and for solutions in thin pores of a glass membrane (*1*, *3*), $r =$ 1.8 nm.

action of these ions is more distinct (Fig. 3.9). If the ion concentration is rather high, the effect of their electric field on the water structure in thin interlayers surpasses the action of the solid surfaces. The weaker the impact of the surface force on water structure, the lower the electrolyte concentration that is capable of destroying the modified layer.

The rise in viscous resistance with increasing solution concentration (Fig. 3.10, curve *1*) observed in sucrose solutions flowing through narrow pores ($r = 1.8$ nm) was more distinct than in the bulk solutions (curve *2*). As seen from Fig. 3.10, the effect of solute on the viscosity is much more distinct in narrow pores than in the bulk. It is well known that sucrose is a structure-stabilizing additive, and its stabilizing action is more pronounced in the case of water in narrow pores.

Changes in viscosity of KCl solutions of various concentrations in pores of the same membrane are shown by curve *3* in Fig. 3.10. The viscosity of KCl solutions becomes less than that of the bulk water, η_0, as also the viscosity of the bulk KCl solutions (curve *4*). Thus, the influence of sucrose and KCl on solution viscosity is similar for bulk water and solutions in narrow pores. However, both effects, stabilizing (in the case of sucrose) and destroying (in the case of KCl), are quantitatively more pronounced for water in hydrophilic pores. The water structure in fine pores is more sensitive to additions of dissolved molecules and ions than the bulk water. Therefore, data for bulk solutions should not be used in estimations of viscosity of aqueous solutions in fine pores, because the discrepancies may be considerable.

In the experiments with model porous bodies and narrow capillaries discussed above, the liquid flow was purely Newtonian, exhibiting only a change in the shear viscosity. As shown in [61], bulk water also does not manifest non-Newtonian properties (within the limits of experimental error of the yield stress measurement, $\sim 10^{-7}$ N/m^2).

In conclusion we should consider the effect of colloidal particles and polymer macromolecules on solution viscosity in narrow pores. Adsorption of surfactants and polymer molecules on a pore surface leads to the narrowing of bores and consequent growth of viscous resistance to the flow [62]. If the walls do not adsorb the polymer, its concentration in the near-to-wall layer is lower reducing the viscous resistance to flow. Therefore, the viscous resistance to the flow of polymers and their solutions depends on the structure of their boundary layers, its positive or negative adsorption on pore walls. Flow of polymer solutions through narrow bores may be accompanied by macromolecule fractionation by size due to their different potentials of adsorption and different mobilities.

The influence of colloidal particles on the viscous flow of solutions through narrow pores varies depending on how the particles interact with the pore surface. If particles stick to the surface due to attractive surface forces, the pore will be fouled, thereby practically stopping the filtration. If particles are repulsed from pore surfaces (when pore surface and particles are of the same sign of charge), their presence in the central part of the pore impedes the flow the stronger the smaller is the residual gap between the particle and pore walls. For spherical particles of radius R in a cylindrical bore of radius r, a relative decrease in a liquid flow rate v, if the particle is in the center of the bore, can be determined using the established solutions [63]:

$$v/v_0 = 1 - [3rN(K_1 - K_2)/2l], \tag{III.9}$$

where v_0 is the Poiseuille flow rate in the absence of particles; N/l is the number of particles per unit bore length l. The coefficients K_2 and K_1 depending only on the ratio R/r are tabulated in [63]. For example, the change in flow rate will be less than 1% for particles of $R = 3$ µm in a capillary of $r =$ 10 µm ($R/r = 0.3$), if the distance between particles is $l/N = 100$ µm. For particles with greater radius ($R/r = 0.6$), the flow rate under the same conditions will decrease by about 20%. As seen from (III.9), the decrease of flow rate due to the presence of particles in the center of the bore will be greater, the larger the particle and the more particles that are within a unit bore length. If the particles are fixed while the liquid flows in a pore ($K_2 = 0$), the flow rate falls more drastically: by 10% in the first case and more than 100 times in the second.

Equation (III.9) may be also applied, as a rough approximation, to the flow of diluted solutions of large molecules. The rise of viscosity of sucrose solutions $C = 0.1$ mol/l ($\delta = R = 0.48$ nm) in the narrow pores of a membrane ($r = 1.82$ nm), calculated by eq. (III.9), is not more than 1%. This shows that the reason for the viscosity change of sucrose solutions (Fig. 3.10) is not due to hydrodynamic effects, but to the changes in water structure.

If the size of colloidal particles is small compared to pore radius, the flow of the diluted dispersion follows the Einstein equation:

$$\eta = \eta_0(1 + 2.5\varphi), \tag{III.10}$$

where φ is volume fraction of particles in the dispersion and η_0 is the viscosity of dispersion media free of particles.

An increase of viscosity ($\eta > \eta_0$) is connected to energy dissipation due to particle rotation in the flow where there is a velocity gradient. The rotational momentum is created by the difference of the forces applied on the different sides of the particle.

Particle rotation creates a force tending to displace them to the center of the bore. Therefore, near-to-surface layers have a lower concentration of particles and consequently lower viscosity. This leads to the effect of "slippage" similar to that of pure liquids with the difference that this appears here at the colloid rather than the molecular level. The reason for this may be also, as mentioned above, the repulsion of colloidal particles from the walls, their negative adsorption.

Equation (III.10) is applicable at $\varphi \leq 0.1$. For more concentrated systems, where hydrodynamic interaction among particles should be taken into account, only numerical solutions are known [63]. The next approximation is the addition of a quadratic term ($+7.52\varphi^2$) to the series. When the concentration becomes so high that, besides hydrodynamic interaction, surface forces considerably influence approaching and colliding of particles, the system is no longer Newtonian, manifesting a yield stress and a dependence of viscosity on the deformation rate.

§4 Density of Liquids in Fine Pores

The density of liquids in fine pores has been far less intensively investigated experimentally than viscosity. Although numerical methods (§1) give evidence for density changes in boundary layers, they are difficult to detect in a direct physical experiment. On the one hand, this is connected to the relatively far smaller changes than, for example, in viscosity measurements. If water viscosity, say, halves with a temperature increase from 20 to 60°C, the density changes do not exceed 1.5%. Viscosity is a more structure sensitive property than density. That is the reason why viscosity measurements have been used predominantly in the study of water structural changes in thin layers.

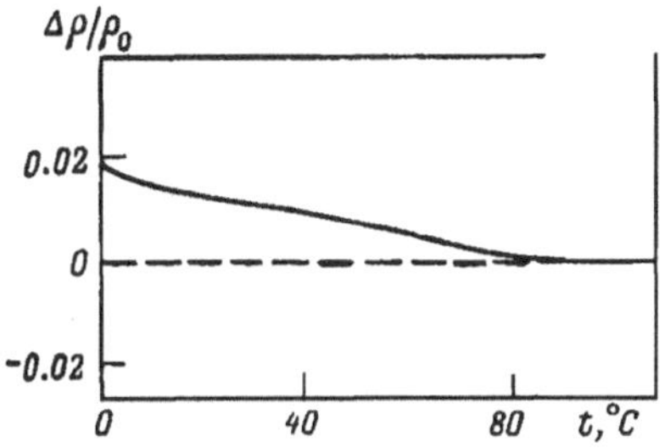

Figure 3.11 Temperature dependence of changes in water density in narrow pores of SiO_2 powder ($r = 5$ nm); ρ_0 is the density of bulk water.

A second difficulty arising in density measurements is that, to measure precisely the volume of a liquid in a pore, it is necessary to establish a geometric boundary between the liquid and the solid phase. Moreover, the position of this boundary may change depending on the size of liquid molecules. Larger molecules may fail to penetrate into some of the cavities of a real solid surface structure, and as a result different liquids may occupy different volumes in the same porous body. This makes it difficult to choose a reference liquid which can be used in pycnometric measurements of density. An ideal reference liquid should occupy the same volume as the liquid under investigation but differ from the latter in that its density in the pore space remains equal to the bulk value. Because of this problem, the data obtained by pycnometry are usually less reproducible.

Results obtained in the measurements of thermal expansion [64, 65] requiring no reference liquid should therefore be considered to be the most reliable. Thermal changes of liquid volume $V(t)$ are related to its density changes. If at some temperature water density in pores is known, then knowing the function $V(t)$ in a wide range of t the density can be determined at any other temperature. First it was established that the thermal expansion of water (H_2O and D_2O) in fine pores of silica (SiO_2) and rutile (TiO_2) powders differs qualitatively from the known dependence $V(t)$ for the bulk water. Measurements of the density are based on the use of V at $t \geq 70$°C as a reference point, when water properties in fine pores and in the bulk are assumed to be the same [65]. Taking water density in fine pores at $t \geq 70$°C equal to that in the bulk, the density dependence $\rho(t)$ may be calculated at lower temperatures, at which water structure in fine pores differs from the bulk.

The results of these calculations are presented in Fig. 3.11. It follows from them that the differences, $\Delta\rho$, between the densities in pores and in the bulk are about 2% at 0°C, decreasing to 1% as the temperature rises to 40°C. Increased water density in narrow ($r \sim 5$ nm) hydrophilic pores of SiO_2 powder qualitatively agrees with the results of computer simulations (§1). In the finer pores of zeolites the density rise achieves 7–20% depending on the method chosen to calculate volume of the pores [66].

Unfortunately, not all porous bodies and liquids can be used for measurements of thermal expansion. Sufficient accuracy can be provided only when the thermal coefficient of expansion of solid phase χ_T is much lower than corresponding values of χ_T for the liquid in pores. In experiments described above, the Aerosil powder ($\chi_T = 0.4 \times 10^{-6}$ 1/deg) was pressed up to 100 MPa (to provide dense packing of particles) into an Invar vessel, an alloy with a very low $\chi_T \sim 10^{-6}$ 1/deg.

Theoretical estimations of liquid density in thin interlayers can be made using methods accounting for the isothermal compressibility χ_P of a liquid within the field of surface forces of the solid substrate. Quantitative estimations were approached in the framework of the theory of dispersion forces [67, 68] restricting application of the method only to nonpolar liquids when electrostatic and structural forces could be neglected. The density of a simple liquid placed in a thin interlayer between two walls is increased $\rho > \rho_0$, if the interaction between its molecules is weaker than that with molecules of the walls. In the framework of the macroscopic theory of dispersion forces this condition can be written as $\varepsilon_1(i\xi) > \varepsilon_3(i\xi)$ and $\chi_P[\partial \ln \varepsilon_3(i\xi)/\partial \ln \rho_3] > 0$, where indices 1 and 3 refer to the wall and the liquid, respectively [68]. The numerical calculations [68] have shown that density of liquid argon in the middle part of a silver pore, $r = 1$ nm, is higher than that in the bulk by 10%. The effect decreases with decreasing pore radii: to 2% at $r = 2$ nm, and to 0.8% at $r = 3$ nm.

In distinction from silica powders and zeolites, measurements with clays gave a 2% lower water density in interlayers of 2.5 nm thick [69]. The method used here consisted in determination of the changes in the volume of a water saturated clay under external pressure with reference to the volume of extruded water. The reasons for a decrease of water density in clays are still not clear, in spite of water viscosity in these pores being significantly increased [70], as in quartz capillaries. It was also revealed that there was a decreased isothermal compressibility c_P of water in clay pores [71] which disagrees with the observed decrease in water density. As known, c_P values for bulk water decrease if it is compressed by external pressure.

Measurements of dielectric relaxation time of water in thin interlayers between lipid membranes have given viscosity values enhanced by 35% at interlayer thickness $H = 5$ nm, and by 90% at 3.6 nm [72]. Similar effects were observed in micelles and microemulsions.

Chapter IV

Flow in Pores Filled with Liquid

Changes in structure and composition of boundary liquid layers influence not only the velocity of their flow in narrow pores under a pressure gradient ∇P, but also result in a number of additional effects: capillary osmosis (flow under a concentration gradient of solutes ∇C), electro-osmosis (flow under a gradient of electrical potential ∇E), thermo-osmosis and thermocrystallization flows associated with a temperature gradient ∇T. These phenomena can reveal themselves either purely or accompany a pressure driven flow, resulting in gradients of concentration, electrical potential and temperature.

Therefore we should account for all the gradients to solve the problem of mass transfer in porous bodies. It is necessary to consider not only mass fluxes of different components of a solution, but also fluxes of energy and electric charge (electric current). Assuming the flow rates to be low, which is usually the case for bodies with narrow pores, a solution to the problem can be found in a framework of the thermodynamics of irreversible processes (TIP) [73, 74]. Fluxes of mass, heat and charge generally disturb a statistical equilibrium. However, if the disturbing processes proceed more slowly than those of relaxation, we can deal with a local equilibrium, that occurs in a physically small volume. TIP is based on the principle of local equilibrium. The approach becomes increasingly more satisfactory as the ratio is reduced between the rate at which the state is changed due to external forces, and the rate at which equilibrium is reestablished due to internal relaxation processes. Because relaxation times in gases and liquids are short (10^{-10} to 10^{-12} s), the application of TIP equations is not restricted to very small deviations from equilibrium of a whole system. This enables TIP equations to be used to elucidate many real processes. In the case of small deviations from equilibrium, one can use the Onsager assumption of a linear relation between fluxes and corresponding gradients. With the exception of chemical reactions, the application of this assumption is rather wide. Applied to mass transfer in porous bodies completely filled with liquid, the appropriate equations of linear TIP for binary solutions are as follows:

$$V = \alpha_{11}\nabla\mu_1 + \alpha_{12}\nabla\mu_2 + \alpha_{13}\nabla E + \alpha_{14}\nabla T/T, \text{ mol/cm}^2\text{ s}; \quad \text{(IV.1)}$$

$$Q = \alpha_{21}\nabla\mu_1 + \alpha_{22}\nabla\mu_2 + \alpha_{23}\nabla E + \alpha_{24}\nabla T/T, \text{ mol/cm}^2\text{ s};$$

$$J = \alpha_{31}\nabla\mu_1 + \alpha_{32}\nabla\mu_2 + \alpha_{33}\nabla E + \alpha_{34}\nabla T/T, \text{ A/cm}^2;$$

$$W = \alpha_{41}\nabla\mu_1 + \alpha_{42}\nabla\mu_2 + \alpha_{43}\nabla E + \alpha_{44}\nabla T/T, \text{ J/cm}^2\text{ s};$$

Here V is the solvent flux (component 1); Q is the solute flux (component 2); $\nabla\mu_1$ and $\nabla\mu_2$ are the corresponding gradients of chemical potential of the components related to the pressure and concentration gradients of the solute as follows:

$$\nabla\mu_1 = -v_m \nabla P; \quad \nabla\mu_2 = -(\partial\mu_2/\partial C)_p \nabla C, \text{ J/mol cm}, \quad \text{(IV.2)}$$

where v_m is the molar volume of the solvent.

Phenomenological coefficients α_{ik} conform to the Onsager relationships of the symmetry $\alpha_{ik} = \alpha_{ki}$, that links cross coefficients. For example, the coefficient α_{12} characterizes capillary osmosis and the cross coefficient α_{21} represents the convective flow of the solute; the coefficient α_{13} represents the velocity of electro-osmosis and α_{31} represents the streaming current in the porous body; α_{14} characterizes thermo-osmosis and α_{41} characterizes the heat of the transfer. The physical identities of the other phenomenological coefficients will be discussed later. The symmetry relations follow from the invariability of the motion equations to time inversion and from the hypothesis of fluctuation decay stemmimg from macroscopic laws.

Symmetrical coefficients α_{ii} characterize the direct effects: filtration (α_{11}), solute diffusion (α_{22}), electrical conductivity (α_{33}), temperature conductivity (α_{44}). The TIP equations do not determine kinetic coefficients α_{ik}. Explicit expressions for these coefficients can be obtained only in a framework of the molecular kinetics approach. The purpose of further consideration is to find relevant relations for the phenomenological coefficients based on the equations of microhydrodynamics and on the theory of long-range surface forces, as applied to different models of porous structure. This consideration will concern basic, practically important mass transfer processes.

§1 Filtration

We will first restrict our consideration to the case of flow under the pressure drop in the absence of other forces. For Newtonian liquids, the flow must obey the Darcy law:

$$V = -\alpha_{11} v_{\mathrm{m}} \nabla P = -K \nabla P/\eta \text{, mol/cm}^2\text{ s,} \tag{IV.3}$$

where K is the filtration coefficient and η is the viscosity. The K value depends, essentially, on pore size and the peculiarities of pore space geometry in a given specimen. For the model of straight cylindrical capillaries of different radii r:

$$K = (m/8v_{\mathrm{m}})\int_0^{\infty} r^2 F(r)dr \text{, mol/cm,} \tag{IV.4}$$

The relevant corresponding expression for α_{11} can be easily obtained from this equation. Here m is the active porosity of a specimen, and $F(r)$ describes the pore distribution by radii (Fig. 2.13). The product $F(r)dr$ characterizes the portion of the total pore area (in parts of unity) within the interval from r to $r + dr$. As seen from eq. (IV. 4), high values of $F(r)$ and r give the major contribution to the integral. Therefore special attention should be paid to the region of the maximum and to the right decreasing branch of the graph $F(r)$. The contribution of narrow pores is generally insignificant. However, they can play an important role in mass transfer processes driven by other gradients.

Specifying the pore shape and the structure of pore space, we can obtain other expressions for K and, correspondingly, for the phenomenological coefficient α_{11}.

In the case where all pores have nearly the same or closely similar radii $r = r_{\mathrm{c}}$, eq. (IV.4) simplifies to

$$K = mr_{\mathrm{c}}^2/8v_{\mathrm{m}}; \quad \alpha_{11} = mr_{\mathrm{c}}^2/8\eta v_{\mathrm{m}}^2. \tag{IV.5}$$

When this expression for K is substituted into eq. (IV.3), the latter transforms into the well-known Poiseuille equation for viscous liquid flow in a single capillary. Equation (IV.5) is often used to obtain the average pore size in porous bodies (Chapter I, §3).

When considering porous bodies with narrow pores, one often has to account for the changed viscosity in boundary layers and its distribution over a pore radius. Solving the Navier–Stokes equations for the variable viscosity $\eta(y)$, that is a function of the distance from the pore wall, gives the following expression for the coefficient α_{11} (at $r = r_c = \text{const}$) [75]:

$$\alpha_{11} = (m/v_m^2 r_c^2)\int_0^{r_c} y\,dy \int_y^{r_c} y\,dy/\eta(y), \tag{IV.6}$$

where y is the radial coordinate.

We should define the function $\eta(y)$ before carrying out further calculations. Statistical treatment of the experimental relationship between the average water viscosity and the capillary radii (Fig. 3.6) showed that the function $\eta(y)$ obeys the following exponential law:

$$[\eta(y) - \eta_0]/\eta(y) = a\exp[-b(r_c - y)]. \tag{IV.7}$$

Here η_0 is the viscosity of bulk water beyond the boundary layer; parameters a and $1/b$ characterize the viscosity near the pore wall (at $y = r_c$) and the length of the region of the changed viscosity (i.e. some effective thickness of the boundary layer), respectively. Substituting eq. (IV.7) into eq. (IV.6) and integrating, we obtain:

$$\alpha_{11} = (mr_c^2/8\eta_0 v_m^2)[1 - 8a(1/2B - 3/2B^2 + 3/B^3 -$$

$$- 3/B^4 + 3/B^4 \exp B)] = mr_c^2/8\eta_c v_m^2, \tag{IV.8}$$

where $B = br_c$. As seen from a comparison of this relation with eq. (IV.5), in the case of boundary layers with increased viscosity one can use the redefined effective value of viscosity $\eta_c > \eta_0$, being a function of parameters a and b and depending on the average capillary radius r_c. Application of this equation is restricted to the values of radii $r_c > (1/b)$, when the boundary layers do not overlap. For water in quartz capillaries, $a = 0.47$ and $b = 6.35 \times 10^5$ cm^{-1}. Here the application of eq. (IV.8) is limited to pore radii $r > 2 \times 10^{-6}$ cm. In the case of the narrower pores, one can use the average values of viscosity η_c given in Fig. 3.6.

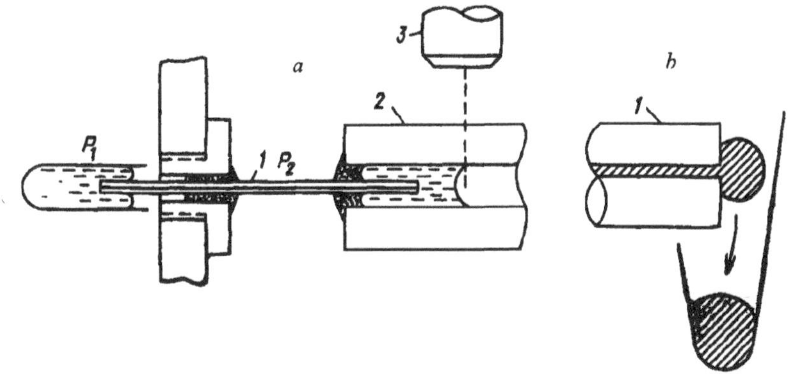

Figure 4.1 Apparatus for measuring liquid slippage in lyophobic capillaries, water (*a*) and mercury (*b*).

The influence of decreased viscosity of water boundary layers (that takes place in the case of a flow in capillaries with hydrophobic surface) can be accounted for by the introduction of the velocity of surface slippage v_0 into the Poiseuille equation. By this approach, the region of decreased viscosity can be reduced to the region of slippage that coincides with the capillary surface. The account of the distribution of the decreased viscosity $\eta(y) < \eta_0$ would be a more strict approach, but this dependence has not yet been experimentally studied. However, using fine hydrophobized quartz capillaries, estimations of the slippage coefficient γ for water were obtained in [47].

The Poiseuille equation adjusted for slippage takes the form:

$$v = r^2\nabla P/8\eta_0 + v_0 = (r^2\nabla P/8\eta_0)(1 + 4\eta_0\gamma/r) =$$
$$= r^2\nabla P/8\eta_c\text{, cm/s.} \qquad \text{(IV.9)}$$

The slippage coefficient $\gamma = v_0/\tau$ characterizes the ratio of the slippage velocity to the shear stress on the capillary wall equal to $\tau = r\nabla P/2$. As seen from eq. (IV.9), we can use again some effective value of viscosity η_c, with the only difference that here $\eta_c < \eta_0$.

The smaller the pore radius, the more pronounced the slippage effect. Quantitative estimations of the slippage coefficient were obtained using narrow quartz capillaries, whose inner surface was made hydrophobic by treatment with trimethylchlorosilane from solution in benzene. Figure 4.1 shows the experimental apparatus. One end of a capillary (*1*) which has been made hydrophobic is dipped into water placed in the ampoule. The ampoule is located in the chamber, where a pressure P_1 can be created exceeding

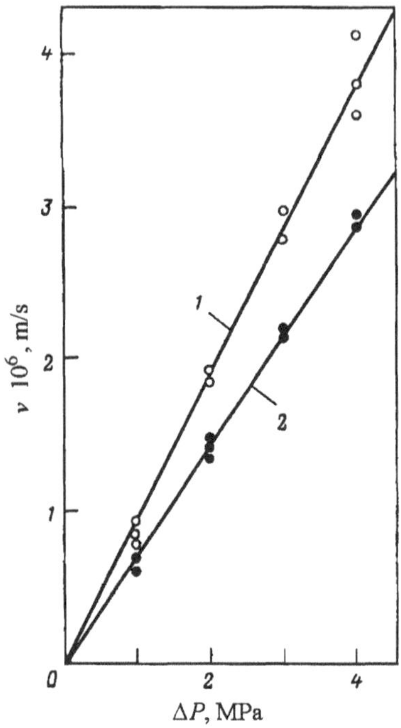

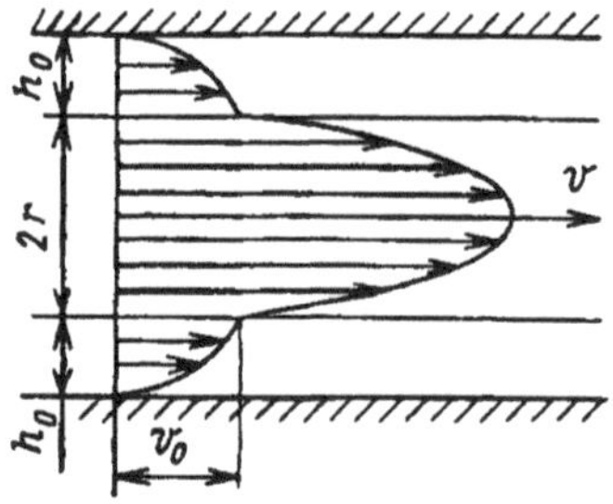

Fig. 4.2 Fig. 4.3

Figure 4.2 Flow rates of water (*1*) and CCl_4 (*2*) in a hydrophobed quartz capillary of radius $r = 0.88$ μm and length $L = 5.71$ cm at 23.7°C.

Figure 4.3 Distribution of flow rates $v(z)$ in the capillary when a fine porous layer of thickness h_0 exists on its surface.

atmospheric pressure P_2. The second end of the capillary is connected to a wider hydrophilic capillary (2). The velocity of water flow in this calibrated capillary is determined by the shift of the meniscus using a microscope (*3*). Direct measurements of the velocity of the meniscus shift in the hydrophobic capillary had low reproducibility because of the wetting hysteresis. For comparison the velocity of mercury flow in quartz capillaries was also measured, where the initial surface is not wetted by mercury. Here the flow rate was measured by collecting and weighing the mercury outflow (Fig. 4.1, *b*).

The slippage velocity was determined by a comparison of the velocity of water flow with that of a completely wetting liquid (carbon tetrachloride) in the same hydrophobic capillary. Using known values of the bulk viscosity of CCl_4, the average capillary radius was determined, so that only the value of slippage coefficient γ remains unknown in eq. (IV.9).

The results of comparative experiments, where the velocity of a meniscus shift in the calibrated capillary was measured for water (curve *1*) and for CCl_4 (curve 2), are presented in Fig. 4.2. The bulk viscosity of these liquids was practically the same within 1% at 23.7°C. Therefore the higher velocity of water flow can only be associated with water slippage.

As a result of tens of measurements, values of slippage coefficient equal to $\gamma = (3 \pm 1)\times10^{-5}$ m^3/N s for water in hydrophobic capillaries were obtained (contact angle θ is from 72° to 100°). This gives the values of $\eta_c/\eta_0 = 0.893$ at $r = 1$ µm, $\eta_c/\eta_0 = 0.455$ at $r = 0.1$ µm and $\eta_c/\eta_0 = 0.077$ at $r = 0.01$ µm. As seen from these estimations, the slippage effect may considerably accelerate water filtration in hydrophobic porous bodies with the average pore radius less than 1 µm. In the case of very narrow pores ($r = 10$ nm), the velocity of flow increases more than ten times compared with truly viscous flow at $\eta_c = \eta_0 =$ const.

The slippage coefficient decreases with improving wetting, and at contact angles $\theta < 70°$ water passes through a capillary with the bulk water viscosity η_0. Further increase in hydrophilicity of a porous body (at $\theta \to 0°$) will lead to increased water viscosity in pores and to flow at an increased effective viscosity $\eta_c > \eta_0$. Thus, the viscous flow of water and other liquids depends considerably on wetting conditions, which in turn depend on a relationship between the forces of intermolecular interaction and those of interaction between molecules and the wall. As seen, the flow rates are governed by the field of surface forces that determines the properties of boundary liquid layers.

For mercury in lyophobic quartz capillaries, higher values for the slippage coefficient were determined: $\gamma = (5 \pm 3)\times10^{-5}$ m^3/N s. The wide scatter of γ values is caused by gas adsorption affecting slippage. At gas pressure (nitrogen) higher than 30–40 atm, the values of γ increase significantly, approaching those previously obtained by Tolstoy for mercury in glass capillaries. According to data obtained by Bogomolov [76], there is a gap of 0.15 nm between mercury and lyophobic walls that drastically reduces the cohesion between glass and mercury surfaces.

In many cases pore walls are not rigid and smooth. It is known that the surface of glass and even quartz, exposed to water for a long time, gradually becomes loosened and forms a surface layer with fine pores, or a gel-layer. By various estimations, the thickness of this layer may achieve several nm. The permeability of the surface gel layer is low, but when the pore radius is commensurate with the thickness of the porous surface layer, the contribution of the latter into the flow may become considerable. As a first approximation, it may be reduced to an effect similar to slippage in lyophobic pores, as seen from the distribution of local velocities shown in Fig. 4.3. The tangential component of the velocity v_0 of liquid flowing relative to the boundary porous layer plays the role of a "slippage velocity". The dependences of such a "slippage coefficient" on the thickness of the porous layer and its filtration characteristics were obtained [77]. Assuming that the porous layer is filled with Newtonian liquid with high viscosity $\eta_p >> \eta$, the increase of the average velocity of flow is:

$$V/V_0 = 1 + (4h_0\eta/r\eta_p), \qquad \text{(IV.10)}$$

where V_0 is the average velocity of flow at $h_0 = 0$. The contribution of the flow in the porous layer into a total flow is greater, the larger its thickness and the lower the viscous resistance η_p.

Microroughness also can play a role similar to the porous layer; in this case h_0 is about a half-height of the roughnesses.

Similar effects may take place in the case of rigid walls, where adsorption layers of surfactants or polymers are formed. The thickness of the adsorption layers of surfactants approaches 3.0–5.0 nm depending on the length of the hydrocarbon chain. Considerably thicker layers form in the case of polymer adsorption; these can achieve hundreds of nm depending on molecular mass. Their characteristic feature is non-uniformity of the adsorption layer structure: more dense near to surface and looser near the periphery. Consequently, the outer part of the adsorption layer can be easily deformed under the shear stress created by the tangential viscous flow; this complicates the effect considerably. The deformation of adsorption layers of polyethyleneoxide (PEO) was experimentally observed during flow of PEO solutions in quartz capillaries [78]. The thickness of a saturated adsorption layer of PEO with molecular mass of 6×10^5 decreased from 20 nm, at rest, to 10 nm when the velocity of flow and, respectively, the tangential shear stress increased up to 300 N/m^2. It was assumed that the deformation occurred only in peripheral regions of the "tails" of the adsorbed molecules.

In the case of viscous-plastic gel layers on a quartz surface, a solution flowing through quartz capillaries broke those outer layers that were rather weakly bound to the surface, as observed in [79]. The detached gel flakes flowing with the liquid through the pore system can stick in pore narrowings and adsorb on pore widenings (where the velocity of flow and shear stresses are lower) or in blind pores changing thereby the structure of the porous space.

All the above mentioned adjustments did not cause deviations from the linear relationship between the pressure gradient and flow velocity according to the Darcy law written as eq. (IV.3). Nevertheless, in a number of cases deviations from this law were experimentally observed in real porous bodies. One of the most often mentioned reasons is the assumption of the existence of a yield stress of bulk water $\tau_0 = 10^{-3}$ N/m^2 [80]. However, more accurate measurements performed under conditions whereby the influence of contact angle hysteresis in a calibrating capillary was considerably reduced, revealed no yield stress in the bulk water within experimental error [61].

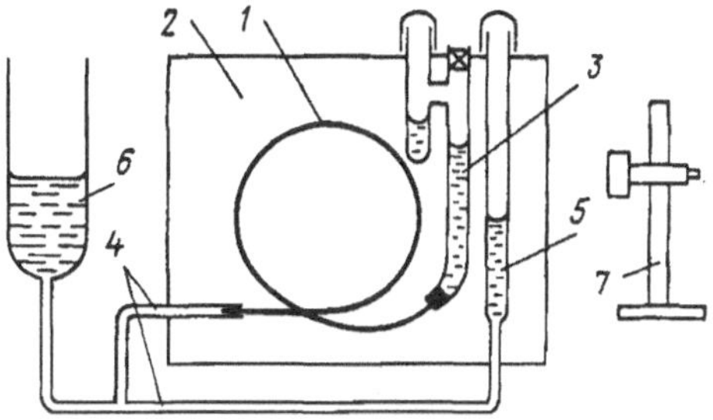

Figure 4.4 Experimental apparatus for measuring flow rates of water at low pressure gradients.

Figure 4.4 shows an experimental apparatus. A 5 m long capillary (*1*) with diameter $d = 13$ μm was stacked in rings and placed in a thermostat (*2*). One end of the capillary was connected to a calibrated capillary (*3*), the other was connected by a polyethylene tube (*4*) to the reference capillary (*5*) and to the vessel (*6*) filled with water. The vessel can be moved in a vertical direction to change the liquid level. The calibrated and reference capillaries were made from the same tube (d = 0.425 mm). The use of the reference capillary excluded the necessity of corrections for capillary pressure of the meniscus in the calibrated capillary. The difference between water levels in the capillaries (*3*) and (*5*) measured by a cathetometer (*7*) controlled the operating pressure drop, and changes of this difference determined the flow rate. All measurements were performed on receding menisci that reduced the influence of the contact angle hysteresis.

Figure 4.5 shows the relationship between the velocity of the meniscus displacement and the pressure drop. It follows that the yield stress, which can be formally determined as the value intercepted by the dependence $V(\Delta P)$ on the ΔP axis, is comparable with the accuracy of experiment characterized by the scatter of points on the graph. This value, $\tau_0 \sim 10^{-6}$ N/m^2, is three orders lower than that expected earlier for bulk water [80].

In further experiments [81], when the capillary length was increased to 92 m ($d = 200$ μm), the possible error in τ_0 determination was reduced by two orders. These experiments revealed that the yield stress in bulk water cannot exceed 10^{-8} N/m^2.

If water possessed the yield stress $\tau_0 = 3\times10^{-3}$ N/m^2, as was earlier supposed [80], the flow in the calibrated capillaries used in experiments should stop when the difference in menisci levels approached 1 m. However, a flow with the bulk viscosity η_0 was observed even when this difference was less than several mm.

At the same time, it would be a mistake to think that liquids in general cannot possess a dynamic yield stress τ_0 and flow with higher viscosity when

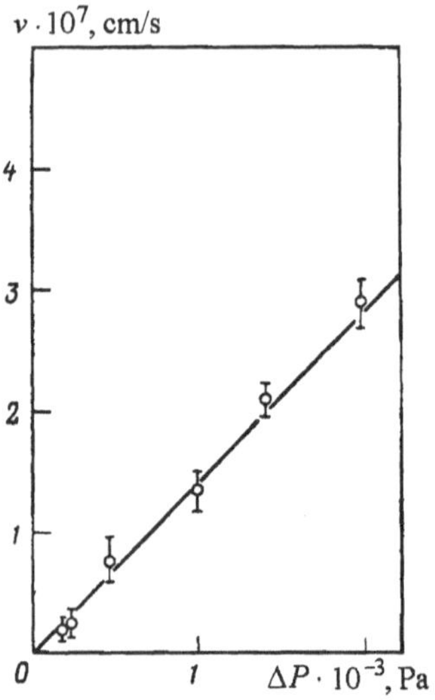

Fig. 4.5

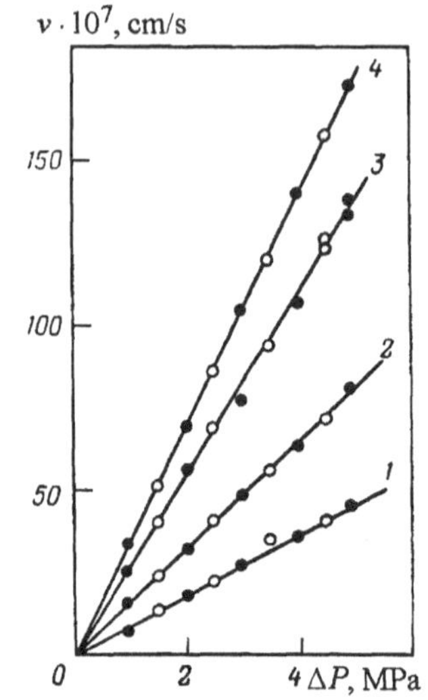

Fig. 4.6

Figure 4.5 Flow rates of water in a quartz capillary of radius $r = 6.5$ μm and length $L = 5.7$ m in the region of low pressure drop ΔP.

Figure 4.6 Dependence of water flow rate through porous glass with mean pore radius $r = 0.8$ nm on the pressure drop ΔP at 20°C (*1*), 40°C (2), 60°C (*3*) and 70°C (*4*).

$\tau < \tau_0$. Computer simulations performed for simple liquids using the methods of molecular dynamics revealed that, in the case of very low deformation rates, liquids possess non-Newtonian properties, and the viscosity depends on the shear velocity. Thus, to determine the real value of τ_0 for bulk water, we should improve the accuracy of our measurements.

The observed deviations from the Darcy law cannot be associated with the non-Newtonian properties of the bulk water. It has been shown experimentally that the Darcy law is obeyed well in pure liquid flows (without solutes) in rigid porous bodies when the structure is not changed under the applied pressure. These measurements were performed on different specimens of porous glass and filters with various average pore radii r in the range from several μm to several nm [45, 49, 50, 55, 60]. The results showed that the Newtonian viscosity of a liquid increases only with decreasing r (Chapter III, §2).

Figure 4.6 shows the results of experiments performed on a glass membrane with fine pores used for the reverse osmosis separation of solutions [50]. As seen from these plots, all curves $V(\Delta P)$ pass through the origin in compliance with the Darcy law of filtration. The dependences

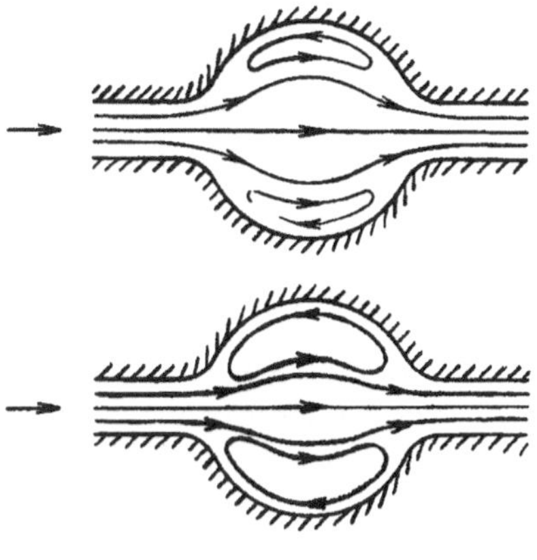

Figure 4.7 Changes of streamlines in local pore widenings during a gradual growth of flow rate.

reproduce well both at increasing (open points) or decreasing (dark points) pressure difference ΔP, i.e. there is no hysteresis. On increasing temperature the flow rate also rises due to the decreasing viscosity. However, in the case of fine pores (as for narrow capillaries, Fig. 3.7), the viscosity decreases more abruptly than in bulk water. This is caused by a destruction of the modified water structure in narrow hydrophilic pores occurring at increased temperature.

In porous bodies with a rigid structure, the deviations from the Darcy law may be caused by non-Newtonian rheological properties of flowing liquid associated with the presence of colloidal particles or polymer molecules building space structures. This takes place in crude oils, water in soils, grounds and porous construction materials, and also with liquids contained in real porous materials used in chemical, food and textile industries and in medical and biotechnology. The threshold gradients were observed, for instance, in clay systems.

The presence of low-molecular weight dissolved substances may lead to deviations from the Darcy law of another type, i.e. caused by capillary osmosis phenomena. This effect will be discussed in further detail in §3.

In solid bodies with wide pores having considerable local widenings, the non-linearity of dependences $v(\Delta P)$ may be caused by the restructuring of stream lines under the change of pressure gradient. This effect was experimentally studied on model porous bodies [82] and has been theoretically justified [83]. This effect may be accompanied by circulating flows in widenings and their configuration and intensity depend on the flow rates (Fig. 4.7).

In non-rigid, easily deformable porous bodies the non-linearity of the dependences $v(\Delta P)$ is caused by reconstruction of the porous structure under the changing pressure drop [84]. This reconstruction may be elastic and fully reversible if the applied pressure does not overcome the force barrier between particles (Fig. 1.4). More often the deformation is followed by a displacement and redistribution of particles within the porous space resulting in irreversible restructuring. A non-linear dependence $v(\Delta P)$ is also characteristic for swelling porous bodies.

The time instability of flow, caused by restructuring of a porous body due to its non-equilibrium state in respect of the composition of the flowing liquid, often prevents precise measurements of filtration rates. In porous bodies, as in colloids, the equilibrium composition of the liquid dispersion medium is established as a result of ionic exchange and adsorption processes. Its replacement with, for example, water or a solution differing in composition from the inherent dispersion medium, will tend to change the potential curve of interparticle interaction (Fig. 1.4) and, consequently, the equilibrium interparticle distances, i.e. cause swelling or shrinkage. In non-rigid porous bodies the displacement of particles proceeds slowly due to high viscous resistance. Relaxation times may achieve several days and sometimes months. As this takes place, the flow removes and separates small and weakly bound particles and transfers them in the flow direction, so making the structure non-uniform. This reconstruction may be most pronounced and lengthy during the filtration of solutions having different composition from the inherent porous solution, for example when carrying out experiments with natural porous bodies using distilled water.

Instability of flow can also be caused by bacterial activity, changes in composition and concentration of dissolved gases [85], bubble formation from dissolved air (at the flow output) [80], that leads to pore obliteration.

All these phenomena should be taken into account in experiments intended both to obtain the structure characteristics of porous bodies and to investigate filtration anomalies.

§2 Electro-osmotic Flow

Many special monographs are devoted to theoretical and experimental investigations of electrokinetic phenomena [48, 57]. Here we will consider this phenomenon briefly, in its applications to the processes of mass transfer in fine-porous bodies.

For diluted electrolyte solutions and in the absence of thermal effects, the set of eqs. (IV.1) can be reduced to the form:

$$V = \alpha_{11}\nabla\mu_1 + \alpha_{13}\nabla E \qquad \text{(IV.11)}$$

$$J = \alpha_{31}\nabla\mu_1 + \alpha_{33}\nabla E. \tag{IV.12}$$

The usual approach for the identification of cross-coefficients is to calculate the flow of a substance (mass, charge, energy, solute) under a pressure gradient using the hydrodynamic equations. In the case of the interrelated fluxes of the solvent and electric charge, the velocity of charge transfer in a convective flow (in the absence of electric field) or streaming current should be determined from:

$$J = \alpha_{31}\nabla\mu_1|_{\nabla E = 0}. \tag{IV.13}$$

Let us specify the model of porous space to apply the hydrodynamic equations. The simplest solution can be obtained for straight cylindrical channels, when EDL (Chapter 1, §1) do not overlap and $\kappa h >> 1$, where h is the characteristic bore dimension. In this case the EDL is located inside the linear region of the velocity profile:

$$v(x) = -xS\nabla P/\Sigma\eta, \tag{IV.14}$$

where x is the distance from the pore surface; Σ is the pore perimeter and S is the area of cross-section. This relation follows from the Newton equation $\tau = \eta(dv/dx) = S\nabla P/\Sigma$, written for the case of $(dv/dx) = \text{const}$.

The streaming current per the cross-sectional area of the porous body equals:

$$J = (n\Sigma/S) = \int_0^\infty v(x)\rho(x)dx, \tag{IV.15}$$

where n is the porosity of the specimen.

The distribution of the charge density $\rho(x)$ in a single EDL is given by the equation:

$$\rho(x) = (\varepsilon_0\kappa^2\psi_1/4\pi)\exp(-\kappa h). \tag{IV.16}$$

Integrating this equation for x from 0 to ∞, we obtain the surface charge density expressed by eq. (I. 7).

Substituting eq. (IV.16) into eq. (IV.15) and using eq. (IV. 17), we get:

$$J = -(\varepsilon_0\kappa^2\psi_1\nabla P/4\pi\eta)\int_0^\infty x\exp(-\kappa h)dx = \varepsilon_0\psi_1\nabla P/4\pi\eta, \tag{IV.17}$$

As seen from this equation, the streaming current at $\kappa h >> 1$ does not depend on the shape of pore cross-section. Because the thickness of the EDL supposed to be much less than the pore size, the flow does not much differ from that near to the plane surface. However, the configuration of the cross-section of a cylindrical pore should have no narrowings and regions of high curvature, where the condition $\kappa h >> 1$ may fail.

A comparison of eq. (IV.17) and eq. (IV.13) gives the expression for the phenomenological coefficient α_{31}:

$$\alpha_{31} = -\varepsilon_0 \psi_1 / 4\pi\eta v_{\mathrm{m}}. \tag{IV.18}$$

Due to the Onsager relation $\alpha_{31} = \alpha_{13}$, and we can obtain the desired equation for the velocity of electro-osmotic flow:

$$V_e = \alpha_{31} \nabla E = -\varepsilon_0 \psi_1 \nabla E / 4\pi\eta v_{\mathrm{m}}, \ \mathrm{mol/cm^2\ s}. \tag{IV.19}$$

Equation (IV.11) allows the calculations of the liquid flow driven simultaneously by gradients of pressure and electrical potential, for which we use α_{11} values determined in §1 in addition to α_{31}. Similarly the streaming current in a porous body driven simultaneously by the external electric field and pressure difference can be determined using eq. (IV.12). In this case, α_{33} is the electrical conductivity of the solution filling the pores. We assume that the skeleton of the porous body does not show considerable electrical conductivity.

Application of the voltage, ΔE, to the porous body possessing a surface charge may create a hydrostatic pressure. Its maximum value determined from eq. (IV.11) at $V = 0$ is equal to:

$$\Delta P = \alpha_{13} \Delta E / \alpha_{11} v_{\mathrm{m}}. \tag{IV.20}$$

The higher the values of electrical potential ψ_1 determining α_{13} and the lower the filtration coefficient of the porous body, the higher this pressure.

For the model of porous body constituted of cylindrical capillaries of equal radii r_{c} we have:

$$\Delta P = -(2\varepsilon_0 \psi_1 / \pi r_{\mathrm{c}}^2 m) \Delta E\,, \tag{IV.21}$$

The cross-effect is represented by generation of a voltage ΔE while the liquid flows under the pressure drop ΔP in the porous body possessing a surface charge. The maximum value of a stationary difference of electrical potentials, achieved at $J = 0$, equals:

$$\Delta E_{J=0} = \alpha_{31}\Delta P/\alpha_{33}\nu_{m}. \qquad (IV.22)$$

The value of $\Delta E_{J=0}$ called the streaming potential, is directly proportional to the applied difference of pressure ΔP. Substitution of the expression for α_{13} from eq. (IV.13) into eq..(IV.22) gives the well-known Helmholtz–Smoluchowski equation:

$$\Delta E/\Delta P = -\varepsilon_0\psi_1/4\pi\eta K_0, \qquad (IV.23)$$

where $K_0 = \alpha_{33}$ is the specific electrical conductivity of liquid (Ω^{-1} cm^{-1}).

The streaming potential generated in a pressure driven flow of the solution in turn generates the counter electro-osmotic flow impeding the filtration process. This phenomenon is called electroviscosity and is often perceived as increased viscosity of a flowing liquid. The role of this effect in the viscosity measurements has been already discussed in Chapter III, §3. The flow rate adjusted for the counter electro-osmosis is equal to:

$$V = -\alpha_{11}\nu_m\nabla P(1 - \alpha_{13}\alpha_{31}/\alpha_{11}\alpha_{33}\nu_m). \qquad (IV.24)$$

For the model of cylindrical capillaries with equal average radii r_c, the substitution of the corresponding expressions for the phenomenological coefficients gives:

$$V = -(K\nabla P/\eta)[1 - (\varepsilon_0^2\psi_1^2/2\pi^2\eta K_0\nu_m m r_c^2)] = -K_e\nabla P/\eta. \qquad (IV.25)$$

As seen from this expression, the effect of electroviscosity does not disturb the linear relationship between V and ∇P, and the Darcy law is formally obeyed, but with the redefined filtration coefficient $K_e < K$. The magnitude of the effect is independent of the sign of potential ψ_1 and increases with decreasing pore radius. However, it should be born in mind that the equations discussed are true only if $\kappa h >> 1$, i.e. when the thickness of the EDL is much less than the half-width h or the pore radius r. Electrokinetic effects in narrow pores, in the case of overlapped EDL at $\kappa h << 1$, will be considered further.

The effect of the streaming potential has various applications. Particularly, it can be used for direct transformation of mechanical energy expended in liquid pumping into electricity. The maximum efficiency of such a transformation ranging up to 17–20% is achieved at $\kappa h \approx 1$ [86, 87]. Pressure transducers are also based on measurements of the streaming potential [88]. Such a transducer consists of a membrane under pressure and a porous diaphragm, through which an electrolyte solution flows, when the membrane deforms under the applied pressure difference ΔP. The difference of potentials,

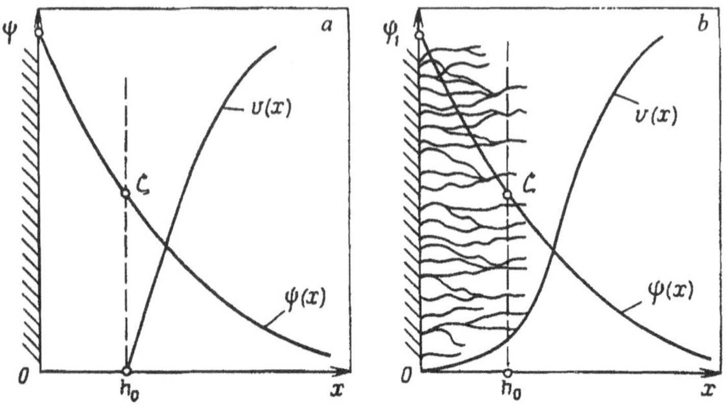

Figure 4.8 Distribution of electrical potential $\psi(x)$ and local flow rates $v(x)$ in the presence of the stagnant liquid layer (a) and in the case of permeable polymer adsorption layers.

ΔE, generated due to the solution flow and proportional to ΔP is measured by electrodes placed in the solution. Using the value of streaming potential, one can determine the direction and filtration rate in subsoils [89]. Electroosmosis is used for dehydration of porous bodies, electromelioration, intensification of immiscible displacement and capillary impregnation, cloth washing [90].

Equations (IV.12) and (IV.23) are used to determine electrokinetic, or ζ-potential assigned to some slippage plane (Fig. 4.8, a) offset by distance h_0 from a solid wall. Previously it was assumed that the liquid layer of thickness h_0 remains stagnant because it has rather high yield stress τ_0. For example in a porous body with pore radius r = 10 μm and thickness l = 5 cm, the tangential shear stress on pore surface will be $\tau_0 = r\Delta P/2l$ = 1 N/m^2 in the case of flow driven by the pressure drop $\Delta P = 10^{-3}$ MPa. The yield stress should be higher than 1 N/m^2 for the layer to remain stagnant. Such τ_0 values may be expected for gels and polymer adsorption layers possessing viscous-plastic properties. As shown above (Chapter III, §3), boundary layers with modified water structure retain Newtonian properties and do not show any measurable yield stress.

The reason for the assumption of a hydrodynamically stagnant layer h_0 was the experimentally found inconsistency between values of surface potential ψ_0, measured by adsorption methods, and electrokinetic potential, with $\psi_0 \geq \zeta$. Since low-molecular weight liquids retain their fluidity near to a solid substrate, the difference between ψ_0 and ζ in real porous bodies is mostly caused by surface gel layers, microroughness (including micropores and surface microfissure), adsorption of polymer molecules and surfactants that impede liquid flow (Fig. 4.8, b). In this case ζ is deemed as some effective value obtained from electrokinetic measurements, that characterizes the position of the plane, beyond which the flow velocity abruptly decreases.

The measurements of ζ-potential are particularly used to determine the hydrodynamic thickness h_0 of adsorption polymer layers on the basis of the Gouy equation:

$$h_0 = (1/\kappa)\ln[\tanh(\psi_1 F/4RT)/\tanh(\zeta F/4RT)],$$

where ψ_1 is the Stern surface potential under the absorption polymer layer. This method was used to estimate h_0 for polyethyleneoxides (PEO) of various molecular mass (from 2.1×10^4 to 2.7×10^5) adsorbed on the surface of quartz capillaries including an investigation of the dependence of h_0 on the equilibrium concentration of PEO solutions (from 10^{-4} to 10^{-1} M) and concentration of the background electrolyte [78].

Electrokinetic measurements in cylindrical quartz capillaries with rigid and molecularly smooth surface showed [79] that in the absence of surface porous formations the electrokinetic potential is equal to the Stern potential and there is no hydrodynamic stagnant layer ($h_0 = 0$).

A gel layer of thickness up to 2 nm can form on the surface of the same capillaries after several weeks of exposure to water and aqueous solutions. Initially the absolute values of ζ-potential increase with time due to the increasing adsorption capacity of the gel layer with respect to OH-ions, but then decrease as the gel-layer grows and its periphery loosens. Higher values of ζ-potential could be restored if the liquid flow driven by high pressure drops tear away peripheral regions of the gel-layer. The yield stress of the outer parts of gel-layer is about 2–3 N/m^2. Deformation of the internal part of the gel-layers requires considerably higher shear stresses.

Figure 4.9 shows the results of measurements of the streaming potential ΔE for a capillary exposed to water for a long time. When the pressure ΔP was applied for the first time (the lower curve), the values obtained for $\Delta E/\Delta P$ correspond to a ζ-potential equal to −94 mV, according to eq. (IV.23). However, at $\Delta P \geq 0.5$ MPa the values of ΔE change nonlinearly and irreversibly. With decreasing ΔP, the linear changes of ΔE now follow the higher curve corresponding to $\zeta = -102$ mV. Higher absolute values of ζ-potential correspond to the break-away of a part of the gel-layer and approach of the slippage plane to the rigid surface of the capillary. At the break-away pressure $\Delta P = 0.5$ MPa we can estimate the yield stress of a peripheral part of the gel-layer: $\tau_0 = \Delta Pr/2l$, where l is the capillary length. This experiment gave the value of $\tau_0 = 30$ N/m^2 close to the yield stress of silica gel at the concentration of solid phase about 30% by weight. Assuming that the electrical potential decays exponentially inside the gel-layer and using the values of ζ-potential before and after the break-away, one can determine the thickness of the removed part of the gel-layer as about 2 nm.

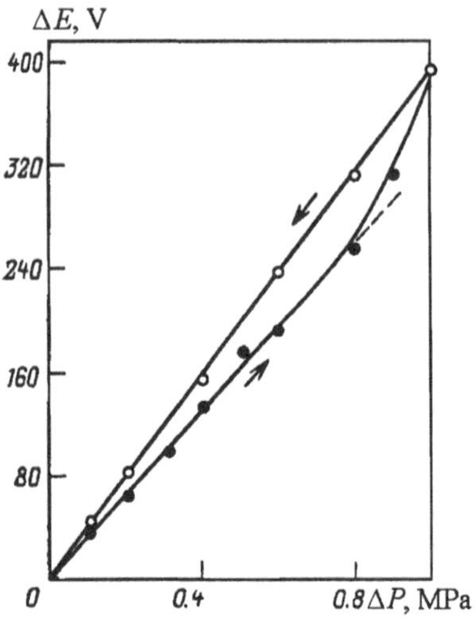

Fig. 4.9

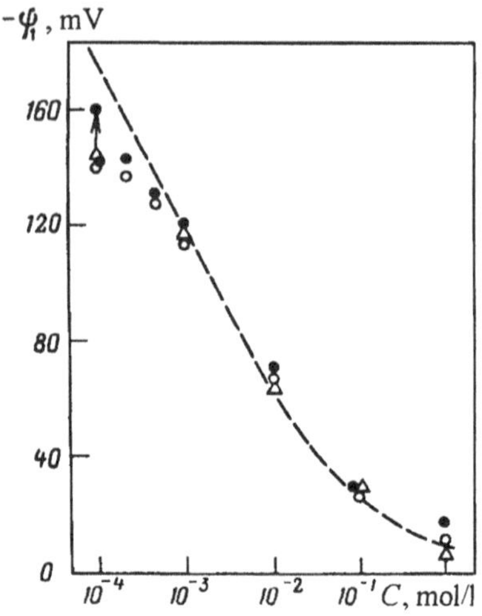

Fig. 4.10

Figure 4.9 Changes in streaming potential ΔE in a capillary of radius r = 10 μm, covered with a surface gel layer, with increasing pressure drop ΔP.

Figure 4.10 Corrections to the measured values of ψ_1-potential for elevated viscosity of water boundary layers.

For capillaries with a rigid solid surface without a gel-layer, we found deviations from the Gouy equation relating the surface charge σ to the Stern potential, provided the concentration of potential-determining ions is constant (for quartz this condition corresponds to constant pH):

$$\sigma = (2\varepsilon_0 RTC/\pi)^{1/2} \sinh(zF\psi_1/2RT) = \text{const}, \tag{IV.26}$$

where z = 1 the ion charge; F is Faraday; C is the concentration of an indifferent electrolyte, KCl. As seen from Fig. 4.10, at $C < 10^{-3}$ mol/l, the experimental points deviate from the dashed theoretical curve plotted according to eq. (IV.26). These deviations can be removed by the introduction of corrections for increased viscosity of the boundary water layers into the Helmholtz–Smoluchowski equation (see eq. (IV.7)). The theory of electrokinetic phenomena, taking into account the increased viscosity of the boundary layers [91], leads to the following modified expression for the streaming potential:

$$\Delta E/\Delta P_{J=0} = -\varepsilon_0[\psi_1 - (4af\gamma_0 RT/zF)]/4\pi\eta_0 K_0. \tag{IV.27}$$

Here

$$f = [\exp(zF\psi_1/2RT) - 1] / [\exp(zF\psi_1/2RT) + 1]$$

$$\gamma_0 = \sum_{n=0}^{\infty} f^{2n}/[1 + 2n + (b/\kappa)], \qquad \text{(IV.28)}$$

where a and b are parameters of viscosity distribution in the boundary water layers from eq. (IV.7).

The arrow and the black point in Fig. 4.10 show the results of calculations based on eqs. (IV.27), (IV.28). The correction for the increased viscosity enables the agreement between the experiment and theory to be improved. The values of a and b obtained for water were used in these calculations. At higher concentration no correction for the increased viscosity is needed (as seen from Fig. 4.10). At electrolyte concentration higher than 10^{-3} mol/l the structure of boundary water layers is partially destroyed and the viscosity of the solution does not differ much from that of the bulk. As was shown in [55], for fine capillaries with radius of 0.073 μm, the viscosity of KCl solutions is equal to that of the bulk at $C > 10^{-3}$ mol/l. Accordingly, at concentrations $C > 10^{-3}$ mol/l, the experimental dependences $\psi_1(C)$ for quartz capillaries with solid undisturbed surface agree well with the theoretical equation (IV.26).

Near to hydrophobic surfaces, where the slippage of liquids may take place (Chapter III, §2), the correction of the opposite sign should be introduced into the Helmholtz–Smoluchowski equation. In the case of low surface potentials, the following modification to this equation is given [92]:

$$\Delta E/\Delta P_{J=0} = -\varepsilon_0\psi_1(1 + \gamma\eta_0\kappa)/4\pi\eta_0 K_0, \qquad \text{(IV.29)}$$

where γ is the slippage coefficient and $1/\kappa$ is Debye radius. Using the measured value for water of $\gamma = 3\times10^{-5}$ m^3/N s and assuming that $\eta_0 = 10^{-3}$ N s/m^2 and $\kappa = 10^5$ l/cm, we obtain the result that $\alpha = \gamma\eta_0\kappa$ is about 0.3. This means that the correction to eq. (IV.23) is about 30% for water. The value of the correction sharply increases with electrolyte concentration. At $C = 10^{-2}$ mol/l for a 1–1 electrolyte ($\kappa = 3\times10^6$ cm^{-1}), the values of ψ_1 were overestimated by one order of magnitude ($\alpha = 9$) without a correction for slippage. Corrections for slippage resulted in significant reduction of values of ψ_1-potential and specific surface charge for a hydrophobic (methylated) silica surface measured in [79].

Electro-osmotic flow can be caused not only by a gradient of electrical potential but also by that of concentration in the absence of an external electric field. In this case the system of equations at $\nabla P = 0$ and $\nabla T = 0$ takes the form:

$$Q = \alpha_{22}\nabla\mu_2 + \alpha_{23}\nabla E$$

$$J = \alpha_{32}\nabla\mu_2 + \alpha_{33}\nabla E. \qquad \text{(IV.30)}$$

Here $\alpha_{22} = D_e/v_m RT$, where $D_e = D_+D_-(z_+ + z_-)/(D_+z_+ + D_-z_-)$ is the effective coefficient of electrolyte diffusion; α_{33} is, as previously, the electrical conductivity. Cross-coefficients $\alpha_{23} = \alpha_{32}$ characterize electrolyte transfer driven by the electric field in the absence of concentration gradient and the reverse effect, the current generated by the gradient of electrolyte concentration at $\nabla E = 0$. These effects occur only if ions have different mobility ($D_+ \neq D_-$).

In general, it is necessary to take into account the so-called diffusion potential ΔE_d, that appears if a stationary concentration drop of electrolyte ΔC is maintained. It follows from eq. (IV.30), at $J = 0$ and being aware that $\Delta\mu_2 = RT\Delta C/C$, that:

$$\Delta E_d/\Delta C = -\alpha_{32}RT/\alpha_{33}C = RT(D_+ - D_-)/FC(D_+z_+ + D_-z_-), \qquad \text{(IV.31)}$$

where z_i are the ion charges.

A diffusion potential may cause an electro-osmotic flow driven by the gradient of electrolyte concentration:

$$V_e = \alpha_{13}\nabla E_d = -(\varepsilon_0\psi_1/4\pi\eta v_m)[RT(D_+ - D_-)\nabla C/FC(D_+z_+ + D_-z_-)]. \qquad \text{(IV.32)}$$

In wide pores, as seen from (IV.32), the rate of the electro-osmotic flow under the diffusion potential arising when $\Delta C \neq 0$ depends on the difference between ion mobilities and surface potential ψ_1.

So far we have studied the electrokinetic phenomena in wide pores and capillaries where the EDL did not overlap. Their overlapping in narrow pores at $\kappa h < 1$ leads to different expressions for phenomenological coefficients $\alpha_{ik} = \alpha_{ki}$.

The differences are mainly associated with a change of expression for the charge density distribution $\rho(x)$ and nonlinear distribution of flow velocities across a pore. Analytical solutions can be only obtained for low ψ_1-potentials and a symmetrical electrolyte. For a plane slit with half-width h, the distribution of charge density is given by the equation [93, 48]:

$$\rho(x) = \varepsilon_0\kappa^2\psi_1\cosh[\kappa(h - x)]/4\pi\cosh(\kappa h). \qquad \text{(IV.33)}$$

The charge density is maximum near to the walls (at $x = 0$ and $x = 2h$), where $\rho_0 = \varepsilon_0\kappa^2\psi_1/4\pi$, and minimum in the middle of the slit (at $x = h$), where $\rho = \rho_0/\cosh(\kappa h)$.

For a plane slit, eq. (IV.15) for streaming current now is written as follows:

$$J = (1/2h)\int_0^{2h} v(x)\rho(x)dx\,. \tag{IV.34}$$

Integration of this equation after the substitution of eq. (IV.33) and of parabolic distribution of the Poiseuille flow velocity

$$v(x) = \nabla P(x-2h)x/2\eta \tag{IV.35}$$

gives at $\Delta E = 0$ the following expression for the coefficient α_{31}:

$$\alpha_{31} = -\varepsilon_0\psi_1[1-(\tanh\kappa h)/\kappa h]/4\pi\eta v_m. \tag{IV.36}$$

As seen, this differs from eq. (IV.18) by the factor in square brackets. When $\kappa h \to \infty$, its value tends to 1 and the equation transforms into that for wide pores. At $\kappa h \to 0$, the ratio $(\tanh\kappa h)/\kappa h \to 1$ and the value of the factor approaches 0. Thus, in microporous bodies the electrokinetic effects are much weaker. For the model of a plane slit pore, the coefficient α_{11} obtained by integrating eq. (IV.35) for a velocity profile, is equal to:

$$\alpha_{11} = -h^2/3\eta v_m^2, \tag{IV.37}$$

Knowing α_{11} and α_{31}, one can obtain the corresponding expressions for the streaming potential and electro-osmotic velocity.

In the light of electroviscosity influence on the flow in porous bodies discussed above, let us write eq. (IV.24) for narrow pores taking into account the effect of EDL overlap. Substituting α_{11}, α_{31} and α_{33} we obtain the following equation, instead of eq. (IV.5):

$$V/V_0 = 1 - 3B(\kappa h - \tanh\kappa h)^2/\{\kappa h(1+K_s)$$

$$+ [\tanh\kappa h - (\kappa h/\cosh^2\kappa h)](\kappa h)^3(B/2)\}, \tag{IV.38}$$

where $V_0 = -(K/\eta)\nabla P$ defines the flow rate in the absence of electrokinetic effects ($\psi_1 = 0$).

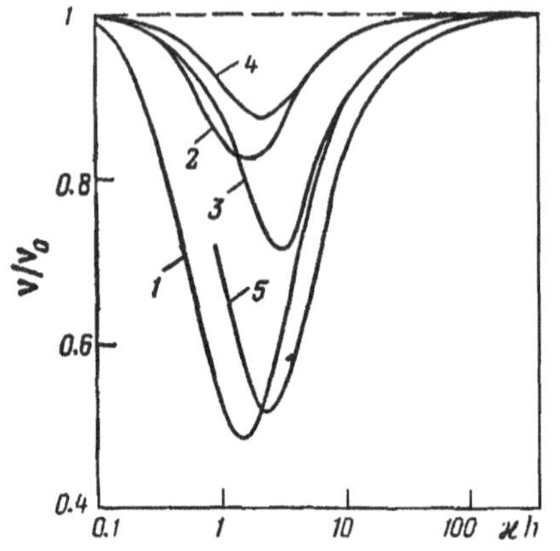

Figure 4.11 Calculated dependences V/V_0 vs. κh characterizing the influence of electro-viscous effects: $B = 5$ (curves *1*, *3*, *5*); $B = 0.5$ (curves *2* and *4*). The results were obtained using eq. (IV.38) (curves *1* and *2*) and according to [48] (curves *3* and *4*) and [94] (curve *5*).

Here B is a dimensionless parameter characterizing the surface potential:

$$B = (\varepsilon_0 \psi_1 \kappa / 4\pi)^2 / \eta K_0, \qquad \text{(IV.39)}$$

where $K_0 = \alpha_{33}$ is the specific electrical conductivity of the solution. The K_s value in eq. (IV.38) characterizes surface conductivity.

Figure 4.11 (curves *1* and *2*) shows the results of calculations of the ratio V/V_0 as a function of κh for the suggested model of the plane slit. Curves *3* and *4* show results obtained by Dukhin [48] for the same model, curve *5* shows the results obtained by Rice and Whitehead [94] for cylindrical capillaries. In the latter case, it was assumed that $h = r$. As seen from Fig. 4.11, the form of the dependences obtained is similar and the electroviscosity effect is the most pronounced in the range of κh from 1 to 3. The fact that the dependences pass through a minimum is associated with the following phenomena: the streaming current generated by a convective transfer abruptly decreases with pore size due to the increased viscous resistance to the hydrodynamic flow (eq. (IV.37)), whereas the bulk electrical conductivity of the solution regulating the value of streaming potential (see eq. (IV.23)) remains constant and is dependent on the total effective cross-section of pores rather than on their size. For real porous bodies, the additional influence of the surface electrical conductivity may result in much more significant decrease of V/V_0 values at $\kappa h \to 0$. In microporous bodies ($r \to 0$) the contribution of the surface conductivity is much more pronounced due to advancing growth of the specific surface at the same porosity.

In conclusion, let us consider one of the thermoelectrical effects, i.e. the generation of a thermoelectromotive power due to the different heat of transfer (J/mol) for anions Q_- and cations Q_+ in the electric field. In this case, the system of electrodynamic equations is written as follows:

$$J = \alpha_{33}\nabla E + \alpha_{34}\nabla T/T$$

$$W = \alpha_{43}\nabla E + \alpha_{44}\nabla T/T. \tag{IV.40}$$

If the heat of dissolution of ions is different ($Q_+ \neq Q_-$), then ionic transfer will arise through a porous membrane between two solutions of equal concentration but at different temperatures. The electric current $J = \alpha_{34}\nabla T/T$ is determined by the coefficient α_{34}. The stationary state of the system is established at $J = 0$, when the ionic thermodiffusion is balanced by the reverse current driven by the thermodiffusion potential that in a symmetrical electrolyte is equal to:

$$\Delta E_t/\Delta T = -\alpha_{34}/\alpha_{33}T. \tag{IV.41}$$

To obtain the expression for α_{34}, one should determine the value of heat flux or heat of transfer, $W = \alpha_{43}\nabla E$, resulting from the difference between the values Q_- and Q_+ for anions and cations transferring by the electric field ∇E at $\nabla T = 0$:

$$W = \alpha_{33}(Q_+ - Q_-)\nabla E/2F. \tag{IV.42}$$

Comparing this equation to the previous one, we obtain:

$$\alpha_{43} = \alpha_{33}(Q_+ - Q_-)/2F. \tag{IV.43}$$

This enables us to determine the electro-osmosis velocity caused by the temperature difference in the absence of a concentration gradient of electrolyte solution and of an external electric field:

$$V_t = \alpha_{13}\nabla E_t = -\alpha_{13}\alpha_{34}\nabla T/\alpha_{33}T. \tag{IV.44}$$

For the model of wide cylindrical capillaries, the substitution of corresponding expressions for α_{13} from eq. (IV.18) into eq. (IV.44) and $\alpha_{34} = \alpha_{43}$ from eq. (IV.43) gives:

$$V_t = \varepsilon_0\psi_1(Q_+ - Q_-)\nabla T/8\pi\eta F v_m T. \tag{IV.45}$$

Thermoelectro-osmotic flow depends neither on pore size (at $\kappa h >> 1$) nor on the electrical conductivity of the solution. This can be explained by the fact that the heat of transfer and reverse electric current driven by the emerging potential difference both depend linearly on α_{33}.

Generally, the values of thermodiffusion potential are not high. For aqueous solutions of KCl and NaCl (when $Q_+ - Q_- \approx 3\times10^3$ J/mol) the $\nabla E_t/\nabla T$ value is about 5×10^{-5} V/deg. This means that even when the temperature drop is about 100°C, the thermodiffusion potential difference does not exceed 5 mV. The effect can be more pronounced when the difference between the heats of ionic dissolution is larger.

As seen from eq. (IV.36), the values of the α_{13} coefficient decrease with decreasing pore size. Because of this, thermoelectro-osmotic effects are less pronounced in microporous bodies. However, this applies to all the electrokinetic phenomena reviewed here.

§3 Capillary Osmosis

Capillary osmosis describes the flow of a solution in pores or in a capillary driven by a concentration gradient in the absence of a pressure gradient. We have already discussed one such phenomenon in the previous chapter. When a constant gradient of concentration of electrolyte solution is maintained, the difference of ion mobilities gives rise to the diffusion potential (see eq. (IV.31)) which resulted in liquid flow. However, this effect was reduced to one of the manifestations of the electro-osmotic flow determined by eq. (IV.32).

Generally capillary osmosis is associated with the existence of near-to-surface adsorption layers of ions or neutral molecules, the diffuse outer part of which can participate in the liquid flow. As will be shown below, the capillary osmosis in electrolyte solutions appears even if the ion mobilities are equal as in the case of the concentration driven electrokinetic effect discussed above. The experimental detection of the capillary osmosis not only in electrolytes but in non-ionic solutions provides direct evidence of the diffuse structure of molecular adsorption layers in solutions.

Capillary osmosis was first revealed and investigated by Derjaguin *et al.* in [95]. The theory of capillary osmosis was developed by Derjaguin, Dukhin *et al.* in [57, 96–98].

To provide a unified approach to the phenomena discussed, let us derive the equations of capillary osmosis on the basis of thermodynamics of irreversible processes. For application to diluted solutions, the system of eq. (IV.1) at $\nabla T = 0$ and $\nabla E = 0$ is as follows:

$$V = \alpha_{11}\nabla\mu_1 + \alpha_{12}\nabla\mu_2$$

$$Q = \alpha_{21}\nabla\mu_1 + \alpha_{22}\nabla\mu_2. \quad \text{(IV.46)}$$

Here the coefficient α_{12} defines the velocity V_c of the capillary osmotic flow driven by the gradient of chemical potential $\nabla\mu_2$ of a solute at $\nabla\mu_1 = 0$. As usual, let us calculate the cross-coefficient $\alpha_{21} = \alpha_{12}$ defining the flux of solute mass $Q = \alpha_{21}\nabla\mu_1$ driven by the pressure gradient $\nabla P = -\nabla\mu_1/v_m$ to determine α_{12}.

In wide pores, where the diffuse adsorption layers do not overlap and the velocity profile $v(x)$ can be taken as linear within the diffuse layer, we have by analogy with eqs. (IV.14), (IV.15) and (IV.17):

$$Q = -(\nabla P/\eta)\int_0^\infty [C(x) - C_0]x dx, \quad \text{(IV.47)}$$

where $C(x)$ is the concentration within the diffuse adsorption layer and C_0 is the bulk concentration of the solution beyond the adsorption layer. Here the concentration is expressed in mol/cm^3. A comparison of eq. (IV.47) with $Q = -\alpha_{21}\nabla P v_m$ gives the expression for α_{21}:

$$\alpha_{21} = \alpha_{12} = (1/\eta v_m)\int_0^\infty [C(x) - C_0]x dx. \quad \text{(IV.48)}$$

Using the known equation for the gradient of chemical potential of a solute $\nabla\mu_2 = RT\nabla C/C_0$, we obtain the following equation for the velocity of capillary osmosis:

$$V_c = \alpha_{12}\nabla\mu_2 = -(RT\nabla C/\eta v_m)\int_0^\infty [(C(x) - C_0)/C_0]x dx$$
$$= -RT\xi_0\nabla C/\eta v_m, \quad \text{(IV.49)}$$

where $\xi_0 C_0$ is the moment of adsorption. The value of $(|\xi_0|)^{1/2}$, having the dimensions of length, characterizes the thickness of the diffuse adsorption layer. At $C(x) = C_0$, when there is no adsorption, capillary osmosis does not take place, $V_c = 0$. In the case of positive adsorption of the solute ($\xi_0 > 0$), the velocity of capillary osmosis is directed to the region of lower concentration.

Further calculations can be performed using the equations of adsorption to derive the concentration distribution $C(x)$ near to a surface and then calculate

ξ_0. One of the options is to use the Boltzmann equation (III.5) and that for the energy of interaction, $U(x)$, between a solute molecule and the substrate. If this interaction is caused by dispersion forces only, then [99, 100]:

$$U(x) = -(\hbar/4\pi x^3)\int_0^\infty [(\varepsilon_1 - \varepsilon_3)\alpha/(\varepsilon_1 + \varepsilon_3)\varepsilon_3]d\xi = -A/x^3, \qquad \text{(IV.50)}$$

where $\varepsilon_1(i\xi)$ and $\varepsilon_3(i\xi)$ are frequency-dependent dielectric permittivities of the material forming the walls and the solvent (see Chapter I), respectively, $\alpha(i\xi)$ is the molecular polarizability of the solute in the solvent. According to Pitaevski [101], the latter can be expressed through the radius of molecules, δ, and their dielectric function $\varepsilon_2(i\xi)$ in the solution:

$$\alpha(i\xi) = \delta^3(\varepsilon_2 - \varepsilon_3)/(\varepsilon_2 + 2\varepsilon_3). \qquad \text{(IV.51)}$$

Using eqs. (III. 5) and (IV. 50), we obtain for ξ_0:

$$\xi_0 = \int_0^\infty [\exp(A/x^3kT) - 1]x\,dx. \qquad \text{(IV.52)}$$

Generally, values for the constant of molecular interaction, A, in eq. (IV.50) are about 5×10^{-35} Jm^3. Exponential index in eq. (IV.52) does not exceed 0.5 at distances $x \geq x_0 = 0.3$ nm from the substrate, where the diffuse part of the adsorption layer is located. This enables the exponent in eq. (IV.52) to be expanded as a series which gives an expression for ξ_0 that can be easily integrated:

$$\xi_0 = (A/kT)\int_{x_0}^\infty dx/x^2 = A/x_0kT. \qquad \text{(IV.53)}$$

Substitution of this equation into eq. (IV.49) gives:

$$V_c = -AN_A\nabla C/x_0\eta\nu_m, \qquad \text{(IV.54)}$$

where N_A is the Avogadro number. Quantitative estimations (for $A = 5\times10^{-35}$ Jm^3, $\eta = 10^{-3}$ Ns/cm^2 and $x_0 = 0.3$ nm) give the following value of the capillary osmosis velocity $v_c = V_c v_m = 10^{-3}\nabla C$, cm/s. This makes up 10^{-5} cm/s or 0.36 mm/h at the concentration gradient of about 10^{-2} mol/cm^4. For reverse-osmosis membranes, the concentration gradient may be much higher (up to 1 mol/cm^4) and the velocity of capillary osmosis also increases. Capillary osmosis is of great importance in biological systems providing the mechanisms of mass transfer through thin cellular walls and water transport in plants.

It follows from eq. (IV.53) that ξ_0 values are 4×10^{-15} cm^2, if the adsorption is caused by dispersion forces, giving the thickness of the diffusion layer equal to ≈0.6 nm. In the case of wide pores and low absorption energies this enables capillary osmosis to be treated as an effect of solution slippage along the walls, because the shear stresses causing the flow are formed in a very narrow near-to-wall region of molecular size.

Capillary osmosis is more pronounced in the case of polar solute molecules or if the solution contains ionic components. It was shown experimentally that for aqueous solutions of alcohols the solute flux driven by the capillary osmotic flow (at $\nabla P = 0$) exceeded the simultaneous diffusion flux by several times. The transfer was driven by the difference of solution concentrations maintained on the porous glass membrane with average pore radius $r_c \approx$ 10 μm [102].

In the case of electrolyte solutions, one should take into account the contributions of the two types of species that form the diffuse layers – anions and cations – to determine the velocity of capillary osmosis. For dilute electrolyte solutions, the expression for the velocity of capillary osmotic flow takes the form [96, 97]:

$$V_c = -(z_-\xi_+ + z_+\xi_-)RT\nabla C/\eta\,, \qquad \text{(IV.55)}$$

where ξ_+ and ξ_- are the values of ξ_0 for cations and anions, respectively.

Estimation of ξ_+ and ξ_- values for relatively low surface potentials ψ_1 (less than 60–70 mV) and symmetrical ($z_+ = z_- = z$) and non-symmetrical electrolytes (1–2 and 2–1) results in the following equation for the capillary osmotic flow [98]:

$$V_c = -\varepsilon_0\psi_1^2\nabla C/32\pi\eta C_0. \qquad \text{(IV.56)}$$

As seen from this equation, the velocity of capillary osmotic slippage of electrolytes depends on the square of the surface potential and its direction (in distinction to the velocity of the concentration driven electro-osmosis, see eq. (IV.32)) does not depend on the sign of ψ_1. Concentration driven electro-osmosis either impedes or accelerates the capillary osmotic flow of electrolyte solutions depending on the relationship between the ion mobilities (D_+ and D_-). In solutions with similar values of ion mobility ($D_+ \approx D_-$) the only effect that occurs is capillary osmotic slippage.

The capillary osmotic flow is associated with tangential shear stresses occurring in the liquid in the region of diffuse adsorption layers. These stresses also act on pore surfaces and are balanced by their reaction. When the particles are not fixed in a porous body or if the individual particles are in the solution, the shear stresses generated by the concentration gradient cause the particle motion: diffusiophoresis. As a first approximation, the equations for the velocity of capillary osmosis are applicable to that of diffusiophoresis but with the opposite sign. Experimental investigations of diffusiophoresis for spherical particles of polystyrene latex ($d = 0.2–2$ μm) showed good agreement with the theory [103, 104].

The case of fine pores and overlapping diffuse adsorption layers requires a separate discussion. For non-ionic solutions this corresponds to narrow pores, of dimension commensurate with $\sqrt{|\xi_0|}$, for example, in silica gels, glasses with fine pores and reverse-osmosis membranes, where the average pore radius r_c varies from 1 to 3 nm. The same method can be used to determine the α_{12} coefficient. However, the calculation of $Q = \alpha_{21} \nabla \mu_1$ requires, first, the usage of a parabolic profile of the velocities; and, second, the determination of the concentration profile in a pore taking into account the interactions of the solute molecules with both surfaces of the slit pore. This approach was realized in [105], where the following expression for the average velocity of capillary osmotic flow was derived:

$$v = -(RT\nabla C/2\eta h)\left[\int_0^h (C(x)/C_0)(h-x)x\,dx - (h^3/6)\right], \text{ cm/s,} \qquad \text{(IV.57)}$$

where h is the width of the slit pore. When $h \to \infty$, this equation turns to eq. (IV.49) derived for wide pores. The function $C(x)/C_0$ is determined from the Boltzmann equation (III.5) and from eq. (III.2) expressing the total energy of interaction of solute molecules with the walls. The substitution of the polarizability $\alpha(i\xi)$ from eq. (IV.51) into (III.2) gives:

$$C(x)/C_0 = \exp\{(\hbar\delta^3/4\pi kT)[1/x^3 + 1/(h-x)^3] \times \int_0^\infty [(\varepsilon_1-\varepsilon_3)(\varepsilon_2-\varepsilon_3)/(\varepsilon_1+\varepsilon_3)(\varepsilon_2+2\varepsilon_3)\varepsilon_3]d\xi\}, \qquad \text{(IV.58)}$$

where $\varepsilon_1(i\xi)$, $\varepsilon_2(i\xi)$ and $\varepsilon_3(i\xi)$ are frequency-dependent dielectric permittivities of pore walls, the solute and the solvent, respectively.

Equation (IV.57) shows that at $C(x) = C_0$, when the solute concentration in a fine pore is the same as in the bulk solution beyond the pores, $v = 0$. Thus, capillary osmosis is connected with concentration variations across a fine pore, not only in the near-to-surface layer, as in the case of wide pores, but throughout the pore space. In this case the capillary osmosis cannot be reduced to the slippage. For $C(x) > C_0$, when adsorption is positive, the expression in the square brackets in eq. (IV.57) is also positive and the capillary osmotic flow is directed towards lower solution concentration. For $C(x) < C_0$, when the solvent but not the solute is predominantly adsorbed, the flow has the reverse direction. When the interactions are controlled by dispersion forces only, the value of $C(x)/C_0$ depends on the sign of the differences $(\varepsilon_1 - \varepsilon_3)$ and $(\varepsilon_2 - \varepsilon_3)$, as seen from eq. (IV.58). Positive adsorption of the solute takes place at $\varepsilon_1 > \varepsilon_3$ and $\varepsilon_2 > \varepsilon_3$ or $\varepsilon_1 < \varepsilon_3$ and $\varepsilon_2 < \varepsilon_3$. This means that the solvent must be more polar than the solute and the pore material; or, quite the reverse, be less polar than them. As discussed in Chapter III, §3, the polarizability, to a first approximation, can be characterized by the value of refraction index $n \approx \sqrt{\varepsilon}$ in the optical range.

To estimate the velocity of capillary osmosis in narrow pores, one can use the value of energy of molecules of the solute averaged over the cross-section:

$$U_c = -[1/(h-2\delta)] \int_{\delta}^{h-\delta} U(x)dx \cong -A_0\delta/h, \tag{IV.59}$$

where δ is the molecular radius and

$$A_0 = (\hbar/4\pi)\int_0^{\infty} [(\varepsilon_1-\varepsilon_3)(\varepsilon_2-\varepsilon_3)/(\varepsilon_1+\varepsilon_3)(\varepsilon_2+2\varepsilon_3)\varepsilon_3]d\xi\}. \tag{IV.60}$$

Introducing the concentration of the solution C_c, averaged over the pore cross-section, we obtain from eq. (IV.58):

$$C_c/C_0 \approx \exp(A_0\delta/hkT) = \text{const.}$$

Then for the velocity of the capillary osmotic flow we have:

$$v = -(RTh^2\nabla C/12\eta)[\exp(A_0\delta/hkT) - 1]. \tag{IV.61}$$

This equation shows that the velocity of capillary osmosis reduces quadratically with decreasing pore width h. However, when h approaches δ, the decrease slows down due to the increase in the exponent. In the case of high A_0 values, the impact of the exponent can prevail, and the velocity of capillary osmosis will increase as the pore size decreases. Thus, the effect of capillary osmosis depends on the energy of interaction between the solute and pore surfaces.

One can derive the value of pressure gradient, necessary to balance the capillary osmotic flow, from eq. (IV. 46) at $v = 0$:

$$V_c = \alpha_{12}\nabla\mu_2 = V_p = -\alpha_{11}\nabla\mu_1 = \alpha_{11}v_m\nabla P. \qquad \text{(IV.62)}$$

Using the expression (IV.61) obtained for $V_c = v/v_m$ and the value of $\alpha_{11} = h^2/12\eta v_m^2$ for a slit pore, we get:

$$\Delta P/\Delta C_{V=0} = RT[1 - \exp(A_0\delta/hkT)]\,. \qquad \text{(IV.63)}$$

Thus, the concentration gradient can lead either to the retardation of flow of the solution (if the directions of flow and capillary osmosis are opposite) or to an acceleration of filtration. This may be one of the reasons for filtration anomalies discussed in Chapter III, §2. If $\exp(A_0\delta/hkT) \approx 0$, that is possible at high negative A_0 values, when a solute cannot penetrate a pore (the case of semi-permeable membranes), eq. (IV.2) turns into the usual expression for the osmotic pressure of the solution $\Delta P = RT\Delta C$.

Due to the complexity of the calculations, more accurate estimations can be provided only by numerical methods. Capillary osmosis of electrolyte solutions in narrow pores, where there is EDL overlap, has still not been studied. It was shown in [106] that in NaCl solutions the velocity of capillary osmosis in a cylindrical pore with $r = 3$ nm increases with ΔC, but remains below that feasible for a semi-permeable membrane. An increase in salt concentration from 0.01 to 0.05 mol/l results in a reduction of the velocity of capillary osmosis due to the decreasing thickness of ion diffuse layers. More exact calculations of capillary osmosis in fine pores with charge-bearing surfaces are complicated because the effects resulting from structural changes in polar liquids have not yet been quantified.

§4 Reverse Osmosis

The coefficient α_{12} in eq. (IV.46) characterizes an effect that is the reverse of capillary osmosis, a flux of solute driven by the gradient of chemical potential of the solvent (pressure gradient) in the absence of a concentration gradient:

$$Q_c = \alpha_{21}\nabla\mu_1 = -\alpha_{21}v_m\nabla P. \qquad \text{(IV.64)}$$

In the case of wide pores, the coefficient α_{12} is defined by eq. (IV.48). Note that the mass flux can be positive (at $C(x) > C_0$) as well as negative: when $C(x) < C_0$ primarily the solvent passes through a pore or a capillary, whereas the solute is retained.

The change in solution concentration generated by pressure driven flow through rather narrow pores is called the reverse osmosis. Because of the wide practical applications of reverse osmosis, especially in separation of liquid mixtures and water desalination [58], let us consider the theory of the effect using, as an example, the reverse-osmosis membranes [54, 107]. Film membranes 100–200 μm thick are produced from different polymer materials and have pore radii of the order of nm.

Membranes with hydrophilic pore surfaces that negatively adsorb the solutes are used to separate aqueous solutions. Water structuring in fine hydrophilic pores makes it impossible to calculate the changes in the energy $U(x)$ of solute molecules and their distribution $C(x)$ over a pore cross-section due to the absence of the relevant theory. Because of this, we will use a value averaged over the pore cross-section of the concentration C in place of the unavailable detailed description of the solution state in a pore. This value is related to the energy of interaction of a molecule with pore walls averaged over the pore cross-section by the Boltzmann equation:

$$C = C_0 \exp(-U_c/kT) = C_0 \exp(-\Phi), \qquad \text{(IV.65)}$$

where $\Phi > 0$ is the effective potential set by the surface force field. In pores of the membrane, where the solute is in equilibrium with the bulk solution of concentration C_0, a concentration is established which is γ times lower than that in the bulk. Here $\gamma = C_0/C$ is the equilibrium distribution coefficient equal to $\gamma = \exp\Phi$.

In the framework of this approach we should also use the averaged velocities of flow in pores $v = V_c v_m$.

With these assumptions and at a constant potential Φ, it follows from eq. (IV.61) that the capillary osmotic flow in membranes equals:

$$\begin{aligned} V_c &= -(RTh^2 \nabla C/12\eta v_m)[\exp(-\Phi) - 1] \\ &= \alpha_{11} RT v_m \nabla C[1 - (1/\gamma)], \end{aligned} \qquad \text{(IV.66)}$$

where the known expression for a slit pore $\alpha_{11} = h^2/12\eta v_m^2$ was used. A comparison of eq. (IV.66) with the definition of $V_c = \alpha_{12} \nabla \mu_2$, from eq. (IV.46), gives:

$$\alpha_{12} = \alpha_{11} C_0 v_m (1 - 1/\gamma). \qquad \text{(IV.67)}$$

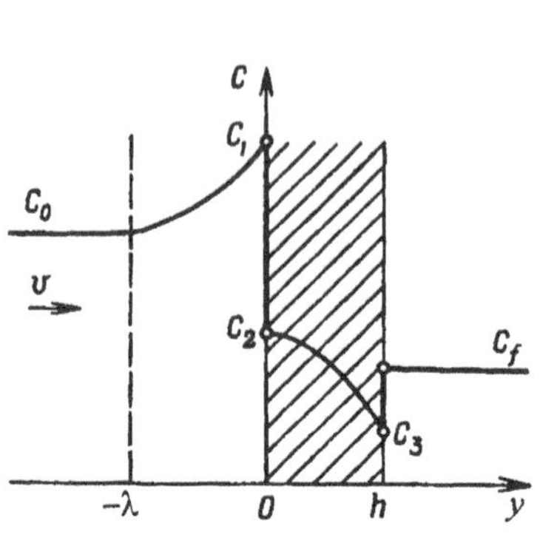

Figure 4.12 Distribution of solution concentration $C(x)$ in the direction of flow through a reverse-osmotic membrane.

At $\gamma = 1$ (when $\Phi = 0$), $\alpha_{12} = 0$ and the effect of capillary osmosis is not observed. In the case of negative adsorption $\gamma > 1$ and $\alpha_{12} > 0$, the flow is directed towards the higher concentration. As $\gamma \rightarrow \infty$, $\alpha_{12} = -\alpha_{11} C_0 \nu_m$, i.e. the solute is completely rejected by the membrane as for semi-permeable membranes when $\Phi = \infty$. When $\gamma < 1$, corresponding to positive adsorption, the capillary osmotic flow is directed towards the lower concentration.

Figure 4.12 shows the schematic distribution of solution concentration for a stationary flow through a membrane. The y-axis coincides with the direction of the flow; $y = 0$ corresponds to the inlet surface of the membrane of thickness h. Since the solution with concentration of $C < C_0$ passes through the membrane, part of the rejected solute accumulates in front of the inlet surface in the region $y < 0$. The excess diffuses back from the membrane creating a region of concentration polarization, where $C > C_0$. The thickness of the region λ depends on the intensity of stirring of the feed solution subjected to separation and maintained at a constant concentration $C_0 = \text{const}$.

Because the solution concentration in pores is γ times lower than in the feed solution contacting the pore inlets, the concentration changes step-wise at the surface of the membrane, $y = 0$, and becomes equal to $C_2 = C_1/\gamma$ at pore inlets. The outflow solution has lower concentration than that of solution under separation $C_f < C_0$. Therefore some concentration drop is established in the membrane that causes solute diffusion. In this case both terms in eq. (IV.46) should be taken into account to calculate the flux Q. To find the distribution of concentration $C(y)$ in the membrane and in the region of concentration polarization, the equation of convective diffusion accounting for both convective (α_{12}) and diffusion (α_{22}) mass transfer should be solved.

By the conditions of eq. (IV.65), the concentration should increase γ times passing the permeate region ($y > h$), i.e. $C_f = C_3\gamma$. The use of the concentration discontinuity at the membrane surfaces is only a convenient mathematical approach that enables an analytical solution to be obtained. In practice the changes in concentration take place in some transition zone of the finite thickness h_0. However, the solution obtained for this more complicated case at $h >> h_0$ does not differ much from the simplified version [107]. In the case of reverse-osmosis membranes the condition $h >> h_0$ is met because h_0 depends on the radius of action of surface forces, being about 10^{-6} cm, and the thickness h of the active layer of the membrane is about 10^{-4} cm.

The purpose of the calculations is to determine the so-called rejection coefficient:

$$\varphi = 1 - (C_f / C_0) \tag{IV.68}$$

controlling the rejecting ability of a membrane. Modern reverse-osmosis membranes possess high rejection coefficients φ up to 0.97–0.98, together with high efficiency characterized by the coefficient α_{12}.

The stationary distribution $C(y)$ can be found by solving a system of equations of convective diffusion written for the regions of concentration polarization, the membrane and the outflowing solution:

$$J = vC_i(y) - D_i \partial C_i / \partial y, \tag{IV.69}$$

where D_i is the respective diffusion coefficient of the i-th component. The solutions of this equation can be obtained by using the condition of the flow continuity J = const in all the regions. Boundary conditions are represented by concentration steps at $y = 0$ and $y = h$ and the constancy of concentrations C_0 and C_f in the bulk solutions.

As a result, we have the following equation for the rejection coefficient:

$$\varphi = 1 - \{1 + (\gamma - 1)[1 - \exp(-Pe_m)]\exp(-Pe_0)\}^{-1}. \tag{IV.70}$$

Here $Pe_m = vh/D_m$ is the internal Peclet number; $Pe_0 = v\delta/D_0$ is the external Peclet number, where D_m and D_0 are the diffusion coefficients of a solution in the pores of the membrane and in the bulk, respectively.

The dependence of φ on the velocity of solution flow in the membrane v, calculated by eq. (IV.70), is presented in Fig. 4.13. Values of φ pass a maximum at some value of v, that evidently determines the optimal velocity of separation. At higher values of v, the separation is worse due to the increasing influence of concentration polarization (increase in Pe_0). At lower values of v, the diffusion flux increases (decrease in Pe_m) affecting the separation. The maximum value $\varphi = \varphi_m$ is linked with the distribution coefficient γ by the following relation:

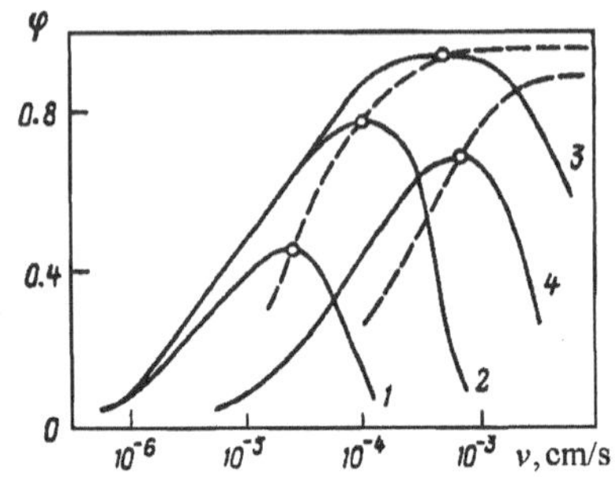

Fig. 4.13

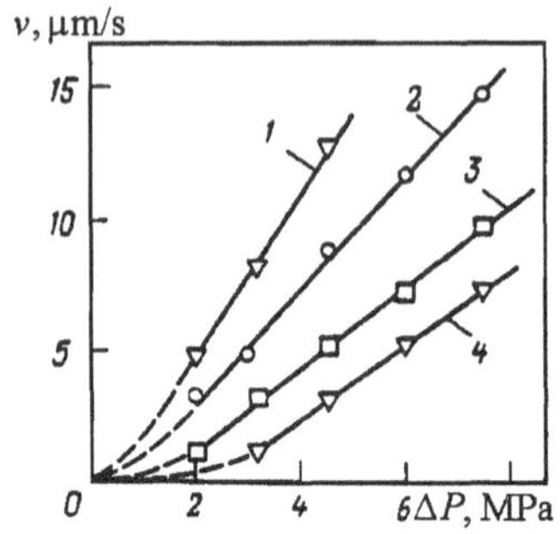

Fig. 4.14

Figure 4.13 Theoretical dependence of the selectivity φ of a reverse-osmotic membrane on the velocity of solution flow v. 1) $\gamma = 25$; $\omega = Pe_m/Pe_0 = 0.1$; 2) $\gamma = 25$; $\omega = 0.5$; 3) $\gamma = 25$; $\omega = 8$; 4) $\gamma = 10$; $\omega = 8$.

Figure 4.14 Effect of the capillary osmosis on the flow of electrolyte solution through the reverse-osmotic membranes: 1) KCl, $C_0 = 10^{-1}$ mol/l; 2) KCl, $C_0 = 1$ mol/l; 3) $BaCl_2$, $C_0 = 1$ mol/l; 4) $LaCl_3$, $C_0 = 1$ mol/l;

$$\varphi_m = 1 - (1/\gamma). \qquad \text{(IV.71)}$$

Many experimental data obtained for reverse-osmosis membranes support the theory developed (eq. (IV.70)). This equation also enables a reverse problem to be solved, that of determining the distribution coefficient γ and the diffusion coefficients of solutes in fine pores of membranes by a comparison with experimental dependences $\varphi(v)$.

Because, in the reverse-osmosis separation, a difference $(C_0 - C_f)$ of solution concentration arises (Fig. 4.12), the influence of capillary osmosis should appear together with diffusion. It follows from eqs. (IV.49) and (IV.57) that in the case of negative adsorption the capillary osmotic flow must be directed towards the higher concentration, i.e. in opposition to the flow causing the solution separation.

To account for the capillary osmotic counterflow, eq. (IV.46) should be used, which after the exchange of chemical potential for those of pressure and concentration gradients takes the form:

$$V = -\alpha_{11} v_m \nabla P + \alpha_{12} RT \nabla C / C_0. \qquad \text{(IV.72)}$$

Replacing $\alpha_{12} = \alpha_{21}$ with its value from (IV.67) and using the relation for the filtration coefficient $K = \alpha_{11} v_m \eta$, we obtain:

$$V = -(K/\eta)\{\nabla P - [1 - (1/\gamma)] RT \nabla C\}. \qquad \text{(IV.73)}$$

The second term in this equation can be interpreted as a dynamic osmosis pressure [55] appearing only when the solution flows through the membrane. When $\gamma \rightarrow \infty$, the dynamic osmosis pressure approaches the static value equal to $\Delta\pi_0 = RT\Delta C$, as in the case of a semi-permeable membrane completely rejecting the solute. In the case of low velocities of flow, the diffusion smoothes out the concentration fall across the membrane: $C_f \rightarrow C_0$ leading to $\nabla C = 0$. Thus, in reverse-osmosis membranes the concentration gradient ∇C is caused only by the solution flow.

It follows from eq. (IV.73) that the Darcy law is not obeyed in the case of filtration through porous bodies rejecting the solute. At a reasonably high velocity of flow, when the values of φ weakly depend on v, the extension of linear part of the dependence $V(\Delta P)$ must intercept on the pressure axis a portion $\Delta\pi$ equal to:

$$\Delta\pi = [1-(1/\gamma)]RT(C_0 - C_f)\,. \tag{IV.74}$$

Replacing $1-(1/\gamma)$ with its value from eq. (IV.71) and using eq. (IV.68), we obtain:

$$\Delta\pi \cong \varphi^2 RTC_0 = \varphi^2 \Delta\pi_0\,. \tag{IV.75}$$

Thus, the higher the rejection coefficient of the membrane selectivity ($\varphi \rightarrow 1$), the closer the dynamic values $\Delta\pi$ approach those obtained under static conditions. In the case of low selectivity ($\varphi \rightarrow 0$) $\Delta\pi$ also approaches zero.

Figure 4.14 shows the experimental data obtained for the reverse-osmosis porous glass membranes [55]. Here measured values of the flow velocity v are plotted in dependence on the pressure drop ΔP. As seen from the graphs, for solutions with reasonably high concentration (when $\Delta\pi_0$ values are high) and for highly selective membranes ($\varphi = 0.8-0.9$) the intercepted values of $\Delta\pi$ range up to 10–20 atm. The dependences of $\Delta\pi$ on φ obtained comply with eq. (IV.75). The deviations from the Darcy law are associated here with the influence of capillary osmotic counterflow that can explain the anomalies observed for solution flow in fine pores (see §1). The decrease in osmotic pressure $\Delta\pi$ compared to the static value $\Delta\pi_0$ explains the ability of reverse-osmotic membranes to reject concentrated electrolyte solutions at pressure drops ΔP less than $\Delta\pi_0$. For example, a 2 M NaCl solution ($\Delta\pi_0$ = 9.1 MPa) was concentrated using a membrane with the rejection coefficient $\varphi = 0.25$ at a pressure drop in the membrane of only 5 MPa [108].

The above-mentioned results were obtained for nonionic solutions and symmetrical electrolytes like KCl, with similar ion mobilities. In general the influence of surface charge and differences in electrical potential appearing on surfaces of the membrane due to different distribution coefficients γ for cations and anions ($\gamma_+ \neq \gamma_-$) should be taken into account. Moreover, the diffusion electrical potential arises (see eq. (IV.31)) when $D_+ \neq D_-$. One should determine the distribution of electrical potential along the y-axis to derive the distribution of ion concentration. The calculations showed that eq. (IV.70) holds its general form but with the redefined parameters [109, 110]. The effective diffusion coefficients are used as D_m and D_0, and the distribution coefficient for a binary solution takes the form:

$$\gamma = \gamma_+^{z_-/(z_+ + z_-)} \gamma_-^{z_+/(z_+ + z_-)}, \tag{IV.76}$$

where z_- and z_+ are ion charges.

For membranes with high density of surface charge the calculations are more complex. In many cases results were obtained by numerical methods [109, 110].

The basic purpose of the theory is to find the effective potential of surface forces Φ responsible for the membrane separation. For non-ionic solutions, the values of $\exp\Phi = C/C_0$ may be calculated on the basis of the theory of dispersion forces (see eq. (IV.58)). The results obtained using this approach agreed well with the experimental data [111]. For electrolyte solutions, the calculations of changes of ion energy in a pore by eqs. (III.4) and (III.5) may be used. As shown in [56, 112, 113], the account of dielectric exclusion and image forces provides, in many cases, good agreement with experimental data. However, in the case of extremely narrow pores there are problems associated with the need to account for the surface charge, ion size and non-uniformity of dielectric permittivity simultaneously. In extremely fine pores one also has to take into account the discrete molecular structure of solutions. The current state of the theory of reverse osmosis in electrolyte solutions is reviewed in [111, 113] and in the monograph [114].

§ 5 Thermo-osmosis

The phenomenon of thermo-osmosis, flow of liquid in pores under a temperature gradient, was first investigated and theoretically justified by Derjaguin and Sidorenkov [115].

Let us consider a system consisting of a porous body filled with liquid and contacting reservoirs between which constant pressure and temperature differences are maintained. Neglecting for the moment the diffusion and electrical phenomena we arrive at the following equations for the thermodynamics of irreversible processes:

$$V = \alpha_{11}\nabla\mu_1 + \alpha_{14}\nabla T/T, \tag{IV.77}$$

$$W = \alpha_{41}\nabla\mu_1 + \alpha_{44}\nabla T/T. \tag{IV.78}$$

For a one-component liquid $\nabla\mu_1 = -v_m\nabla P$, and α_{11} is expressed by (IV.3) as $\alpha_{11} = K/\eta v_m$, where η is the viscosity, v_m is the molar volume of liquid and K is the filtration coefficient.

A temperature gradient, ∇T, must generate a liquid flow according to eq. (IV.77) and a pressure gradient must generate a heat of transfer according to (IV.78). The intensity of these processes is characterized by the phenomenological coefficient $\alpha_{14} = \alpha_{41}$. According to Derjaguin, a heat flux due to the liquid flow in a porous body may be caused by structural changes in boundary liquid layers (Chapter III) changing the specific enthalpy ΔH (J/m^3) of the liquid compared to the bulk. This is associated with the changes in the energy of intermolecular bonds in the liquid, for example weakening or strengthening of hydrogen bonds between water molecules. In the first case, $\Delta H > 0$, and the transition of the boundary layer to bulk state is accompanied by heat absorption. In the second case, $\Delta H < 0$, and the heat is released when the inflow bulk phase forms a boundary layer. In particular, methods for the estimation of surface hydrophilicity are based on the measurement of heat of wetting [53].

When $\Delta H < 0$, the flowing liquid releases heat at the inlet and absorbs at the outlet of a porous body. The reverse situation occurs at $\Delta H > 0$. In both cases (at $\Delta H \neq 0$), the flow driven by the pressure drop generates a heat flux, the heat of transfer $W_0 = \alpha_{41}\nabla\mu_1$. The cross-phenomenon, a mass flux driven by the temperature gradient $V_t = \alpha_{14}\nabla T/T$, was called the thermo-osmosis. From the physics standpoint, the phenomenon is associated with gradients of thermal tension in boundary layers that tends to develop tangential shear stresses in this region.

The analogy to this effect is, for example, the thermal dependence of surface tension connected with restructuring of transition layers between the liquid and its vapor. A temperature gradient generates a shear stress $\tau = \partial\sigma/\partial T$ at the liquid surface which, in turn, generates the thermocapillary film flow (see Chapter V).

To determine the coefficients $\alpha_{14} = \alpha_{41}$, one must derive the expression for the heat of transfer based on some model for the pore structure. If pores are reasonably wide and the thickness, h_s, of the boundary layers is much lower than the pore radius, r, we can employ the approach used for electro-osmosis at $\kappa h >> 1$ (see §2) to calculate the heat of transfer. The velocity distribution near to the surfaces of straight cylindrical channels may be considered linear.

The heat of transfer is determined by analogy with eq. (IV. 15) as

$$W_0 = -(n\Sigma/S)\int_0^r v(x)\Delta H(x)dx, \tag{IV.79}$$

where x is measured along a normal to the pore walls. Because the thickness of near-to-wall layers, where $\Delta H(x) \neq 0$, is far less than r, the upper limit can be changed to ∞ in integration.

The velocity distribution is specified by eq. (IV.14). The distribution of enthalpy changes $\Delta H(x)$ is generally unknown. Some data on these changes can be obtained from computer simulations. For example in [34, 38, 116], the effects of changes in both numbers and energies of intermolecular hydrogen bonds in boundary layers were calculated for water near to a solid surface.

However, these changes essentially depend on surface hydrophilicity, allocation and a number of centers capable of forming hydrogen bonds with water molecules. In [117] the Monte-Carlo simulation was used to obtain the average potential energy U_m of water molecules in the film near to a solid surface on which N active centers, able to form hydrogen bonds with water molecules, were located at the points of a square lattice. Due to the effect of the solid surface, the values of U_m varied from $U_m = -4.2\times10^4$ J/mol at $N = 0$ (the model of a hydrophobic surface) to $U_m = -6.3\times10^4$ J/mol at $N = 25$ (the model of a hydrophilic surface). As N is increased, the water dipoles are re-oriented from predominantly parallel, at $N = 0$, to predominantly normal to surface direction, at $N = 25$; this was accompanied by a change of the order parameter $s = 0.5\,(3 < \cos\phi > - 1)$ from $s = -0.18$ to $s = +0.1$.

The isotropic state of the boundary layers, $s = 0$, was associated with the number of centers ($N = 13$) for which the distances between centers were close to the intermolecular spacing in the bulk water. The corresponding value of U_0 was close to that calculated for the bulk water by the same approach $U_0 = -4.9\times10^4$ J/mol. Thus, the change of the number of active surface centers results in marked structural changes and the thermodynamic properties of the boundary water layers.

Simple analytical expressions for the thermo-osmotic velocity can be approached in two cases. For wide pores, an appropriate model is that presented by a thin boundary layer of thickness $h_s \ll r$ with some effective, average value of $\Delta H_s = \text{const}$. Beyond the boundary layer, at $x > h_s$, $\Delta H_s = 0$. Integration of eq. (IV.79) provides the following expression for the heat of transfer:

$$W_0 = n\Delta H_s h_s^2 \nabla P / 2\eta. \tag{IV.80}$$

Whence it follows:

$$\alpha_{41} = \alpha_{14} = -n\Delta H_s h_s^2 \nabla P / 2\eta v_m. \tag{IV.81}$$

Thus, the thermo-osmotic flow is equal to

$$V_t = -n\Delta H_s h_s^2 \nabla T / 2\eta v_m T = \chi \nabla T / v_m T, \tag{IV.82}$$

where χ is the so-called thermo-osmosis coefficient, in cm^2/s.

Since $h_s << r$, the thermal stresses occur only in a thin near-to-wall layer and we can discuss thermo-osmotic slippage of liquid in pores, as in the case of electro-osmosis at $\kappa h >> 1$. The most significant changes in ΔH occur in the case of polar liquids. However, since the structures of all liquids change at a plane boundary (Chapter III), the thermo-osmotic flow is a general phenomenon, as was recently shown by the methods of statistical physics [118]. In polar liquids, especially in water, this effect is more pronounced (due to sizable changes of boundary layer structure) and, therefore, can be determined in a direct experiment.

As seen from eq. (IV.82), when $\Delta H_s < 0$, the thermo-osmotic flow is directed towards the hot side, but when $\Delta H_s > 0$, it is directed towards the cold side.

For porous bodies with wide pores, the velocity of thermo-osmosis is low and commensurate with that of viscous flow driven by the occasional pressure variations, that hinders the measurements. Because of this, most experimental data on thermo-osmosis coefficients have been obtained for microporous bodies. However, we cannot use the foregoing solutions, because in this case h_s is commensurate with pore radius.

We can use here another approximation assuming that at $h_s \sim r$ some average value, ΔH_c = const, independent of x is assigned to the liquid in a pore. Then, instead of (IV.79), this can be written

$$W_0 = \alpha_{11} v_m^2 \Delta H_c \nabla P.$$

Whence

$$\alpha_{41} = -\alpha_{11} v_m \Delta H_c = -K\Delta H_c/\eta, \text{ mol/cm s.} \quad \text{(IV.83)}$$

The velocity V_t and the thermo-osmosis coefficient χ are directly related to the filtration coefficient K and average changes of enthalpy ΔH_c:

$$V_t = -K\Delta H_c \nabla T / \eta T = \chi \nabla T / v_m T; \; \chi = -K v_m \Delta H_c/\eta. \quad \text{(IV.84)}$$

Equation (IV.84) makes it possible to determine the changes of enthalpy ΔH_c using the measured values of the thermo-osmosis velocity and filtration coefficient.

A number of effects besides the heat of transfer and the thermo-osmosis follow from eqs. (IV.77) and (IV.78). For example, if the thermo-osmotic flow of liquid is compensated by the pressure driven counterflow, when $V = 0$, the ratio $\Delta P/\Delta T$ characterizing the thermomechanical effect, a stationary pressure drop, ΔP, under the temperature drop ΔT in a porous body, can be determined. At $V = 0$, eq. (IV.77) gives:

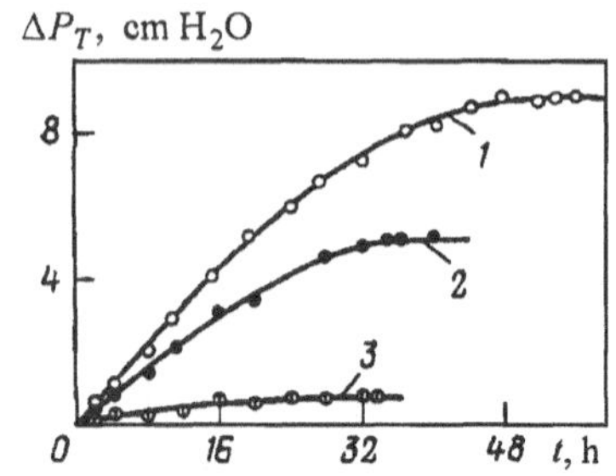

Figure 4.15 Measured change in thermomechanic pressure difference ΔP_T in porous glass at a temperature gradient ∇T = 170 deg/cm and at different average temperatures 20°C (*1*), 40°C (*2*), 60°C (*3*).

$$\Delta P/\Delta T = \alpha_{14}/\alpha_{11}v_m T. \tag{IV.85}$$

For microporous bodies, substituting the relevant expressions for α_{11} and α_{14} from eq. (IV.83), we get:

$$\Delta P/\Delta T = -\Delta H_c/T. \tag{IV.86}$$

Therefore, measurements of the ratio $\Delta P/\Delta T$ provide values of the average changes of enthalpy in fine pores.

Figure 4.15 shows, as an example, the results of measurements of the thermomechanical pressure performed on a specimen of porous glass with the average pore size r_c = 8.3 nm [49]. The values of ΔP are plotted in dependence of time. Stationary values of ΔP were achieved after 30–60 h. Calculations by eq. (IV.86) give the values of $\Delta H_c = 1.5\times10^4$ J/m^3 at 20°C. Consequently, the values of ΔH_c are positive since the pressure increases at the colder end of the specimen, and $\Delta P/\Delta T < 0$.

Increase in the average temperature of the specimen results, as seen from Fig. 4.15, in a decrease in ΔP values at the same temperature gradient ∇T = 170 deg/cm. At 40° $\Delta H_c = 8\times10^3$ and $H_c = 1.3\times10^3$ J/m^3 at 60°. Such a decrease is caused by the destruction of a specific water structure in fine pores and the approach of its properties to those in the bulk (see Chapter III).

At $W = 0$, when the heat of transfer is balanced by the heat conductivity, the expression for a stationary temperature drop, ΔT, arising due to viscous flow, can be obtained from eq. (IV.73):

$$\Delta T/\Delta P = \alpha_{41}v_m T/\alpha_{44}. \tag{IV.87}$$

In application to microporous bodies we obtain the following expression for a thermomechanical effect:

$$\Delta T/\Delta P = -Kv_m \Delta H_c T/\alpha_{44}\eta, \tag{IV.88}$$

where α_{44} is the coefficient of thermal conductivity.

However, the temperature differences arising are very small and masked by the heat effects associated with energy dissipation in a viscous flow. Thus, to determine the enthalpy changes, ΔH_c, measurements of $\Delta P/\Delta T$ or thermo-osmosis velocity are usually employed.

Values of the thermo-osmosis coefficient obtained in experiments with hydrophilic microporous bodies with $r_c < 10$ nm (porous glass, membranes, clays, silica gels), range from -10^{-6} to -10^{-8} cm^2/s [49, 119–122]. These χ values are negative, because the thermo-osmotic flow is directed towards the cold side. An increase in temperature and pore size results in a reduced thermo-osmosis velocity. Hydrophobization of pore surface changes the sign of χ: the flow in pores of acetate-cellulose membrane without surface hydroxyls is directed towards the hot side [123]. This agrees with the above-mentioned changes of U_m occurring when turning from hydrophilic surfaces, when $U_m > U_0$, to hydrophobic ones, when $U_m < U_0$ [117].

Experiments with porous bodies show that in narrow hydrophilic pores the changes in enthalpy are positive and, for example, equal to $\Delta H_c = 10^4 - 10^3$ J/m^3 for microporous glass depending on pore size. For comparison, the specific heat of ice melting is four orders higher than the maximum obtained value for ΔH_c. This means that even small changes in water structure in fine pores can generate measurable thermo-osmosis velocities.

In porous glass, the thermo-osmotic flow directed towards the cold side was also observed for another polar liquid: methyl alcohol. However, in this case χ values are one order lower [49].

In porous bodies with wider pores ($\kappa h > 1$), besides the thermo-osmosis, some thermoelectrical effects associated with heat polarization of EDL can be detected. In this case, the electric current should also be taken into account in addition to heat and mass fluxes. Then the system of equations (IV.1) takes the form:

$$V = \alpha_{11}\nabla\mu_1 + \alpha_{13}\nabla E + \alpha_{14}\nabla T/T,$$

$$J = \alpha_{31}\nabla\mu_1 + \alpha_{33}\nabla E + \alpha_{34}\nabla T/T,$$

$$W = \alpha_{41}\nabla\mu_1 + \alpha_{43}\nabla E + \alpha_{44}\nabla T/T. \tag{IV.89}$$

Two new phenomenological coefficients $\alpha_{34} = \alpha_{43}$ describing two cross thermoelectrical effects appear here. As has been shown in [124], the heat of transfer generated by the gradient of electrical potential $W_e = \alpha_{43}\nabla E$ includes the heat of transfer of electro-osmotically flowing water polarized in the

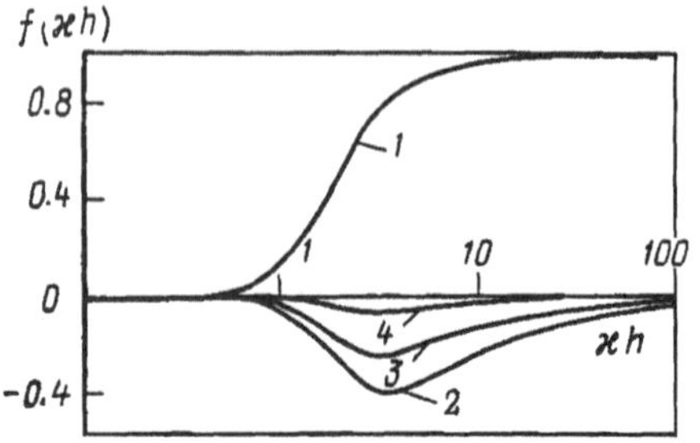

Figure 4.16 Dependences of the dimensionless thermo-osmotic coefficients $\chi_{1e} \sim f_2(\kappa h)$ (curve *1*) and $\chi_{2e} \sim f_1(\kappa h)\alpha_{13}/\alpha_{33}$ (curves *2–4*) on κh. $\psi_1 = 150$ mV (*2*); 100 mV (*3*); 50 mV (*4*).

EDL, and the difference between the heat of transfer of anions Q_- and cations Q_+ [74]. For the model of plane slit pore of half-width h the coefficient α_{43} is equal to:

$$\alpha_{43} = [\varepsilon_0^2\psi_1^3\kappa^2 a f_1(\kappa h)/64\pi^2\eta] + (\alpha_{33}(Q_+ - Q_-)/2F), \qquad \text{(IV.90)}$$

where $f_1(\kappa h) = (\tanh \kappa h/\kappa h) + \tanh^2 \kappa h - 1 + (2\tanh^3 \kappa h/\kappa h)$; ψ_1 is the potential of surface of the slit; F is the Faraday and $a = [1 + (T/\varepsilon_0)(\partial\varepsilon_0/\partial T)] \cong -0.4$ for water.

The cross coefficient $\alpha_{34} = \alpha_{43}$ defines the relevant value of the electric current J generated by heat polarization of the EDL and ionic thermo-diffusion.

Polarization of water inside the EDL changes its specific enthalpy ΔH_e. In the case of charged surfaces, this effect is "added" to the changes in enthalpy ΔH of water near a hydrophilic surface discussed above. Local changes in ΔH_e will occur under an external temperature gradient. Due to this, the liquid flow driven by the pressure gradient will generate a heat of transfer $W_{0e} = \alpha_{41}\nabla\mu_1$. In application to the same model, this gives the following expression for the kinetic coefficients:

$$\alpha_{41} = \alpha_{14} = \varepsilon_0\psi_1^2 a v_m f_2(\kappa h)/16\pi\eta, \qquad \text{(IV.91)}$$

where $f_2(\kappa h) = (\tanh \kappa h/\kappa h) - \tanh^2\kappa h - 1 + (4/3)\kappa h^2[1 - \tanh^2\kappa h]$. The coefficient $\chi_{1e} = \alpha_{14}v_m$ determines the thermoelectric component of thermo-osmosis.

Thus, in charged porous bodies with pore sizes or electrolyte concentration, that meet the condition $\kappa h \sim 1$, the heat of transfer is determined by the sum of two components:

$$V_T = \alpha_{13}\nabla E_T + \alpha_{14}\nabla T/T. \quad \text{(IV.92)}$$

The second term specifies the thermo-osmotic flow generated by water polarization in the EDL. The first term is associated with electro-osmotic flow driven by the gradient of electrical potential ∇E generated by the external temperature gradient. It follows from eq. (IV. 89) that at a steady state, when $J = 0$ and $\nabla\mu_1 = 0$:

$$\nabla E_T = -\alpha_{34}\nabla T/\alpha_{33}T. \quad \text{(IV.93)}$$

Substituting this value into (IV.92), for the thermoelectric flow we have:

$$v_T = v_m[(-\alpha_{13}\alpha_{34}/\alpha_{33}) + \alpha_{14}]\nabla T/T = (\chi_{2e} + \chi_{1e})\nabla T/T, \text{ cm/s}, \quad \text{(IV.94)}$$

The expressions for coefficients α_{34} and α_{14} were given above: eqs. (IV.90), (IV.91). The coefficients α_{13} and α_{33} are well-known from the theory of electrokinetic phenomena in slit pores (see §2) and represent the coefficients of electro-osmosis and electrical conductivity, respectively.

Analysis of the results obtained shows that, if $Q_+ = Q_-$, the coefficients χ_{1e} and χ_{2e} always have different signs and are proportional to the square of the potential of the pore surface ψ_1^2. Consequently, $\chi_{1e} > 0$ and the corresponding flow is directed towards the cold side. Figure 4.16 shows the dependence of dimensionless values of χ_{1e} and χ_{2e} calculated as a function of κh for various values of ψ_1. When $\kappa h < 1$, i.e. in microporous bodies and electrolyte solutions of low concentration, the thermoelectric phenomena are weak. Here the flow under the temperature gradient is associated only with liquid structuring, i.e. with pure thermo-osmosis.

When $\kappa h \geq 5$, the major contribution is given by the component of thermoelectric flow associated with water polarization in the EDL (curve *1*). In the range of κh between 2 and 10, the second component, the thermo-electro-osmotic term, manifests itself (curves *2* to *4*). The contribution of this component increases with rising potential of the pore surface ψ_1.

The effect of ion thermodiffusion can also show up and may be the more pronounced the greater the difference between values of heat transfer $\Delta Q = Q_+ - Q_-$. As estimations show [124, 74], this effect can exceed other components of thermoelectric transfer when $\kappa h > 0.1$, if the surface potential is sufficiently high. The difference in heat of transfer of ions leads to the thermodiffusion difference of potentials causing the thermo-electro-osmotic flow of liquid. The direction of this component of thermoelectric flow depends on the sign of the difference ΔQ and that of ψ_1.

Therefore, to calculate the heat of transfer in charged porous bodies at $\kappa h \geq 1$, a number of contributions, depending not only on structural changes in

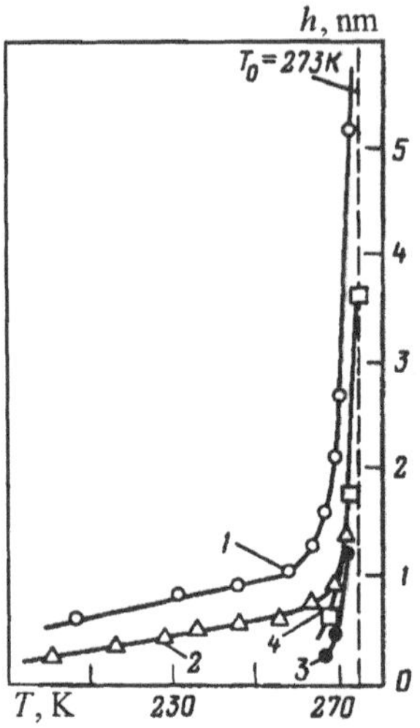

Fig. 4.17

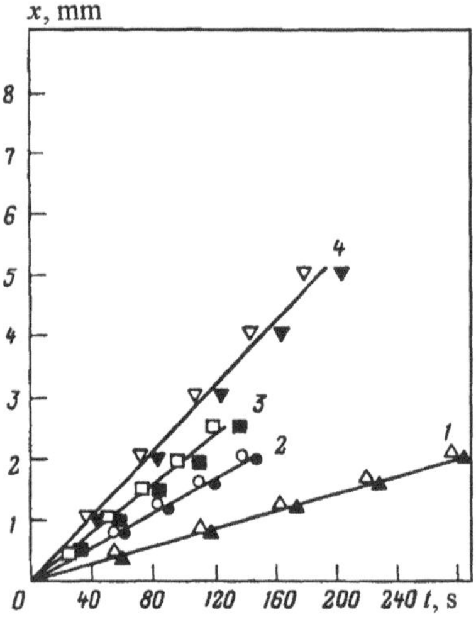

Fig. 4.18

Figure 4.17 Temperature dependence of thickness of non-freezing layers near to the surface of Aerosil particles (*1*), Na-montmorillonite (*2*), Teflon (*3*) and ice–air interface (*4*) [134].

Figure 4.18 The rate of displacement $x(t)$ of an ice column of length l = 1.67 mm in a capillary of r = 9 μm at −2.05°C under the pressure drop ΔP = 0.02 (*1*); 0.04 (*2*); 0.06 (*3*); 0.08 (*4*) MPa.

liquid but also on surface charge, heat of ion transfer and dielectric properties of media, should be taken into account.

§6 Flow of Non-freezing Water in Porous Bodies

Water transfer in freezing porous bodies (soils, grounds and construction materials) involves three mechanisms: diffusion of vapor, flow of non-freezing water films on particle surface and flow of non-freezing interlayers between ice and solid surface. At constant temperature, the non-freezing interlayers and films are in equilibrium with ice and vapor for which the pressure is governed by the temperature of the bulk ice.

The influence of surface forces on the structure of the water boundary layers provides the reason why films do not freeze (Chapter III). Non-freezing interlayers represent a boundary phase of ice, and the structure of this phase is changed (due to the action of the adjacent boundary) to the extent that a transition takes place from the crystalline to the amorphous liquid state. A distinguishing feature of this phenomenon is an interface between the bulk ice and the liquid non-freezing interlayer indicating a definite thickness h of the interlayer depending on temperature and pressure.

Liquid films at the ice–vapor interface [125–131] and liquid interlayers between ice and solid surfaces [132–135] were experimentally observed. In the latter case, the thickness h of a non-freezing interlayer is greater, the higher the hydrophilicity of a solid substrate. Figure 4.17 shows the temperature dependences of the thicknesses h of non-freezing interlayers. For hydrophobic Teflon the thickness of the interlayer does not exceed 0.5 nm at −3°C, but for the hydrophilic surface of Aerosil particles it approaches 3 nm at the same temperature. The values of h are somewhat lower for clays. The lowest values correspond to the interlayers between ice and hydrophobic surface. The measurements, shown in Fig. 4.17, were performed on frozen Aerosil dispersions by the NMR method [134]. The thickness h was determined as the volume of liquid phase in a frozen system divided by total surface of particles.

It is apparent that the non-freezing interlayers must have a structure different from that of bulk water that opposes the transition to the ice phase with which they are in direct contact. A decrease in temperature results in gradual freezing out of the interlayers [125]. At temperatures lower than −50°C their thickness approaches the thickness of a monolayer, that hardly can be regarded as a liquid phase.

Non-freezing water interlayers between ice and molecularly smooth surface of quartz capillaries were also detected by the observation of displacement of ice columns under an applied gas pressure ΔP [136] and by the method of the capillary dilatometer [135].

Figure 4.18 presents the measured displacement, x, of the ice column as a function of time t at various values of ΔP. Black and white points correspond to the ice column moving back and forth inside the same segment of a capillary. Linearity of the functions obtained $x(t)$ and their independence of the direction of displacement of the ice column prove the cylindrical shape of the capillary bore and the absence of hysteresis. The rate of displacement of the ice columns at different temperatures, $v = dx/dt$, was calculated from the slope of the graphs $x(t)$. Figure 4.19 shows the calculated dependences $v(\Delta P)$ that fit the equation of viscous flow for non-freezing water interlayers:

$$v = rh\nabla P/2\eta, \qquad \text{(IV.95)}$$

where $\nabla P = \Delta P/l$ is the pressure gradient, l is the length of the ice column, r is the radius of the capillary and η is the viscosity of the interlayer.

Such experiments, which enabled the ratio h/η to be determined, were performed for water and aqueous solutions of different concentrations [136, 137]. The experimental scheme and the main results are shown in Fig. 4.20, where the values of h/η are plotted on a logarithmic scale.

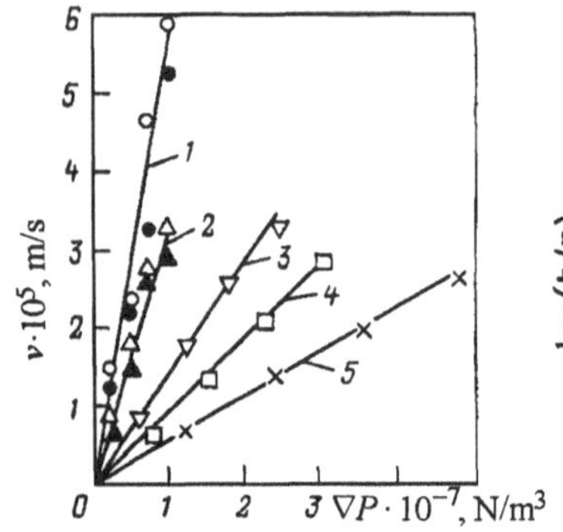

Fig. 4.19

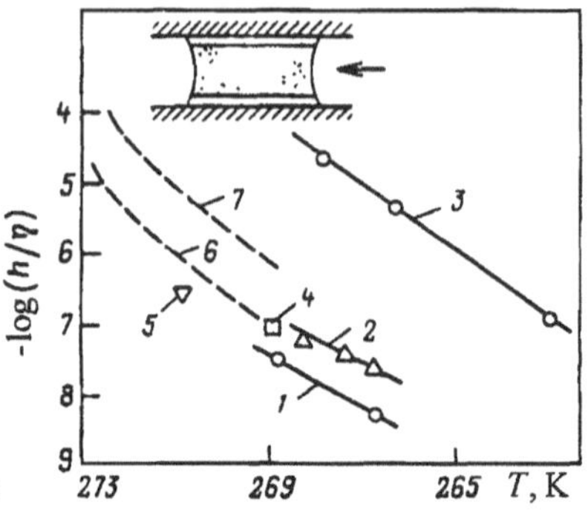

Fig. 4.20

Figure 4.19 The rate of displacement of ice columns $v = dx/dt$ in a capillary $r = 9$ μm vs. gas pressure gradient ∇P at various temperatures: –0.25°C (*1*), –0.5°C (*2*), –1°C (*3*), –1.5°C (*4*) and –2.05°C (*5*).

Figure 4.20 Ratio h/η vs. temperature T for water (*1*) and aqueous solutions: 10^{-1} mol/l KCl (*2*); 1 mol/l KCl (*3*); 10^{-2} mol/l KCl (*4*); 10^{-2} mol/l H_2SO_4 (*5*). Dashed lines *6* and *7* are plotted using the data from [136] for water (*6*) and 10^{-1} mol/l KCl (*7*). Thickness, in cm; viscosity, in poise.

The ratio of the thickness of non-freezing interlayers to their viscosity always decreases with temperature. An increase in solution concentration causes h/η values to grow. Separate estimations of the thickness and viscosity were obtained in only two cases: for pure water (curve *6*) and for concentrated KCl solution (curve *3*). For water, one can use the reported data on the thicknesses of non-freezing interlayers between ice and Aerosil particles (Fig. 4.17). Using the calculated values of h/η, we find that at –2°C the viscosity of interlayers of thickness $h = 5$ nm is 0.04 Ns/m^2, that is 20 times higher than the viscosity of the bulk water at the same temperature. As the temperature decreases to –5°C, the thickness of the non-freezing interlayers reduces to 2 nm but the viscosity rises to 0.2 Ns/m^2. However, such an increase in viscosity can be associated not only with structural changes of water, but rather with some influence of the roughness of ice surface. Friction between peaks of ice crystallites and the capillary surface can also impede the motion of the ice column in the capillary bore.

For concentrated solutions, we can use another assumption based on the known effect of destruction of the specific structure of a boundary layer. We can assume that the viscosity of non-freezing interlayers in concentrated solutions does not differ much from that of the bulk solution. Using the data provided by curve *3* (Fig. 4.20), we find that at –5°C the thickness of the non-freezing interlayer of the solution is equal to 6 nm exceeding by three times the thickness of a non-freezing water interlayer at the same

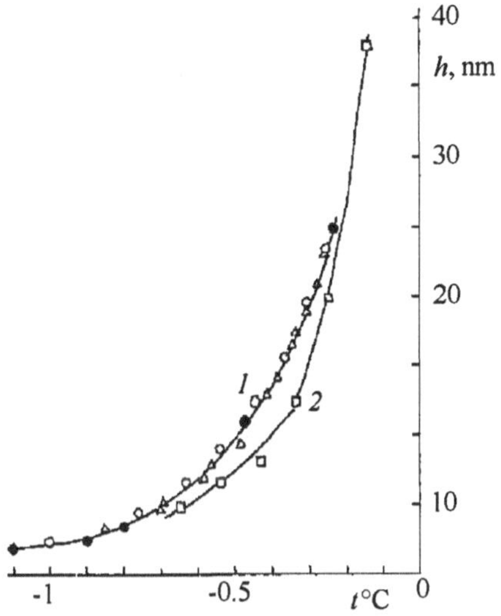

Figure 4.21 Thickness of non-freezing water interlayers h vs. temperature for the particles of silica hydrosol (curve *1*) and for water filling a separate thin capillary of $r \approx 1$ μm (curve 2).

temperature. In the case of solutions of lower concentration (0.1 mol/l KCl: curves 2 and 7) the values of thickness and viscosity of interlayers range between those for water and for concentrated solutions.

Water and aqueous solutions crystallize only after a considerable supercooling (by −30°C to −40°C) following the mechanism of homogeneous nucleation. Heterogeneous crystallization is not possible due to the high hydrophilicity of the capillary surface. Non-equilibrium partially amorphous structures form (on the boundaries of ice crystallites) in the course of rapid crystallization which transform irreversibly into an equilibrium state of the ice-I, if the temperature rises. Equilibrium branches of the graphs $V(T)$, where V is the volume of the system, are characterized by the independence of V on the path through which it was achieved: by increasing or decreasing the temperature. Equilibrium between the ice and the solution in the capillaries is established for 0.1 mol/l KCl solution at $t \geq -10$°C and for 1 mol/l at $t \geq -7.5$°C. Pure ice in capillaries with $r = 1-10$ μm melts exactly at 0°C.

More precise measurements of the thicknesses of non-freezing interlayers were performed at temperature higher than −1°C by the method of the capillary dilatometer [135]. The basic assumption in the method is to use narrow exactly cylindrical quartz capillaries with molecularly smooth surfaces to measure the volume of the ice. This enables the friction between the ice and the capillary surface to be excluded at least at $t \geq -1$°C. Capillaries with r about 10 μm were filled with monodisperse silica sol and the changes in the

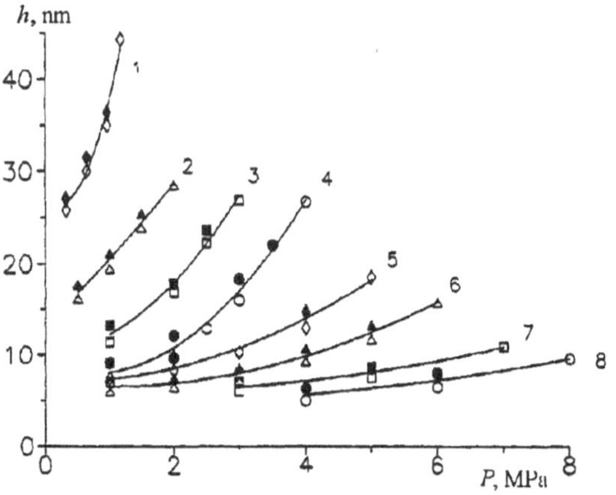

Figure 4.22 Thickness of non-freezing water interlayers h between ice and surface of the narrow quartz capillary (r = 1.14 μm) vs. pressure P and temperature t = −0.14°C (*1*), −0.25°C (*2*), −0.33°C (*3*), −0.42°C (*4*), −0.55°C (*5*), −0.64°C (*6*), −0.82°C (*7*), and −1°C (*8*).

ice volume caused by temperature changes were observed. The ice formed after supercooling was transformed into the equilibrium phase of ice-I by being maintained a long time at −0.2°C. Volume changes of the frozen dispersion, due to partial transition of water from non-freezing liquid interlayers into the phase of bulk ice, were followed by the measurements of the length of the frozen column in the capillary by a comparator.

Figure 4.21 shows the experimentally obtained dependences of the interlayer thickness on temperature. Curve *1* corresponds to the results of measurement for silica hydrosol with particle radii of 68 nm. Curve *2* was obtained for non-freezing interlayers formed directly on the internal surface of a narrow ($r \approx 1$ μm) quartz capillary. When the temperature approaches the melting point of ice, the thickness of non-freezing interlayers increases up to 30–40 nm near −0.15 to −0.2°C. At temperatures below −0.6°C the thickness reduces abruptly to 5–6 nm.

Observations of the displacements of small ice columns were carried out in the same narrow quartz capillary ($r \approx 1$ μm) by the method considered above. Since in this case the thickness of interlayers is known (Fig. 4.21, curve *2*), more accurate estimations of viscosity were obtained for the non-freezing interlayers. From these, the measured values of η/h proved to be equal to those obtained earlier for water (Fig. 4.20, curve *6*). It was shown that in the range of temperatures from −0.2°C to −0.6°C the viscosity is increased compared to bulk water at the same temperature by at least two times [135, 138]. Formally, the higher values of viscosity measured at lower temperature, at which the mean thickness of the interlayer h fell to 10 nm, might be caused

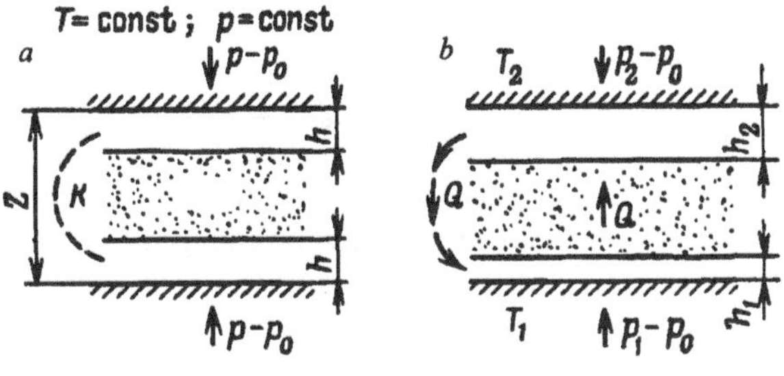

Figure 4.23 Ice and non-freezing water interlayers in a slit pore: *a*, in equilibrium state at T = const and p = const; *b*, under pressure or temperature gradients.

by the influence of the ice roughness which increased (at the same mean h value) the viscous resistance to interlayer flow.

The capillary dilatometer was also used to obtain the dependences of interlayer thickness on the external pressure P applied to ice. The results of measurements performed at various temperatures are shown in Fig. 4.22. The interlayer thickness grows as P approaches the pressure of phase equilibrium between liquid water and ice, $P_*(t)$. When P was increased up to P_*, complete ice melting in a narrow capillary was observed. The isotherms of disjoining pressure for non-freezing interlayers $\Pi(h)$ were calculated on the basis of these data [135]. The shape of the isotherms obtained agrees well with the well-known exponential relation (I.16) for structural forces. The temperature dependences of parameters K_s and λ for non-freezing layers were also obtained. Thermodynamic analysis of the results obtained proved the conclusion reported earlier [139] that the exponential shape of the isotherm leads to a logarithmic dependence of the thickness of non-freezing interlayers on temperature:

$$h = \lambda \ln(\Delta v K_s T_0 / L \Delta T),$$

where $\Delta v = v_w - v_i$ is the difference between molar volumes of water and ice; L is the heat of ice melting; $\Delta T = T_0 - T$; T_0 is the melting point of ice; K_s and λ are the parameters of the isotherm.

It has been shown [135], that when $h > 7$ nm the results of calculations carried by this equation agree with experimental data, assuming $K_s = 7\times10^6$ N/m^2 and $\lambda = 17$ nm. The values of K_s are of the same order as those for the liquid water (Table 1.2). However, the values of λ are higher by one order, that may be associated with the growth of correlation length in colder non-freezing interlayers adjacent to the ice.

To solve the problem of transfer of the non-freezing water driven by temperature and pressure gradients, one should consider the interrelated mass and energy fluxes using the thermodynamics of irreversible processes [140]. To identify the main physical mechanism of the phenomenon, let us consider the slit model of porous space shown in Fig. 4.23. Here the ice plate is located between parallel solid walls near to which there exist non-freezing interlayers of thickness h. The model corresponds to a deformable porous body: the distance $z >> h$ between the pore walls can be changed by external pressure. The upper and lower interlayers are connected with each other by non-freezing communications K constituted by narrower pores. The real situation corresponds, for example, to an ice lens in a pore widening.

The system showed in Fig. 4.23 is in equilibrium at constant temperature $T < T_0$, where T_0 is the melting temperature, and at a constant pressure applied from the side of the walls (neglecting the influence of gravity). The thickness of the non-freezing interlayers being in equilibrium with the ice depends, as shown above, on the temperature and pressure.

The excess external pressure $(p - p_0)$ applied to the interlayers is balanced by their disjoining pressure Π that depends on their thickness and temperature: $\Pi(h, T) = p - p_0$, where p_0 is the pressure in the bulk supercooled liquid phase capable of being in equilibrium with the interlayer. Let us disturb the equilibrium of the system applying a small temperature difference $\Delta T = T_2 - T_1$, or small pressure difference $\Delta P = p_2 - p_1$ between the interlayers, or both of these simultaneously (Fig. 4.23, b). This will cause heat and mass transfer accompanied by ice melting on one surface and water crystallization on another.

Consider a stationary process established after a time, when the mass flux Q, g/cm^2 s, in the liquid phase in communications is exactly equal to the counterflux in the ice phase $Q_s = \rho_s q_s$, where ρ_s is ice density and q_s is the velocity of vertical displacement of the ice plate. The force applied to its unit area to achieve mechanical equilibrium must be equal to the pressure drop Δp and directed oppositely. When $p_2 > p_1$ the liquid flows inside communications from the upper plate to the lower one and the ice plate moves upwards completing the mass circulation.

The thermodynamics of irreversible processes provide the following kinetic equations for the heat and mass transfer:

$$q_s = \alpha_{11}\Delta p + \alpha_{12}\Delta T/T, \text{ cm/s}, \tag{IV.96}$$

$$W = \alpha_{21}\Delta p + \alpha_{22}\Delta T/T, \text{ J/cm}^2\text{ s}. \tag{IV.97}$$

Here the coefficient α_{11} characterizes the resistance to ice displacement depending on the hydrodynamic resistance of communications and interlayers. When $h << z$, the coefficient α_{22} can be considered equal to the thermal conductivity of ice.

The cross-coefficients α_{12} and α_{21}, equal to each other according to the Onsager principle, specify the thermocrystallization mass flux $q = q_0$ (at $\Delta p = 0$), driven by the temperature difference, and the isothermal heat flux (or heat of transfer) $W = W_0$ (at $\Delta T = 0$) driven by the pressure drop.

The values of the coefficients α_{12} and α_{21} are governed by the physical properties and geometry of the model system. To determine $\alpha_{12} = \alpha_{21}$, let us write the relation for the heat of transfer W_0. Obviously, it is equal to a product of the velocity of the isothermal mass flux and the heat of the water–ice phase transition L, J/g:

$$W_0 = \alpha_{11}\Delta p\rho_s L. \tag{IV.98}$$

Since the specific enthalpy of thin interlayers differs from its bulk value by ΔH, the liquid flow from one interlayer into another is accompanied by the isothermal heat of transfer. Due to this, in general, we should add the value of ΔH to L, i.e. to account for the effect of thermo-osmotic flow. However, since $L >> \Delta H$ ($L \sim 3\times10^2$ J/g), the influence of the thermo-osmosis can be neglected.

It follows from eq. (IV.97) that $W_0 = \alpha_{21}\Delta p$. Comparing this relation with eq. (IV.98), we obtain: $\alpha_{12} = \alpha_{21} = \alpha_{11}\rho_s L$, that gives after substitution into eq. (IV.96):

$$q_s = \alpha_{11}(\Delta p + \rho_s L\Delta T/T). \tag{IV.99}$$

In particular, it follows from this equation that the velocity of thermo-crystallization flow:

$$q_0 = \alpha_{11}\rho_s L\Delta T/T \tag{IV.100}$$

is governed by the heat of the phase transition, L, and this justifies the name given to the flow. For each given temperature and pressure, the only possible thickness of the equilibrium interlayer is established. If one of these conditions is not fulfilled, the equilibrium is disturbed and the ice plate will either grow or melt.

An increase in temperature tends to increase the thickness of non-freezing interlayers and to decrease the developing disjoining pressure. Absolute value of interlayer thickness depends on the shape of isotherm $\Pi(h)$. Because the coefficient α_{11} (at low hydrodynamic resistance to communication) depends on h, the kinetics of mass transfer processes in real frozen bodies depends considerably on the shape of the isotherm $\Pi(h)$.

The model described for a single slit pore being filled with ice (Fig. 4.23) can be easily extended to another important case corresponding to the mass transfer between thawed and frozen regions of a porous body, for example to the case of grounds. To regard this case, we should only assume that for one of the interlayers the following conditions are valid: $\Pi_2 = p_2 - p_0 = 0$, $T_2 = T_0$, $h \to \infty$. These conditions correspond to the real situation shown in Fig. 4.24. Here the temperature T_0 separates the thawed region I from the region II placed above, where the ice lenses co-exist with non-freezing interlayers at $T < T_0$. Since the lenses are thin compared to their distance from the melted region, the temperature of interlayers bounding the lens may be considered equal. This enables the mass transfer between the melted and frozen regions to be calculated, applying eq. (IV.99):

$$q_s = \alpha_{11}[-\Pi_i + \rho_s L(T_0 - T_i)/T_0], \qquad \text{(IV.101)}$$

where T_i and Π_i are the temperature and disjoining pressure in the i-th layer of the ground.

As temperature falls, h_i reduces but the pressure, $\Pi_i > 0$, acting on the pore walls bounding the lens increases. If the positive disjoining pressure Π_i developing by interlayers exceeds the hardness of the porous body, it will deform and the pores will be filled with ice due to water inflow through the non-freezing communications. While the counter-pressure of the ground grows when it contracts, the velocity of ice lens growth decreases and approaches zero when some new equilibrium conditions are achieved. From this it follows that the ice lenses are likely to form in weakened parts of a porous body confirming the experimental results. Ground packing, that causes defective structure regions to liquidate, also reduces the probability of ice lens formation and the effect of frost heaving.

If the interlayers generate positive disjoining pressure capable of causing frost destruction, the first and the second terms of eq. (IV.101) have different signs, since $T_0 - T_i > 0$ and $\Pi_i > 0$. This means, that the thermocrystallization mass inflow ($q_0 > 0$) is accompanied by the counterflow driven by the disjoining pressure $\Pi_i > 0$ of the interlayers. Obviously, the maximum value $\Pi_i > 0$ is achieved at $q_s = 0$, when the flows are balanced and the growth of ice lenses is stopped. The maximum disjoining pressure generated by the interlayers can be obtained from eq. (IV.101), assuming $q_s = 0$:

$$\Pi_{max} = p_{max} - p_0 = \rho_s L(T_0 - T_i)/T_0. \qquad \text{(IV.102)}$$

This equation coincides with that known for the maximum crystallization pressure (the pressure applied to a closed face of a growing crystal) derived from the equations of equilibrium thermodynamics [141].

The applicability of the theory developed for a thermocrystallization flow was confirmed by comparisons of the calculated results with experimental

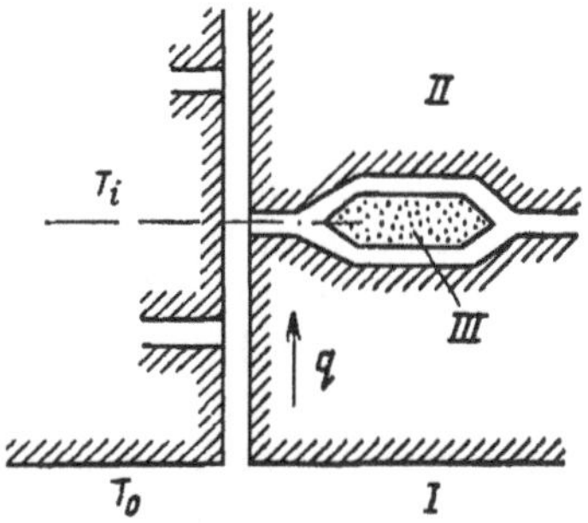

Figure 4.24 Mass transfer between the thawed I and frozen II zones of the ground, III is the ice lens in the i-th ground layer at $T = T_i < T_0$.

data obtained for model systems [142, 143]. In these experiments, the velocity of ice growth due to inflow of melt through non-freezing communications (a narrow slit or microporous filter) was measured. Quantitative agreement between the results calculated by eq. (IV.102) and the data of the experimental work [143] was achieved. In this work the pressure p has been measured, under which the inflow of water to ice ($q_s = 0$) terminates, and it has been shown that this pressure equals $p = \Delta T/K$, where $K = 0.083$ degree per atm. K value remained constant, as $\Delta T = T_0 - T_i$ changed from 0.02 to 0.05°C. According to eq. (IV.102), the coefficient K is equal to $T_0/\rho_s L$. Substituting $L = 3.34\times10^2$ J/g, $\rho_s = 0.91$ g/cm^3 and $T_0 = 273$ K into the equation, we obtain $K = 0.09$, that is close to the experimental value given above.

Therefore, frost destruction of porous bodies and frost heaving of grounds are caused by disjoining pressure generated by thin non-freezing water interlayers and their thermocrystallization flow. The theory developed describes the phenomenon of frost heaving better than the theory of Everett and Haynes [144] which associated this effect solely with pressure drop at a curved water–ice interface. Loch and Miller [145] have shown that the effective pressures developed are far beyond those calculated using this theory. In the case of flat lenses, the theory of Everett and Haynes fails to explain the origin of the pressure that causes a frost heaving at all.

From the thermocrystallization theory it follows that an increase in hydrodynamic resistance of non-freezing communications (Fig. 4.24), that reduces α_{11} values, must diminish the velocity of ice formation. Indeed, the observations show, that the effect of frost heaving reduces with reducing permeability of soils and grounds. In compliance with the theory developed, frost heaving also reduces with increasing hardness of a porous body or, as in the case of grounds, under the action of a packing stress.

The existence of non-freezing interlayers contacting ice also allows a number of other phenomena to be explained: for example, motion of solid particles and air bubbles, frozen in ice, driven by a temperature gradient [146, 147]; repulsion or capture of particles by the moving crystallization front [148, 149]. Frozen particles obviously move in the opposite direction to the crystallization flow. The velocity of such thermophoresis falls with temperature due to the decreasing thickness of non-freezing interlayers. The velocity of thermophoresis is also reduced by increasing thermal conductivity of particles, that lowers the local values of a temperature gradient. Experimental verification of the thermophoresis phenomenon associated with non-freezing water interlayers flowing along the surface of glass particles frozen in ice, was obtained in [146, 147].

It should be mentioned that the formation of thin liquid layers contacting crystal phase at a temperature close to the melting temperature of the bulk crystal, is observed not only for water, but also for a number of other liquids, for example naphthalene and stilbene [150], diphenyl [151] and liquid helium [152]. Hence, the phenomena discussed for transfer of non-freezing interlayers and films may be of more general importance.

Chapter V

Film Flow

As shown in Chapter II, at relative vapor pressure $p/p_s \geq 0.2-0.3$ a capillary condensate is in equilibrium with adsorption or wetting films. Free porous space is filled with vapor being in equilibrium with liquid and adsorbed phases. When capillary bridges formed around the contact zone of particles are separated by solid surface regions covered with adsorption films (see Fig. 2.15), mass transfer in porous bodies is usually considered in the framework of space and surface diffusion. The area of the film regions depends on the relative vapor pressure (p/p_s) and the peculiarities of the pore structure. When $p/p_s \leq 0.2-0.3$, vapor and surface diffusion represent the main mechanisms of mass transfer in the absence of condensate. Mass transport of vapors and gases in dry porous bodies is considered in the next chapter.

When $p/p_s > 0.9$, wetting films may exist instead of the adsorption monolayers. The description of mass transfer in wetting films is based on hydrodynamic equations. The thickness of the forming wetting films is governed by the corresponding isotherms of disjoining pressure $\Pi(h)$ (Chapter II). We will show in this chapter how the isotherms may be included into the equations of the hydrodynamics of wetting films. When a porous body is partly saturated by capillary condensate and wetting films, the liquid and vapor fluxes cannot be considered separately due to interaction between the fluxes: internal evaporation and condensation. The more complicated problem of joint liquid and vapor fluxes will be considered in Chapter VII.

§1 Hydrodynamics of Wetting Films

Let us consider the flow in a planar slit of a wetting film between liquid menisci that takes place under a difference of capillary pressures. At equi-

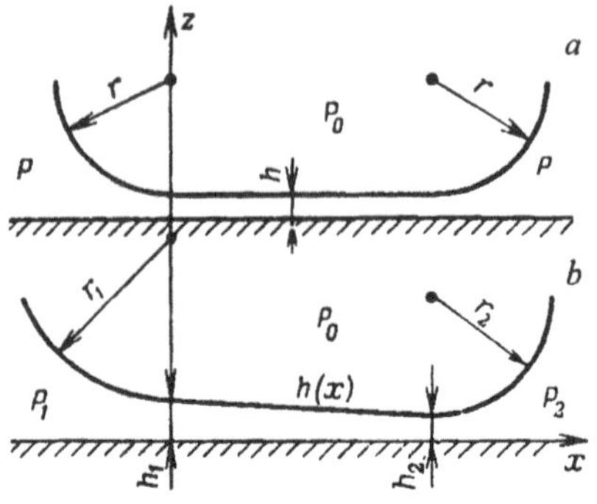

Figure 5.1 Equilibrium (*a*) and flow (*b*) of a wetting film between the menisci in a slit pore.

librium (Fig. 5.1, *a*), a constant film thickness, h, in the region between the menisci is determined by the equality of pressure of liquid in the film and in the capillary condensate. Pressure under the menisci is lower than the pressure, P_0, in the gas phase by the value of the capillary pressure drop P_c given by: $P = P_0 - P_c = P_0 - (\sigma/r)$, where r is the curvature radius of the cylindrical surface. The same decrease in pressure should take place in the film to be in equilibrium with menisci. On the other hand, the disjoining pressure, by definition (Chapter I), is equal to $\Pi = P_0 - P$, because in this case P_0 is the pressure acting on the film surface and P is that in the bulk liquid being in equilibrium with the film. Therefore, the pressure inside the film equals $P = P_0 - \Pi$, and the capillary pressure is equal to the disjoining pressure acting in the wetting film: $P_c = \Pi$.

It follows from the theory of surface phenomena that in thin wetting films and interlayers, as in a transition layer between liquid and vapor, the pressure is non-isotropic. The normal component of the pressure tensor P_N does not change when transversing a plane boundary, but the tangential component $P_T(z)$ varies across the film.

As will be shown below, the introduction of an isotropic pressure inside the film, P, that is equal to the difference between $P_0 = P_N$ and the disjoining pressure Π, enables the processes of film flow in the framework of hydrodynamics of isotropic fluids to be described. A positive disjoining pressure acting from the liquid film side of the boundary surfaces reduces the external pressure P_0 to the value of P. Particularly, this was the basis from which Derjaguin and Nerpin [153] considered the decrease of pressure inside the liquid film by Π, similar to the decrease of pressure in liquid under a concave meniscus by the value of capillary pressure P_c. In the first case the decrease is caused by long-range surface forces, and in the second by capillary forces. Application of the isotropic pressure P makes it possible to use a common hydrodynamic approach to describe the flow of liquid in a film, in capillary condensate and in the transition layers between them.

If there is a capillary pressure drop (Fig. 5.1, *b*), then within the regions where the film is in equilibrium with menisci the pressure will change to the value $P_1 = P_0 - P_{c1} = P_0 - (\sigma/r_1)$ at the left meniscus side and to $P_2 = P_0 - P_{c2} = P_0 - (\sigma/r_2)$ on the right. Because $r_1 > r_2$, it follows that $P_1 > P_2$ and this makes the liquid flow from the left to the right, to a region of decreased pressure. At the same time, the pressure in the gas phase, P_0, remains constant. The change in capillary pressure will also cause the disjoining pressure and consequently the film thickness to change. The film thickness will be greater at the left meniscus than at the right ($h_1 > h_2$), because $\Pi_1 < \Pi_2$, and the increased film thickness corresponds to a decrease in disjoining pressure according to the isotherm $\Pi(h)$ (Fig. 2.6, *a*). Obviously, this refers to stable films when $d\Pi/dh < 0$ and $\Pi > 0$. These films flow to a region of smaller thickness h. When $d\Pi/dh < 0$, random local perturbations of the film surface spontaneously resolve and the film reaches a stable plane state. At the same time the film surface is not ideally smooth due to thermal fluctuations of its thickness, for which the amplitude is of molecular dimensions under normal conditions, far from critical.

If the film is thermodynamically unstable and its thickness corresponds to the region of the isotherm where $d\Pi/dh > 0$ (e.g. the region of thickness between h_1 and h_2 in Fig. 2.6), then the film flow will be directed to greater thickness. This will result in the appearance of thickened regions, and drop formation on a solid substrate. In this case a more stable film, wetting or adsorption, will remain on the substrate behind the receding liquid that corresponds to the α-branch of the isotherm.

Turning back to Fig. 5.1, let us write the equation for film flow of a non-volatile liquid. With films of gently sloping surface of modest curvature, when $dh/dx << 1$, the equation of viscous flow in a plane slit may be used at each local vertical section z–z. Because the tangential stress on the film surface is $\tau = 0$, the equation of the flow takes the same form as that for a half of a plane slit $2h$ wide. The approximate distribution of the local velocities $v(z)$ and shear stresses $\tau(z)$ across the film are shown in Fig. 5.2.

The stationary flux across a given film section of the coordinate x can be written as [1, 26, 63]:

$$q(x) = -[\rho h^2(x)/3\eta]\nabla P(x), \text{ g/cm}^2\text{ s}, \tag{V.1}$$

where ρ is the density and η is the liquid viscosity. For $P = P_0 - \Pi$, we can pass from the pressure gradient to that of disjoining pressure $\nabla\Pi = -\nabla P$:

$$q(x) = [\rho h^2(x)/3\eta]\nabla\Pi(x). \tag{V.2}$$

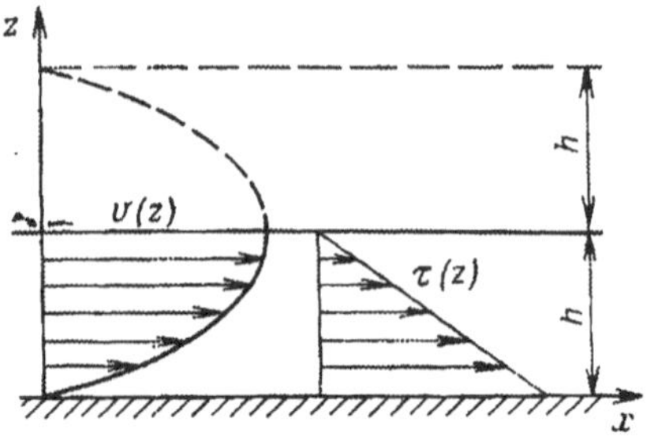

Figure 5.2 Distribution of local velocities v and shear stresses τ across the wetting film.

To quantify the film flow, the isotherm of disjoining pressure $\Pi(x)$ should be known to provide the link between the values of h and Π involved in this equation.

In contrast with the bulk liquid flowing in a plane slit, the thickness of a flowing film changes in the direction of flow to fulfil (at steady state) the condition of flow continuity $q = \text{const}$. Using this condition, it is easy to find the film profile $h(x)$ in a region between the menisci (Fig. 5.1, *b*) and the pressure distribution along the film, $\Pi(x)$. For example in the case of the isotherm of molecular forces $\Pi = -A/6\pi h^3$ and for a one-dimensional flow we get from eq. (V.2):

$$h(x) = \rho h_1 A/(\rho A + 6\pi\eta h_1 qx), \tag{V.3}$$

where h_1 is the film thickness at $x = 0$ and x is the distance from the reference point; $A < 0$ is the constant of molecular forces.

Expressing h in terms of Π, we can obtain the relevant equation for the pressure distribution inside the film $P(x) = P_0 - \Pi(x)$. In contrast with the flow in a plane slit filled with liquid, the pressure in the wetting film and film thickness, as seen from eq. (V.3), changes in a non-linear manner. It should be mentioned that these calculations refer to the case of non-volatile liquids. The mass transfer between the film and the vapor diffusing to the more concave meniscus will change both the thickness distribution $h(x)$ and that of pressure $\Pi(x)$. Joint flow of the liquid film and of its vapor will be considered in Chapter VII.

An example of a flow obeying eqs. (V.1) and (V.2) could be film flow on the walls of a conical capillary (Fig. 5.3). In porous bodies this may represent a local narrowing of the pore. As experiments have shown [154], in the case of rather thick films the mass transfer between menisci is caused mainly by a film flow along capillary walls. If the walls are completely wetted with water,

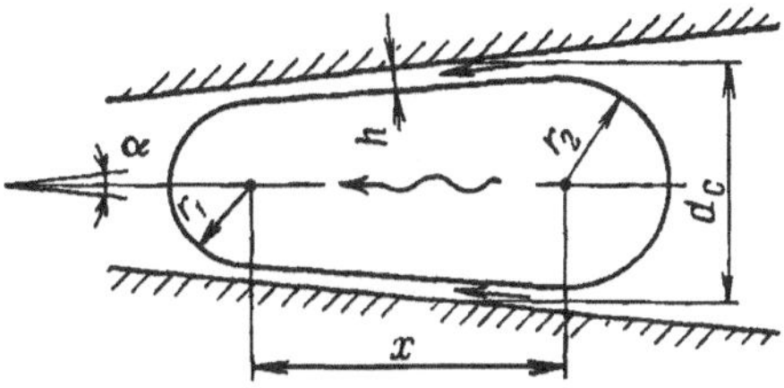

Figure 5.3 Film flow in a conical capillary.

the contribution of vapor diffusion to the mass flux does not exceed 1% in capillaries of $r = 3$ to 30 μm and with cone angles between $\alpha = 10^{-4}$ to 10^{-2} rad.

Assuming that at a low pressure difference between the menisci the film thickness h may be considered approximately constant, we get for a conical capillary instead of eq. (V.2):

$$q = \alpha\rho\sigma h^2 / 12\eta r_c^2,$$

where σ is the surface tension and r_c is the mean capillary radius in the region between the menisci. The larger the angle of conicity α determining the pressure gradient and the smaller the mean radius, the greater the film flow q (g/cm^2 s).

In porous bodies, film flow towards the narrower pores results in the displacement of a gas bubble to the wider part of the pore. This displacement will persist until capillary pressures of both menisci equilibrate or the bubble merges with the bulk gas phase at the boundary of the porous body. Conical pore formation is possible as a result of shrinkage of surface layers of the sample during drying. In the case of air retention within narrowed pores, spontaneous film flow towards sample surfaces may take place when the drying is stopped, for example in soils at night, when evaporation does not occur.

Studying the schemes in Figs. 5.1 and 5.3 more strictly, it should be borne in mind that there is a transition zone between the meniscus of constant curvature and a flat equilibrium film where both surface and capillary forces manifest themselves (Fig. 2.9). The length of the transition zone is about $l \sim \sqrt{hH}$, where H is a characteristic linear pore size and h is the film thickness [13]. At $H = 0.1$ cm and $h = 50$ nm, l is about 10 μm. The surface of the transition zone has a variable curvature $K(x)$. The non-uniform thickness of the transition zone should (at equilibrium) satisfy the condition of pressure constancy:

$$P = P_0 - \sigma K(x) - \Pi(h, x) = \text{const}, \tag{V.4}$$

where P_0 = const is the pressure in the gas phase. If the surface is convex, $K >$ 0. Due to the action of disjoining pressure $\Pi > 0$ in this zone, its surface should be less curved than the meniscus: $K < K_0$, where $K_0 = 1/r$ is the meniscus curvature in a slit pore. The surface of the transition zone becomes flatter with growing disjoining pressure and decreasing local thickness of a transition layer h, and ultimately transforms into an equilibrium plane film of constant thickness h_0. In this case $P = P_0 - \Pi(h_0)$ and $K = 0$. In contrast, when the local film thickness h is greater than the surface forces radius, $\Pi = 0$ and $P = P_0 - \sigma K_0 = P_0 - P_c$, which corresponds to a meniscus of constant curvature K_0.

The profile of the transition zone $h(x)$ in a plane slit can be determined from eq. (V.4) using the equation of the isotherm of disjoining pressure and the expression for the curvature of a cylindrical surface [155]. Therefore knowing that $P_0 - P = P_c$, for the isotherm of molecular forces, we get:

$$P_c = (\sigma h''/[1 + (h')^2]^{3/2}) - (A/6\pi h^3) = \sigma/r = \text{const},$$

where $h'' = \partial^2 h/\partial x^2$ and $h' = \partial h/\partial x$ are the second and the first derivatives, respectively. The solution of this differential equation is given in the monograph [26].

For a flow in the transition zone eq. (V.1) is still applicable, but instead of eq. (V.2) we get:

$$q(x) = (\rho h^2/3\eta)\nabla[\Pi(x) + \sigma K(h)]. \tag{V.5}$$

For gentle profiles of the transition zone, when $K \cong h''$, we have for one-dimensional flow along the x-axis:

$$q(x) = (\rho h^2/3\eta)(\partial\Pi/\partial x + \sigma\partial^3 h/\partial x^3). \tag{V.6}$$

This equation is widely used in calculations of droplet spreading and film drainage.

When wetting films are not placed horizontally, the influence of the gravitational force must be taken into account [153]. Therefore, the pressure gradient should be introduced into eq. (V.2) $\nabla P = -\nabla\Pi + \rho g \sin\omega$, where g is the gravity acceleration and ω is the angle of inclination. At $\omega = 0$, the additional term vanishes, and it has the maximum value when the film is placed vertically, at $\omega = \pi/2$. The film equilibrium on a vertical wall obeys the condition $q = 0$, when $\nabla P = 0$ and $\rho gH - \Pi = 0$. Here the height $H(z)$ is measured from the level of the bulk liquid, where $\Pi = 0$. The measurement of film thickness h at different height H makes it possible to obtain the isotherm of disjoining pressure, when the magnitude of the disjoining pressure $\Pi = \rho gH$ corresponds to each value of $H(z)$. For example Ingram [156] has obtained the isotherm of disjoining pressure of hydrocarbon films on quartz measuring ellipsometrically the distribution of film thickness along the height $H(z)$ above the liquid level. When $\Pi = 30$ N/m^2, values of $h = 15-20$ nm were obtained for octane, hexane and heptane.

The method of suction of liquid through a filter (Fig. 2.7, *b*) employs the same approach to determine Π from the value of hydrostatic pressure ρgH. The only difference is that in this case a horizontal flat wetting film is formed at the height H, which considerably simplifies the measurement of its thickness.

In porous bodies with pore radius of about 10 μm, the influence of gravity is small. The capillary pressure existing here corresponds to H values of about 1.5 m that is much greater than the usual linear dimension of porous bodies. Exceptions include massive porous systems such as grounds and soils, where processes of mass transfer occur in layers of considerable thickness.

In zero gravity conditions, all mass transfer processes (not only in porous bodies) are governed by capillary effects and disjoining pressure of films. Here they are of fundamental primary importance.

At high values of gravity, created artificially, for example using a centrifuge, the effect of inertia may become equal to or even exceed the action of capillary and disjoining pressure. The rate of liquid film flow on a surface of a sealed capillary placed into a centrifuge was measured in [157]. In this case, at $\rho g >> \nabla P$, the stationary flow of liquid could be expressed in the form:

$$q = g\rho h^2/3\eta \text{, cm/s.} \tag{V.7}$$

At 2000–3000 rpm, water films of thickness $h \sim 100$ nm flowed at the rate of 1 μm/s.

The action of a non-uniform electric field results in pulling a liquid dielectric into the region of higher electric field strength, E, creating pressure gradient in a wetting film. As known from electrostatics, the force acting on a unit volume of dielectric is:

$$f = [(\varepsilon_0 - 1)/8\pi]\nabla(E^2) \text{, N/cm}^3\text{,} \tag{V.8}$$

where ε_0 is the dielectric constant. The action of the force, f, may be reduced to the corresponding pressure gradient, thus eq. (V.1) takes the form:

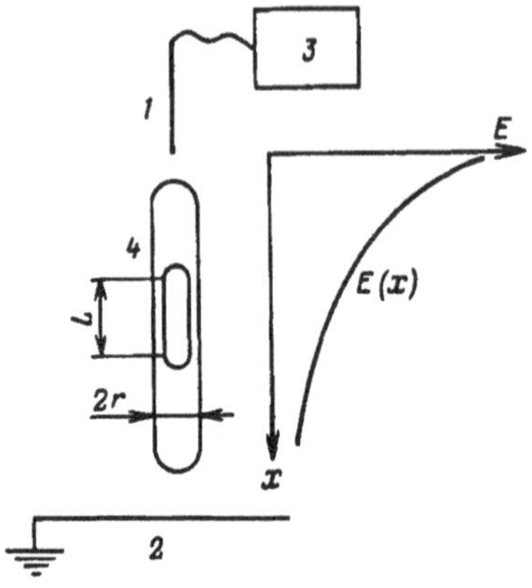

Figure 5.4 Apparatus for studying the flow of wetting films in a non-uniform electric field.

$$q = -[\rho h^2(\varepsilon_0 - 1)/24\pi\eta]\nabla(E^2), \text{ g/cm}^2\text{ s}. \qquad \text{(V.9)}$$

The scheme and the results of experimental research for wetting film flow of various liquids on a surface of quartz capillaries [158] are presented in Figs. 5.4 and 5.5. A non-uniform electrostatic field was created by two electrodes (of a type needle *1* – plane 2) and regulated by changing the voltage on the needle from a high voltage supplier *3*. The rate of displacement v of the bubble of length L, that moves due to the flow of wetting films along the walls of the capillary *4*, was measured by a cathetometer. The thickness of films in capillaries with radii of tens of microns was 150–300 nm, which makes it possible to neglect the influence of vapor electromigration (see Chapter VI). The expression for v can be obtained from eq. (V.9) taking into account the difference between the film cross-section $2\pi rh$ and the capillary bore πr^2:

$$v = [h^3(\varepsilon_0 - 1)/12\pi r\eta]\nabla(E^2).$$

The dependences obtained (Fig. 5.5) are linear in accordance with this equation. However, they do not pass through the origin due to the hysteresis of the contact angle. The values of the intersect $\nabla(E^2)_0$ on the abscissa axis, when the flux is equal to zero, can be used for estimation of the gradient of capillary pressure, that determines the rate of the counter flow:

$$\nabla P_c = 2\sigma(\cos\theta_R - \cos\theta_A)/rL = (\varepsilon_0 - 1)\nabla(E^2)/8\pi. \qquad \text{(V.10)}$$

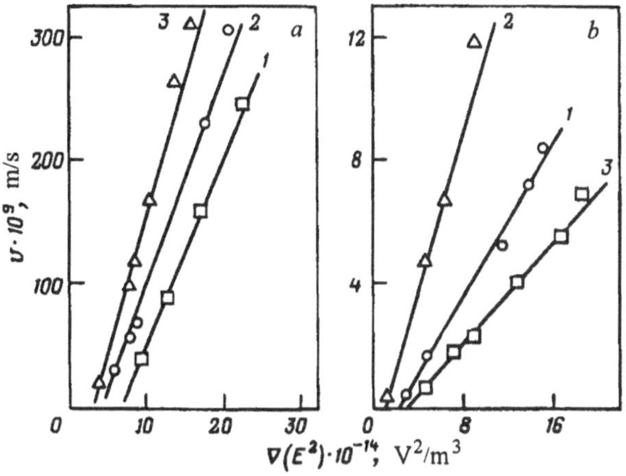

Figure 5.5 Rates of the bubble displacement (Fig. 5.4) due to the flow of wetting films in a non-uniform electric field; *a*) water, *1*) $r = 8.6$ μm, $L = 0.54$ mm; *2*) $r \approx 10.1$ μm, $L = 0.81$ mm; *3*) $r = 37.8$ μm, $L = 0.87$ mm; *b*) ethylene glycol, *1*) $r = 17.8$ μm, $L = 0.47$ mm; *2*) $r = 33.9$ μm, $L = 1.13$ mm; *3*) propyl alcohol, $r = 40.6$ μm, $L = 0.86$ mm.

Assuming that $\cos\theta_R \approx 1$, we get values of the advancing contact angle θ_A ranging from 2° to 9°. These values are close to those determined by the independent method based on the measurement of the rate of displacement of the condensate column in the same capillary. This shows positively that the deviations of experimental data from those of eq. (V.9, a) were related to the hysteresis of the contact angle.

The film thickness could be determined from the slope of straight lines in Fig. 5.5 according to eq. (V.9, a). First the value of $\nabla(E^2)$ in the points of the bubble location in the capillary was found by field calibration [159]. The dependences of h on the disjoining pressure, $\Pi = \sigma/r$, were obtained by this method. For water, the values of β-film thicknesses obtained were close to those calculated from the electrostatic component of disjoining pressure according to eq. (I.13). Values of h decreased from 300 nm to 150 nm with diminution of capillary radius.

The action of the non-uniform electric field is increased by a rise in polarity of the liquid in the film. As seen from the comparison of the graphs in Fig. 5.5, the rates of flow regularly increase in the series: propyl alcohol (ε_0 = 19.7), ethylene glycol ($\varepsilon_0 = 38.7$), water ($\varepsilon_0 = 78.3$).

Examples of flow of wetting films under the pressure gradient created by the various methods considered in this section show that the flow significantly differs from that in liquid-filled channels bounded by rigid walls. The difference is connected with the availability of the free liquid surface to

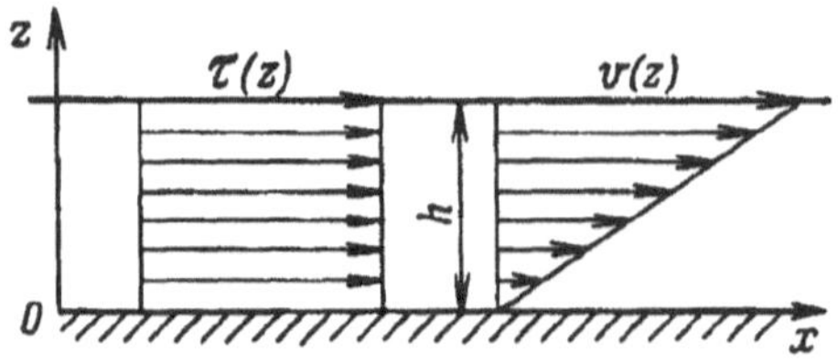

Fig. 5.6

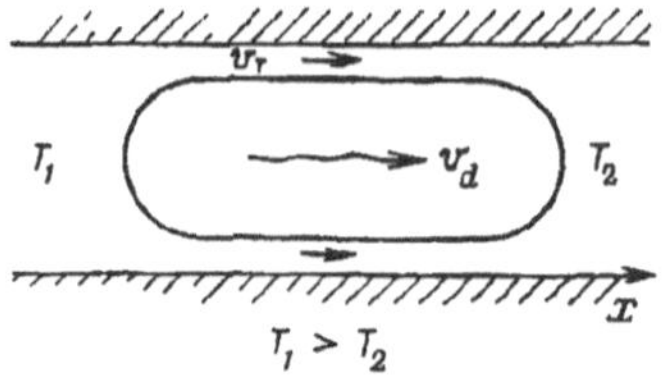

Fig. 5.7

Figure 5.6 Distribution of shear stress $\tau(z)$ and velocities $v(z)$ during the thermo-capillary film flow.

Figure 5.7 Measurement of the rates of thermo-capillary film flow from the rate of an air bubble displacement.

regulate the local film thickness in accordance with pressure distribution in the flowing film.

The assumption that the shear stress is equal to zero at the film surface is no longer applicable if there are gradients of temperature, concentration or electrical potential. This leads to alternative and different mechanisms of mass transfer having no analogues to the cases of flow in pores filled with liquid.

§2 Thermocapillary Film Flow

Consider a plane wetting film on a solid substrate along which the temperature, T, decreases linearly in the direction of the x-axis corresponding to a constant temperature gradient ∇T = const. As the surface tension σ of liquids rises with diminution of temperature, the shear stress $\tau = \partial\sigma/\partial x$ acts on the film surface, that is constant within the film: $\tau(z)$ = const. According to the Newton equation, this will cause a liquid flow at a constant velocity gradient:

$$\partial v/\partial z = \tau/\eta = \text{const.} \qquad \text{(V.11)}$$

The appropriate profiles of $\tau(z)$ and of the flow velocities $v(z)$ are shown in Fig. 5.6. The velocity on the film surface, from the eq. (V.11), is equal to $v(h) = \tau h/\eta$. Thus the average velocity of flow is $v_c(h/2) = \tau h/2\eta$ and the liquid flux in the film is equal to [160]:

$$q = v_c \rho = \rho\tau h/2\eta = (\rho h/2\eta)(\partial\sigma/\partial x)$$

$$= (\rho h/2\eta)(\partial\sigma/\partial T)(dT/dx)\,,\ \mathrm{g/cm^2\,s}. \qquad (V.12)$$

Because $\partial\sigma/\partial T < 0$, the thermocapillary flow of wetting films of pure liquids is always directed towards the colder side.

If the film is bounded by menisci at different temperatures T_1 and T_2 (see Fig. 5.7) then in addition to thermocapillary flow the film will participate in the flow driven by the different capillary pressure of the menisci. The expression for this flow may be found from eq. (V.1) using the relation: $P = P_0 - P_c$ and assuming that grad $P_c = (\sigma_1 - \sigma_2)/rL = (\partial\sigma/\partial x)/r$. Then:

$$q = (\rho h^2/3\eta r)(\partial\sigma/\partial T)(dT/dx)\,, \qquad (V.13)$$

where r is the curvature radius of the menisci.

As seen from the comparison of eqs. (V.13) and (V.12), the flow under the temperature difference of capillary pressures is $(3r/2h)$ times less than that for the thermocapillary flow. Because the film thickness, h, is considerably less than the curvature radius of the menisci, roughly equal to pore radius, then in calculations of thermocapillary film flow the contribution of the "thermomenisci flow" may be neglected. For example for films of nonpolar liquids stable due to molecular forces the relation h/r is 2×10^{-4} at $r = 50$ µm and 2×10^{-3} at $r = 0.5$ µm.

Thermocapillary flow, in the absence of other effects, has been studied on wetting films for the nonpolar liquid *n*-dodecane [161]. This choice was determined by its low volatility, permitting the contribution of vapor flow to be reduced, together with complete wetting of glass surface that excludes the influence of hysteresis of the contact angle. Moreover, in nonpolar liquids neither electric nor structural effects manifest themselves, thereby excluding the involvement of other transfer mechanisms. A capillary with a trapped air bubble (Fig. 5.7) was placed in a hole made in a metal plate, and different temperatures were maintained at the ends of the plate by a thermostat and thermoelements of a semiconductor microcooler. This made it possible to maintain temperature gradients ∇T on the plate up to 20 deg/cm at a mean bubble temperature of 8°C and higher. The rates of the bubble motion V measured by a comparator ranged from 3 to 25 µm/h depending on the experimental conditions. The measurements lasted for 6–8 h, so the total displacement of the bubble was at least 100–250 µm. The instrumental error did not exceed 1 µm.

For cylindrical capillaries, the expression for the bubble velocity V_T due to the thermocapillary flow, derived from eq. (V. 12), takes the form:

$$V_T = 2hq/\rho r = (h^2/\eta r)(\partial\sigma/\partial T)(dT/dx)\,, \qquad (V.14)$$

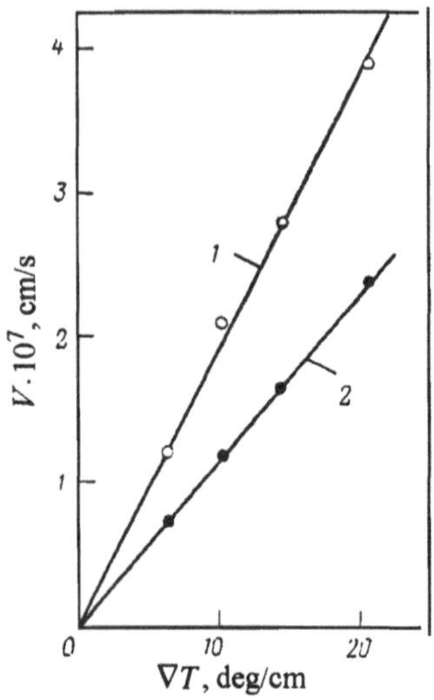

Figure 5.8 The rate V of an air bubble displacement (L = 1.8 mm) in a glass capillary (r = 0.73 μm) filled with n-dodecane under the action of a temperature gradient ∇T. Measured values (curve *1*) and values calculated using eq. (V.17) for vapor diffusion (curve 2) are shown here.

where r is the capillary radius.

Besides the thermocapillary flow in the liquid film, the diffusion vapor flow, displacing the bubble at the rate V_d, should be taken into account:

$$V_d = (v_m D/RT)dp/dx, \tag{V.15}$$

where v_m is the molar volume of the liquid (cm^3/mol), R is the gas constant, D is the coefficient of vapor diffusion through the gas and p is the partial vapor pressure of n-dodecane. Under non-isothermal conditions, the pressure gradient dp/dx (at $r \geq 0.1$ μm) is equal to the gradient of saturated vapor pressure $dp_s(T)/dx$. Using for $p_s(T)$ the Clausius–Clapeyron equation $\partial p_s(T)/\partial T = \lambda_v/Tv_m$ and the equation of state of an ideal gas, $p_s = RT/v_m$, we get:

$$dp/dx = (dp_s/dT)dT/dx = (\lambda_v p_s/RT^2)dT/dx, \tag{V.16}$$

where λ_v is the molar heat of evaporation of n-dodecane. The substitution of this expression into eq. (V.15) gives the following equation for vapor flow:

$$V_d = (v_m \lambda_v D p_s / R^2 T^3) dT/dx = (\lambda_v D / RT^2) dT/dx. \qquad \text{(V.17)}$$

Here the values of v_m, λ_v, D and p_s should refer to the mean temperature of the bubble. Due to its small length (L = 1 to 3 mm), the difference of menisci temperature does not exceed several degrees.

Figure 5.8 shows the results of an experiment with a capillary of $r = 0.73$ μm at the mean temperature of 31°C. The dependences obtained for the shift x of the bubble during time, $x(t)$, enabled the rates of displacement $V = dx/dt$ and their dependence on the temperature gradient to be calculated. As seen from Fig. 5.8, the rate of bubble displacement V (curve *1*) is, as follows from the theory, a linear function of the temperature gradient. Curve *2* shows the rate of bubble displacement, calculated according to eq. (V.17), which could be caused solely by vapor diffusion. The difference between these two curves corresponds to the contribution of the thermocapillary flow in the film. Using V values from eq. (V.14), one can determine the mean film thickness h if the capillary radius r is known. The film thickness obtained in the experiments varied from 2.5 nm in the narrowest capillary (r = 0.4 μm) to 10 nm in the widest (r = 17 μm). As shown above, the capillary radius (at complete wetting) also determines the disjoining pressure in the film $\Pi = \sigma/r$. This allowed the experimental dependence $h(\Pi)$ to be obtained, i. e. the isotherm of disjoining pressure of wetting films of *n*-dodecane. The results obtained appeared to agree with the theoretical isotherm for molecular forces $\Pi = -A/6\pi h^3$. The experimentally obtained values, $A = -(1.5 \pm 0.4) \times 10^{-20}$ J are close to the theoretical value for the Hamaker constant, $A = -1.34 \times 10^{-20}$ J, calculated using spectral data [26].

This agreement shows that the expression obtained from the theoretical isotherm for molecular forces can be used instead of h in eq. (V.14):

$$V = (-A/6\pi\sigma)^{2/3} (1/\eta r^{1/3}) (\partial\sigma/\partial T) dT/dx. \qquad \text{(V.18)}$$

The inclusion of the isotherm $\Pi(h)$ into the equation for thermocapillary flow removes all the unknown variables. The parameter of the isotherm of molecular forces A can be calculated using spectral characteristics of the liquid and the substrate (see Chapter I, §1).

The lower the surface tension and the liquid viscosity and the higher the constant A, i.e. the polarity of the substrate, the more pronounced is the thermocapillary flow. For films on metals, the Hamaker constants are approximately one order higher than those on dielectrics such as glass or quartz. In this case the thermocapillary flow may be several times greater. The smaller the capillary radius, the greater the contribution that is made by thermocapillary film flow to the total mass transfer. Therefore, this mechanism of mass transfer plays an especially important role in fine-porous bodies. As the diffusion flux does not depend on r, the thermocapillary flow becomes the main mechanism of thermo-transfer in incompletely wetted fine porous bodies.

In films of polar liquids the effect of thermo-osmosis may appear in addition to the thermocapillary flow. The equation for completely liquid-filled pores (Chapter IV, §5) is also applicable to film flow:

$$q = (\rho\chi/T)dT/dx, \tag{V.19}$$

where χ is the coefficient of thermo-osmosis. Let us compare the coefficient for dT/dx in this equation with the corresponding coefficient in eq. (V.12). For water this relation is:

$$2\eta\chi/Th(-\partial\sigma/\partial T) \approx 4\times10^{-4}\,\chi/h, \tag{V.20}$$

where χ is expressed in cm^2/s, and h in cm. Because the absolute values of χ do not exceed 10^{-6} cm^2/s for hydrophilic porous bodies (see Chapter IV, §4, 5), the contribution of thermo-osmosis to thermocapillary flow of wetting films may be neglected.

In films of aqueous solutions, as in narrow liquid-filled pores, thermo-electrical fluxes may affect the mass thermotransfer. Consequently, it is worth addressing the analysis of experimental results obtained for model systems, water films on a capillary surface. In the first experiments [160, 162] glass capillaries were used, because much thicker water films are formed on their surface due to the higher hydrophilicity compared to quartz. The hysteresis of contact angle was small, but the thermoelectric effects caused by the surface charge might play a significant role. As a result, after the temperature gradient was removed, a spontaneous reversal of film flow took place [163]. This might be due to a reverse electro-osmotic film flow under the streaming potential that arises. This shows that the thermocapillary flow of aqueous films may be significantly complicated by thermo-electro-osmotic effects.

In another series of experiments [164], narrow quartz capillaries were used where much thinner α-films were formed unlike those with glass. In such thin films electroviscosity effects do not markedly manifest themselves but, because of incomplete wetting of α-films with the bulk water, the hysteresis of contact angle affects measurements of the thermocapillary flow. The difference between contact angles of advancing θ_A and receding θ_R menisci gives rise to a reverse flow in a film under a pressure gradient equal to:

$$\nabla P = -(2\sigma/rL)(\cos\theta_A - \cos\theta_R). \tag{V.21}$$

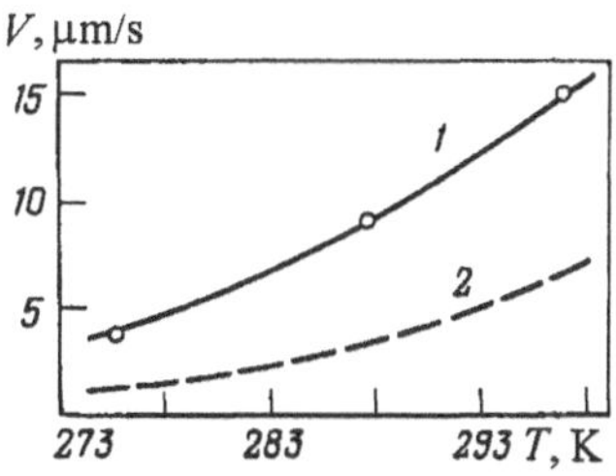

Figure 5.9 Temperature dependence of the rate of displacement $V = dx/dt$ for a bubble filled with nitrogen under the pressure of 4.6 MPa in a glass capillary, $r = 0.73$ μm (curve *1*). The contribution of water vapor diffusion is shown by curve *2*. The temperature gradient $\nabla T = 10$ deg/cm.

Film thickness calculated according to eq. (V.14) ranged from 5 to 12 nm being close to that of water α-films (Chapter 2, §2). The observed scatter of h values might be caused by the changes of θ_A from 0 to 10° (at $\theta_R \approx 0$) in different experiments.

Figure 5.9 shows the temperature dependence of the rate of bubble displacement V for a capillary of radius $r = 0.8$ μm (curve *1*). The contribution of the vapor diffusion to the water transfer is indicated by curve *2*. In this case, the vapor diffusion is significantly suppressed (even at 22°C), because the bubble was filled with nitrogen at pressure $P = 4.6$ MPa during the capillary sealing.

The measurements were performed at approximately equal temperature gradients. The increase in the rate of bubble displacement at higher temperature was caused by the increased velocity of vapor diffusion, according to eq. (V.17), because the saturated vapor pressure and the diffusion coefficient had increased. The contribution of the thermocapillary film flow (the difference between curves *1* and *2*) depends weakly on T, because in the range of temperature from 274 K to 295 K the parameters entered into eq. (V.14) vary less as compared with vapour pressure.

The climbing of wetting films of a nonpolar liquid (polydimethylsiloxane) along a vertical wall caused by thermocapillary flow has been studied in [165, 166]. Expressions for the height of film climb based the gravitational drainage and capillary effects (caused by cylindrical drop formation at the upper end of the film) have been obtained.

§3 Thermocrystallization Film Flow

A thermocrystallization film flow takes place together with thermocapillary movement in thin non-freezing water films on the capillary wall between ice

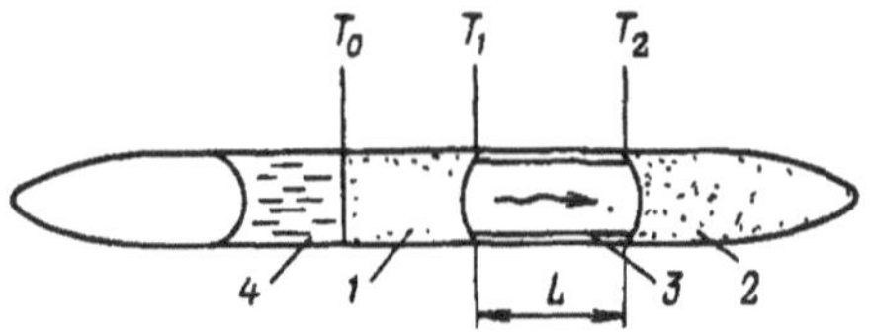

Figure 5.10 Formation of a gas bubble of length L between the ice menisci (*1*) and (*2*) in the capillary when measuring the rates of thermo-crystallization flow of non-freezing water films (*3*); (*4*) is the bulk water at $T > T_0 = 273$ K, the temperature of ice melting.

menisci maintained at different temperatures (Fig. 5.10). The theory of this effect, connected with the heat of the ice–water phase transition, has been presented in Chapter IV, §6.

The expression for the thermocrystallization flux in the film has the form [164, 167]:

$$q = (\rho h^2 \lambda_i / 3\eta T) dT/dx, \qquad \text{(V.22)}$$

where λ_i is the heat of ice melting, J/g.

Let us compare the thermocapillary and thermocrystallization film flows at equal temperature gradients. Dividing the right side of eq. (V.22) by that of eq. (V.12), we obtain a ratio of fluxes equal to:

$$3T(-\partial\sigma/\partial T)/2h\lambda_i .$$

Using the known values of $\partial\sigma/\partial T = -1.7\times10^{-6}$ N/cm deg and $\lambda_i = 3.6\times10^2$ J/g for water, we find that at $h = 10$ nm the value of this ratio is equal to 0.02 and at $h = 1$ nm it is equal to 0.2. Therefore, in frozen porous bodies the thermocrystallization flow is the preferable mechanism of the flow of non-freezing films under temperature gradients. Cryological investigations [168] confirm this conclusion by showing that the effective coefficient of vapor diffusion, $D_i = 0.85$ to 0.9 cm^2/s, through a snow layer at temperatures ranging from −0.2° to −8°C is several times higher than that of vapor diffusion through air under the same conditions ($D_0 = 0.21$ cm^2/s). Although a part of this effect may be explained by vapor transfer due to successive evaporation and condensation on non-frozen water inclusions, it is highly probable that at a temperature of −0.5° to −0.2°C, the main contribution results in the flow of non-freezing films.

Direct measurements of the velocities of thermocrystallization film flow were performed on a section of a capillary between ice menisci maintained at

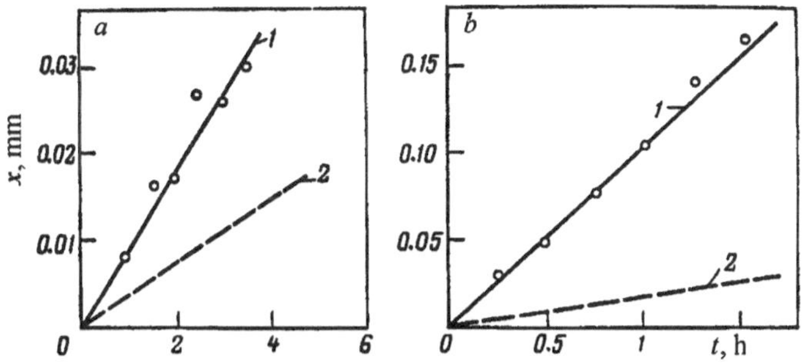

Figure 5.11 Displacement of ice menisci in capillaries as a result of vapor diffusion and thermo-crystallization flow of non-freezing water films (curves 1). Curves 2 show the contribution of vapor diffusion. a) r = 13.4 μm; L = 1.07 mm; t = −0.95°C; ∇T = 12.6 deg/cm; P = 1.05 MPa; b) r = 2.5 μm; L = 0.13 mm; t = −0.56°C; ∇T = 12.4 deg/cm; P = 0.17 MPa; P is the nitrogen pressure in the bubble.

different temperatures (Fig. 5.10). The capillaries were inserted in a hole of a metal plate of a thermogradient chamber [161, 164]. A water film was formed by vapor adsorption on the capillary surface. The position of the ice–water interface in the capillary (T_0 = 273 K) was taken as a reference point when determining the mean temperature of the film T_c in these experiments. A system of thermoregulation enabled different temperature gradients ∇T and mean temperatures T_c to be maintained. To suppress vapor diffusion, the open capillary space was filled with nitrogen at pressures P up to 2 to 3 MPa [167].

Data on the displacements of ice menisci x during the time of the experiment t in a capillary of radius r = 13.4 μm at the mean temperature of the film −0.95°C are presented in Fig. 5.11, *a*, as an example. The rate of menisci displacement $V = dx/dt$ was about 10 μm/h under these conditions, fell to 5 μm/h when the temperature decreased to −2°C, and became 2 μm/h at −5.7°C.

In a capillary of smaller radius (Fig. 5.11, *b*), where the contribution of the thermocrystallization film flow is much higher, at a temperature close to 0°C the rate of menisci displacement rose to 100 μm/h.

The menisci displacement, calculated according to eq. (V.15), that might have been caused by the vapor diffusion is shown in Fig. 5.11 by dashed lines. The values of the diffusion coefficient were determined from the empirical equation:

$$D = 0.217(1/P_T)(T/273)^{1.88}, \text{ cm}^2/\text{s}, \qquad (V.23)$$

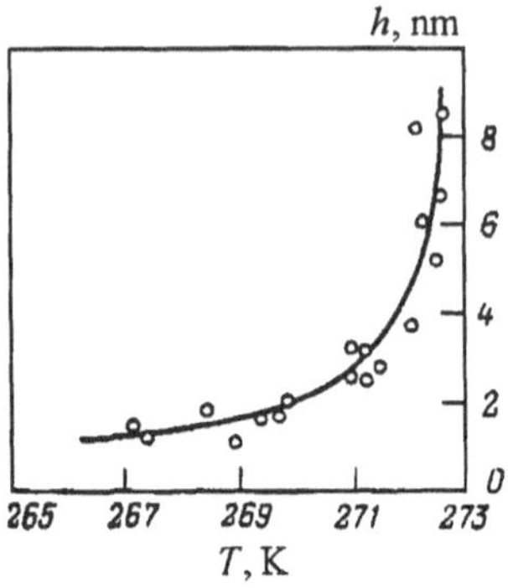

Figure 5.12 Temperature dependence of the thickness h of non-freezing water wetting films on quartz surface.

where P_T is the pressure (in atm) of vapor–gas mixture in the capillary during the experiment and T is the mean temperature of the experiment.

As seen from comparison of the experimental data (lines *1*) in Fig. 5.11 with the dashed lines 2, the film flow makes a substantial contribution to the mass transfer.

The rate of the thermocrystallization film flow was determined from the difference between the measured rate of displacement of ice menisci $V = dx/dt$ and that calculated from the equation of vapor diffusion. The values obtained were compared with these obtained theoretically from eq. (V.22):

$$V_T = (2\rho h^3 \lambda_i / 3\eta r \rho_i T) dT/dx, \qquad \text{(V.24)}$$

where ρ_i is the ice density, η is the film viscosity. Assuming the viscosity of the thin non-freezing films is $\eta = 1.5\eta_0$, where η_0 is the viscosity of the bulk water at the experiment temperature, one can determine the thickness of non-freezing films from eq. (V.24).

The calculated $h(T)$ dependence is presented in Fig. 5.12. Maximum values of h are close to the thickness of the α-films on quartz surface at room temperature (Fig. 2.6, *b*). The non-freezing film thickness decreases with temperature, being 1.5 nm at −6°C. The form of the curve $h(T)$ agrees qualitatively with those of non-freezing interlayers (Figs. 4.17, 4.21). The deviations rise with decreasing temperature: the thicknesses of non-freezing interlayers decrease more abruptly with temperature.

Non-freezing films on quartz are thicker than those on ice [131], although qualitatively the dependences are similar. It is also known that the thicknesses of water films on ice are different for different crystal faces [130]. Therefore, the thickness of non-freezing water layers depends significantly on the properties of a solid phase they contact. This is connected with the different nature of surface forces, binding energies and water molecule orientation near the different surfaces.

As in the case of thermocapillary flow, we may compare the rates of thermocrystallization film flow and vapor diffusion. Using eqs. (V.24) and (V.17), we get:

$$V_T/V_d = 2\rho h^3\lambda_i RT/3\eta r\rho_i D v_m \lambda_v, \qquad \text{(V.25)}$$

where λ_v is the heat of sublimation.

At atmospheric pressure and negative temperatures near to 0°C, the value of the ratio V_T/V_d at $h = 10^{-6}$ cm is approximately equal to $10^{-3}/r$, where r is expressed in cm. Therefore, at $r \approx 10^{-3}$ cm the thermocrystallization and diffusion fluxes give equal contributions into the total mass thermotransfer. At $r < 10$ μm thermocrystallization film flow prevails while at $r > 10$ μm the vapor diffusion dominates. The contribution of vapor diffusion increases with decreasing film thickness. At $h = 5$ nm, the film flux is equal to vapor flux only at $r = 1$ μm.

§4 Wetting Film Flow Driven by a Concentration Gradient

A concentration gradient may also cause a shear stress on a free surface of a film making it flow. In this case $\tau = d\sigma/dx = (d\sigma/dC)(dC/dx)$, where $d\sigma/dC$ expresses the dependence of surface tension on solute concentration. The derivative $d\sigma/dC$ may be found from the Gibbs equation:

$$\Gamma = -(C/RT)(d\sigma/dC), \qquad \text{(V.26)}$$

where Γ is the amount of adsorption, mol/cm^2 and C is the appropriate equilibrium concentration.

The equation for a solution film flow has, according to eq. (V.12), the following form:

$$q = \rho h\tau/2\eta = -(\rho h\Gamma RT/2\eta C)d\ln C. \qquad \text{(V.27)}$$

For positive adsorption ($\Gamma > 0$), the film flow is directed towards lower solution concentration. This is especially pronounced in the case of surfactant solutions that features high $d\sigma/dC$ values (at concentration below the critical micelle concentration) and positive Γ values. This type of flow, called the Marangoni–Gibbs effect, may be of great importance in the case of evaporation of liquid films containing surfactants. For example, the concentration of surfactants on the outer surface of porous bodies in the course of water evaporation will give rise to considerable concentration gradients causing the liquid film flow to be directed from the surface inside the body.

A concentration gradient generated during such solution evaporation may cause a climbing of the liquid film along a surface of semiimmersed electrodes equilibrated by gravity [169, 170]. Films of micron thickness may climb to heights of several cm.

The impact of the Marangoni–Gibbs effect on foam and emulsion film thinning has been well studied both theoretically and experimentally [171–174]. The flow of wetting films has been experimentally studied far less. Marangoni–Gibbs effects have frequently been included in calculations of the kinetics of drop spreading [175–177].

It has been shown in [178] that concentration gradients of surfactants may considerably influence film flow in sands and soils incompletely saturated with water. The velocity of film flow was up to 2 to 3 cm/day.

In electrolyte solutions when ions are negatively adsorbed at the surface of water ($\Gamma < 0$), the flow direction is reversed. However, due to low values of $d\sigma/dC$, the velocity of film flow is small compared to that of other flows, namely, that due to capillary osmosis.

Film flow under a concentration gradient of KCl in water solutions was observed in [179]. However, in these first experiments it was difficult to distinguish capillary-osmotic flow from that driven by electroosmosis.

§5 Electro-osmotic Flow of Wetting Films

When the free film surface has no charge ($\sigma_2 = 0$), the equations obtained in Chapter IV, §2 are applicable to film electro-osmosis. The Navier–Stokes equation for one-dimensional flow of a flat wetting film at $\sigma_2 = 0$ takes the form

$$\rho_0 \nabla U = \eta (d^2 v / dz^2), \tag{V.28}$$

where $U(x)$ is the electrical potential and $\rho_0(z)$ is the charge density. The analytical expression for the velocity of electro-osmosis has been derived for the case of highly overlapping EDL and low surface potentials ψ of the substrate [93]:

$$q = (\rho \varepsilon_0 \psi / 4\pi\eta)[1 - (\tanh \kappa h / \kappa h)] \nabla U, \tag{V.29}$$

where κ is the inverse Debye radius, ρ is the liquid density and h is the film

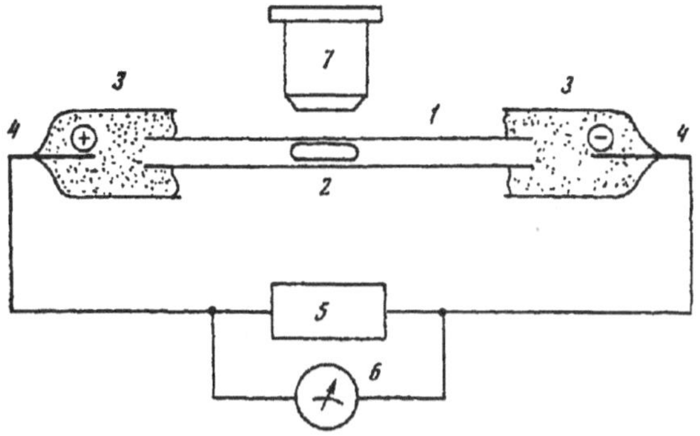

Figure 5.13 Scheme of the capillary cell for studying electro-osmosis of wetting films for aqueous solutions.

thickness. Common boundary conditions $v = 0$ at $z = 0$, $dv/dz = 0$ at $z = h$ were used here.

Electro-osmotic flow of wetting films has been experimentally studied in [180] using the previously described method of the bubble in a capillary (Fig. 5.7). Rates of bubble displacement V as a function of ∇U, composition and solution concentration were measured. Results approached qualitative agreement with the theory of electro-osmosis for rather thick β-films of aqueous solutions stable in the cases when the length of bubbles was less than 1.5 mm. Quantitative examination of the theory will require, together with the measurement of the rate of electro-osmosis, independent determinations of the ψ-potential and the film thickness.

As in the case of porous bodies filled with liquid, the electro-osmotic film flow takes place when the capillary condensate is separated by films in pores [181, 182]. At a low extent of filling of a porous space, when the capillary liquid was separated by film regions, changes in the rate of electro-osmosis correlated with the growth of moisture content. This agrees with eq. (V.29) from which it follows that q values increase with the film thickness h.

The assumption that the surface charge of the air–film interface is equal to zero, when deriving eq. (V.29), is not necessarily applicable in practice. Exerova and Zakharieva [183] have shown experimentally that the negative surface charge of free water films stemming from the predominant hydroxyl ion adsorption may be of 0.1 $\mu C/cm^2$ at pH > 6. The adsorption of ionic surfactants may lead to a much higher surface charge of free films.

The application of eq. (V.28) to this system was considered in [184]. Expressions for the distribution of charge density $\rho_0(z)$ between two film surfaces of different charges σ_1 and σ_2 and different potentials ψ_1 and ψ_2, were used. Here the index 1 refers to the substrate surface and 2 to free surface of the wetting film. The second boundary condition should also be changed,

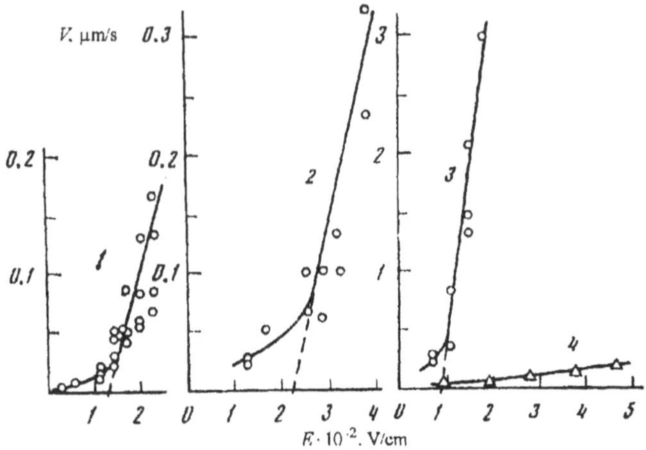

Figure 5.14 Rate of a bubble displacement V in the capillary due to electroosmotic film flow vs. field strength E: *1*) 10^{-4} M SDS, $r = 8.9$ µm; $l = 0.35$ mm; 2) 10^{-3} M KCl, $r = 4.8$ µm; $l = 0.315$ mm; *3*) 10^{-1} M SDS, $r = 8$ µm; $l = 0.275$ mm; *4*) 10^{-4} M SDS, $r = 5.7$ µm; $l = 0.107$ mm.

because the shear stress $\tau(h) = \sigma_2 \nabla U$ caused by the interaction of the surface charge σ_2 with the electric field operates on the free surface of the film.

An analytical solution was obtained for the case of low potentials ψ_1 and ψ_2 which do not vary with the film thickness. The velocity of electro-osmotic flux is expressed by the following equation:

$$q = (\rho\varepsilon_0\psi_1/4\pi\eta)[1-(1+\psi_2/\psi_1)\tanh(\kappa h/2)/\kappa h]\nabla U . \quad \text{(V.30)}$$

Analysis of this expression shows that at $\sigma_2 = 0$, when $\psi_2 = \psi_1/\cosh(\kappa h)$, the second term in square brackets becomes equal to $\tanh(\kappa h)/\kappa h$ and the expression for the electro-osmotic flow transforms into eq. (V.29).

Let us consider the case when the potentials of both film surfaces are negative. Then at $|\psi_2| > |\psi_1|/\cosh(\kappa h)$, the charge σ_2 is negative and grows in absolute value with the difference between ψ_1 and ψ_2. This negative charge impedes the film flow, because the applied electric field acts in opposite directions on the surface charge σ_2 and on the cations in the EDL. At some value of σ_2, the flow in the film becomes equal to zero. The condition $q = 0$ corresponds, as seen from eq. (V.30), to the following ratio of potentials:

$$\psi_2/\psi_1 = \{\kappa h \sinh(\kappa h)/[\cosh(\kappa h)-1]\}-1 . \quad \text{(V.31)}$$

For example at $\kappa h = 1$, this ratio is equal to 1.16, whereas the transition to the flow without surface charge ($\sigma_2 = 0$) corresponds to $\psi_2/\psi_1 = 0.648$, when, obviously, $q \neq 0$.

When $|\psi_2| < |\psi_1|/\cosh(\kappa h)$, the surface charge is positive thereby leading to an acceleration of the film flow due to an additional field action on the surface charge $\sigma_2 > 0$ having the same sign as the counter-ions near the solid surface. Thus the accelerated electro-osmotic flow in wetting films may be expected when $\psi_2 << \psi_1$ that is usual for water films [9]. On the other hand, in the wetting films where $\psi_2 >> \psi_1$ the electro-osmotic flow is weak and may have the opposite direction with respect to the flow of counter-ions in the EDL near to solid surface.

Studies of electro-osmotic flow of thinner α-films formed in quartz capillaries of r = 5–10 μm have been attempted [185]. The experimental method used for this study is shown in Fig. 5.13. A capillary *1* with the air bubble *2* formed in it was dipped in glass ampoules *3* with agar-agar. The voltage from the power supply *5* was applied to platinum electrodes sealed in the ampoules and was controlled by the voltmeter *6*. The displacement of the bubble under the voltage ΔU (up to 12 V) was observed using horizontal comparator *7* with an accuracy of ±1 μm. The equipment was located on the table of the comparator. The lengths of the bubbles l were 100 to 300 μm. The aqueous solutions of KCl (10^{-4}–10^{-2} N) and sodium dodecyl sulfate SDS (10^{-4}–10^{-1} N) were used in the experiments.

Experimental plots for the rates of the bubble displacement V as a function of the field strength E are shown in Fig. 5.14. Because the electric resistance of thin films is much higher than the resistance of the solution in the filled part of the capillary, E values were taken equal to $E = \Delta U/l$. The experimental points are considerably scattered, so we can only discuss approximate results. As in [180], a linear dependence of the electro-osmotic velocity on the field strength was observed only above some threshold value E_0 intercepted on the plots by the dashed extensions of linear regions. In different experiments the values of E_0 were 100 to 200 V/cm.

The thicknesses of wetting films were estimated using eq. (V.30). Taking for the solutions of low concentration (curves *1* and *2*) $\psi_1 = -100$ mV and $\psi_2 = -45$ mV [183], $\kappa = 10^6$ cm^{-1}, $r = 4.8$ μm, we obtain $h \sim 5$ nm close to the reported thickness of α-films (Fig. 2.6, *b*). At concentrations of SDS above the CMC, the velocity of electro-osmosis increases (curve *3*) due to greater film thickness ($h \sim 25$ nm). The effect of CMC is clearly seen when compared with low SDS concentration (curve *4*).

The non-linear growth of the electro-osmotic flux with E can be explained in the framework of the so-called secondary electro-osmosis [186], because in the thin films, as in the fine pores of membranes, primarily one type of ion migrates. According to this concept, the velocity of electro-osmosis in such ion-selective systems should grow as E^2 due to the accumulation of cations at the inlet because they penetrate the thin film less readily. As seen from

Fig. 5.14, the results obtained show a close to quadratic dependence of V on E. Deviations from linearity were observed, as in [186], when the strength of the external electric field exceeded about 100 V/cm.

To investigate this effect in more detail, it is necessary to measure the current through the film together with measurements of film flow rates. Such experiments with aqueous solutions of different compositions and concentrations, including various surfactants, would enable the mechanism of this phenomenon, which can play a significant role in disperse materials and porous bodies, to be verified.

Chapter VI

Vapor Flow in Porous Bodies

Analyses of mass transfer processes in porous bodies, except the case of non-volatile liquids, should take into account vapor transfer. This mechanism prevails in volatile liquids with high pressure of saturated vapors (especially at high temperature) and in porous bodies with a low mass content.

Different mechanisms of vapor and gas transfer will be considered in this Chapter. Free-molecular, viscous and transient regimes are characteristic of one-component vapor and gas flow in porous bodies. In these cases we need to take into account porous space geometry and such supplementary effects as gas slippage at low Knudsen number and surface diffusion. The next section contains the results obtained for vapor diffusion through immobile gas, accompanied by surface diffusion of adsorbed vapor. Temperature gradients cause the effects of thermodiffusion and heat slippage. It is shown that a non-uniform electrical field can accelerate or impede the vapor fluxes. The more complicated problem of mutual diffusion of vapor/gas mixtures in thin capillaries, even at temperatures higher than boiling point is discussed, including the effects of capillary conicity and application of temperature gradients.

The equations related to the cases considered are compared in this Chapter with experimentally obtained data.

§1 Free-molecular, Viscous Flow and a Transient Regime

A vapor flow obeys different physical laws depending on vapor pressure, the presence of other vapors or gases and pore width. For Knudsen numbers $Kn = (\lambda/d) >> 1$ (where λ is the molecular free path length and d is the characteristic cross-section of a pore), a free-molecular Knudsen flow occurs when the vapor molecules collide only with the pore walls. When the molecule reflection from the walls of a cylindrical capillary of radius r is fully diffusive, the vapor flux is equal [187–189] to:

$$q = (8r/3)(\mu/2\pi RT)^{1/2}\nabla p = K_1 \nabla p, \text{ g/cm}^2\text{s}, \quad \text{(VI.1)}$$

where μ is the mass of a mole and p is the vapor pressure. This equation does not include viscosity, because there is no exchange of impulses among molecules. The Maxwell distribution of thermal velocities is specified by the temperature T of the walls.

This flow occurs in a vapor or vapor–gas mixtures for which the pressure meets the condition $Kn \geq 5$. The free path length is related to the pressure p by the following approximate relation: $\lambda \approx 0.057\ kT/\delta^2 p$, where k is the Boltzmann constant and δ is the molecular radius. For air at atmospheric pressure $\lambda \approx 0.06$ µm. Under these conditions, Knudsen flow can take place in microporous bodies with pore radii of $r < 5$ nm. The coefficient of permeability for the Knudsen regime of flow K_1 in eq. (VI.1) decreases linearly with capillary radius.

When deriving eq. (VI.1), Knudsen used the assumption that the molecular reflection from capillary walls is fully diffusive. According to this assumption, a gas molecule after a collision is temporarily retained near to the wall in the potential well created by intermolecular forces (the gas is adsorbed) and then evaporates in an arbitrary direction having no memory about the direction of its motion before the collision. This assumption has proven itself only partially. Generally it should be taken into account that only a proportion f of the molecules reflects diffusely, but another part $(1 - f)$ reflects specularly obeying the law of elastic collision. In this case, as shown by Smoluchowski, the coefficient K_1 must be multiplied by $(2 - f)/f$. As a rule, f values do not differ much from unity and depend on the molecular structure of the substrate and its smoothness. For example, in the case of noble gases, $f = 0.9$ for polished glass and $f \approx 0.8$ for metals. In the case of purely specular reflection ($f = 0$), a capillary does not resist a gas flow, Δp tends to zero as $f \to 0$.

Experiments with molecular gas beams and single capillaries showed that at decreasing angles of incidence a greater proportion of the molecules reflects specularly. For crystal substrates there are some preferential directions for specular reflection. At low velocities of transfer, $f \to 0$. High-speed molecules have greater probability of being reflected specularly. Specular reflection dominates in light gases and gases with lower adsorption energy.

An exact account of all the factors is complicated and requires detailed statistical analysis of interaction between the molecules and the walls.

Equation (VI.1) can be considered to be a good approximation for practical purposes, provided that f values are close to unity.

In fine cylindrical capillaries, the molecules incident at low angles can travel through a capillary reflecting purely specularly. Derjaguin and Fedjakin [190] derived, instead of eq. (VI.1), an expression for a flow independent of capillary radius (due to mainly specular reflection of molecules) and depending weakly on the pressure drop:

$$q = (2\alpha_0^2 p\lambda/L)(\mu/2\pi RT)^{1/2}\ln p_1/p_2,$$

where p_1 and p_2 are the values of pressure at the ends of the capillary of length L; α_0 is the limiting angle of specular reflection. The equation is applicable for $Kn > 1/\alpha_0$. This restricts the application of eq. (VI.1) to Knudsen numbers in the range of $(1/\alpha_0) > Kn > 1$. For water vapor in capillaries of $r < 0.1$ μm, $\alpha_0 \approx 0.1$ rad, or about 6°.

The Dashman interpolation formula is often used, if the capillary length L is commensurate with its radius [188]:

$$q = (\mu/2\pi RT)^{1/2}(p_1 - p_2)/[1 + (3L/8r)]. \qquad \text{(VI.2)}$$

When $L/r \to \infty$, and $(p_1 - p_2)/L = \Delta p$, it transforms into eq. (VI.1), and at $L/r \to 0$ into the equation for a gas outflow through a hole in an infinitely thin wall:

$$q = (p_1 - p_2)(\mu/2\pi RT)^{1/2}.$$

In free-molecular flow, each of the components of a gas mixture passes through a capillary independently, since the molecules do not collide.

In [191] the influence of non-uniformity of porous structure on the Knudsen gas diffusion was analysed by the Monte-Carlo method. The porous structure was modelled by a three-dimensional pore network built from spherical voids connected by segments of cylindrical pores. The coefficient of Knudsen diffusion was calculated by root-mean-square displacement of molecules in unit time. It was shown that in the case of low porosity, the difference from the known solutions for a sinusoidal pore profile does not exceed 15–20%. At high porosity the difference increases.

When $Kn \ll 1$, vapor molecules collide chiefly with each other resulting in viscous vapor flow. The only distinction from the Poiseuille equation for a liquid flow is that the gas compressibility and the dependence of its density on pressure is taken into account. This leads to the following equation:

$$q = \mu r^2(p_1^2 - p_2^2)/16\eta RTL = \mu r^2\langle p\rangle\Delta p/8\eta RT = K_2\Delta p, \qquad \text{(VI.3)}$$

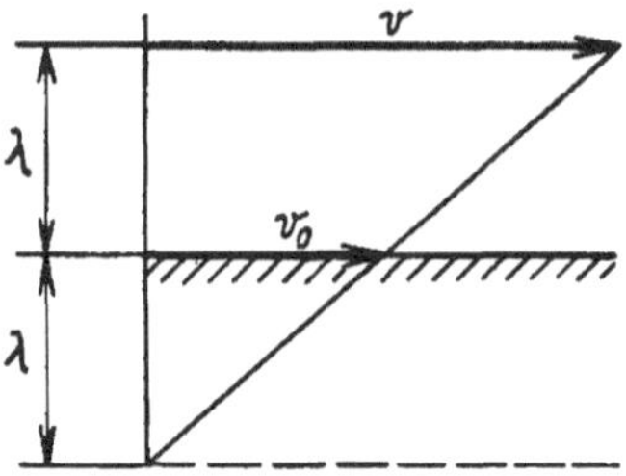

Figure 6.1 Distribution of velocities in the case of gas slippage.

where η is the vapor viscosity and $\langle p \rangle$ is the average pressure. During viscous flow, the permeability coefficient K_2 depends to the second power on capillary radius and is also a function of pressure.

For fine capillaries of radius comparable with λ (at $0.1 > Kn > 0.01$), the slip effect resulting from the diffuse reflection of gas molecules should be taken into account. Since the velocity component of a diffusely reflected molecule parallel to the capillary axis is equal to zero, and other molecules in the near-to-wall layer possess the mean drift velocity v, then, on average, near to the wall $v_0 = v/2$, where v_0 is the slippage velocity. Molecules located in the near-to-wall layer of thickness λ strike and leave the wall without a collision. In this case the distribution of velocities $v(z)$ is as shown in Fig. 6.1. The slip effect results in an apparent expansion of a capillary bore by the value of λ. Evidently, we can neglect this effect when $r >> \lambda$. At $r \sim \lambda$, eq. (VI.3) takes the following form [187]:

$$q = [(\mu r^2 \langle p \rangle / 8\eta RT)][1 + (4\zeta / r)]\nabla p, \qquad \text{(VI.4)}$$

where $\zeta \sim 1.38$ is the slip coefficient. The effect of gas slippage in capillaries is well-studied experimentally [192]. However, a more rigorous treatment again introduces the problems encountered when eq. (VI.1) was extended to consider the imperfect diffuse reflection of molecules from the capillary walls.

For Knudsen numbers of the order of unity, molecules collide with each other and with the walls at equal frequency, corresponding to the transient regime of gas or vapor flow. Only approximate or numerical solutions are known for this regime.

The calculated ratio of the coefficient of permeability, K, for transient and viscous flow regimes to that at the Knudsen regime, K/K_1, plotted against the reverse Knudsen number are shown in Fig. 6.2 by the dashed lines. These calculations were performed for water vapor at $T = 300$ K. For the transient region the data from [193] were used. The curves are plotted for various

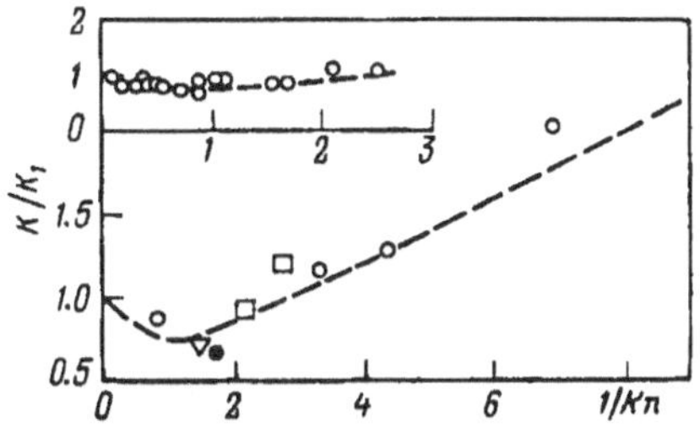

Figure 6.2 Dependences of the ratio of permeability coefficients, K/K_1, on the reverse Knudsen number in the course of water evaporation from quartz capillaries. Theoretical data are shown by dashed lines.

scales to cover a wider range of the Knudsen numbers. The experimental data obtained for water evaporation from narrow quartz capillaries with r ranging from 0.2 to 50 μm [194] are shown by points in Fig. 6.2. To exclude the influence of film flow, the measurements were carried out at low relative vapor pressure, and the capillary surface was previously made hydrophobic.

When $1/Kn > 3$, the dependences obtained correspond to a viscous regime of flow. Here the dependence of K/K_1 on $1/Kn$ is practically linear, since the $1/Kn$ values are proportional to the average vapor pressure $\langle p \rangle$, on which the coefficient K_2 from eq. (VI.3) depends linearly. The values of K/K_1 pass a weak minimum at $Kn \sim 1$, ascribed by Derjaguin and Bakanov [195] to a specific mechanism whereby molecule impulses in collisions are balanced without the formation of a velocity profile. The existence of this effect makes it possible to use the equation of free molecular flow (VI.1) with a sufficient accuracy (~10%) for Knudsen numbers $Kn \geq 0.5$.

At lower Knudsen numbers, the equations with viscous flow work well. As a first approximation, it is possible to use for calculations of water evaporation eq. (VI.1) when $Kn > 0.5$ and eq. (VI.3) when $Kn < 0.5$. The experimental results obtained for water (Fig. 6.2) supported the validity of this approach. For a more volatile liquid, e.g. carbon tetrachloride, the evaporation from capillaries (at $\langle p \rangle = 1.3 \times 10^4$ N/m^2) is well described by the equation of viscous flow [194]. This equation also governs water evaporation from capillaries at elevated temperature (380 K) [196]. However, any further increase in the rate of evaporation produces deviations from eq. (VI.3) towards lower values of K_2 associated with a decreasing temperature of the evaporating meniscus. To calculate the rate of evaporation in this case, one should solve the problem of a combined mass and heat transfer taking into account the heat of evaporation and the thermal conductivity of the porous body.

The applicability of eq. (VI.1) has not only an upper ($r < \lambda$) but also a lower limit. When the pore size is comparable with the radius of attractive forces

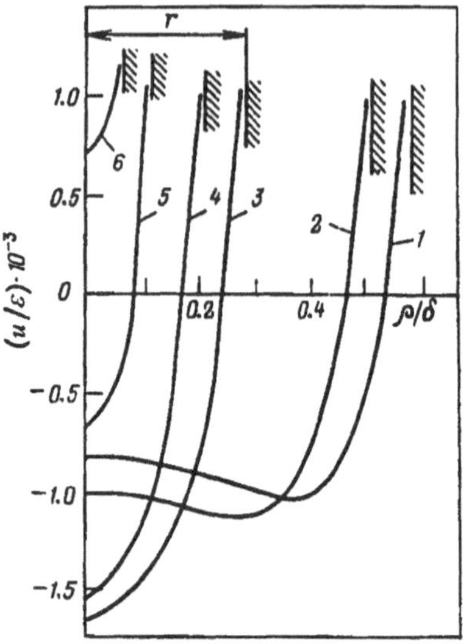

Figure 6.3 Radial distribution of potential energy, $U/\varepsilon = f(\rho)$, of a molecule in capillaries of various relative radius $r/2\delta$ = 0.73 (*6*), 0.882 (*5*), 0.956 (*4*), 1.03 (*3*), 1.25 (*2*) and 1.32 (*1*). Molecular diameter 2δ = 0.34 nm.

between molecules and the wall, the linearity of the trajectories of molecular motion, on which the Knudsen equation is based, breaks down. Molecular forces influence the flow markedly at capillary radii of about 10 molecular diameters. However, at increasing temperature or decreasing energy of molecule–wall interaction, Q, (Fig. 2.2) Knudsen flow is maintained. In this case, when $Q \ll kT$, the applicability of eq. (VI.1) is extended.

To calculate the interaction between vapor molecules and those of the material of the wall, the Lennard–Jones potential (eq. (III.1)) is used, as has been shown in Chapter III. For flat slits, the integration of pair molecular interaction gives a potential of type 3–9 (eq. (III.2)). For cylindrical capillaries the dependence of molecular potential energy on the radial coordinate, $U(\rho)$, is more complex [197]. The potential curves for capillaries of various radii, r, obtained by numerical methods are shown in Fig. 6.3. Here the dimensionless values of energy U/ε, where $\varepsilon = 120\ kT$ is a parameter of the potential (III. 1), are plotted on the y-axis. The calculations were performed for microcapillaries, in which the fields of surface forces overlap and the potential energy is negative, which corresponds to an attraction, even in the centre of a channel. When approaching the capillary walls, the attraction is replaced by repulsion, that sharply increases (approximately as $1/\rho^9$) when $\rho \rightarrow r$. As a result, there is a potential minimum near to each wall, which corresponds to the maximum probability of molecule location (curves

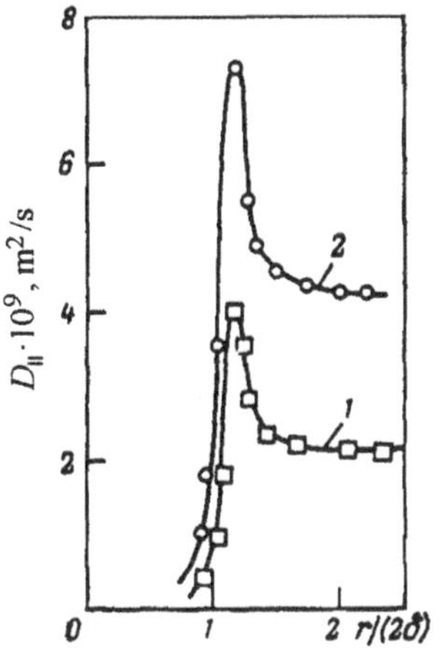

Figure 6.4 Dependence of the coefficient $D_{\|}$ of longitudinal self-diffusion on dimensionless capillary radius $r/2\delta$ at $T = 20$ K (*1*), 40 K (2), $\varepsilon = 120\ kT$.

1 and *2*). This state can be considered as a preferred zone of molecule localization, i.e. the adsorption state (where the probability of occupation increases with *r*). If *r* reduces (curves *3* and *4*), the single (and deeper) potential minimum *U* is found in the center of the channel instead. This corresponds to a transition from a molecular adsorption state near the wall to one in the center of the micropore. In this case the adsorption ceases to be proportional to the area of pore surface. The number of adsorbed molecules is governed by the total volume of micropores and the depth of potential minimum which depends on *r*. Transition from surface adsorption to micropore filling takes place (for selected potentials, Fig. 6.3) at *r* between 1 and 0.5 nm according to the micropore definition suggested by Dubinin [15].

Deriving eq. (VI.1), we account for only a short-range, repulsive component of the potential causing molecules to bounce from the walls. Obviously, if a potential field in a micropore differs from the Knudsen model assumption, it should be reflected in a dependence of coefficients of permeability, *K*, on capillary radius.

The coefficients of longitudinal self-diffusion of molecules in microcapillaries, $D_{\|}$, with potential curves given in Fig. 6.3 were calculated by the Kirkwood method as a ratio of the average molecular displacement to the average intervals without collisions between molecules and walls. This approach gives the following expression for $D_{\|}$ [197]:

$$D_{\|} = kT/(m\,\langle\Delta^2 u\rangle)^{1/2}, \qquad \text{(VI.5)}$$

where k is the Boltzmann constant, m is the molecular mass and $\langle \Delta^2 u \rangle$ is the squared derivative $du/d\rho$ (proportional to the force of interaction between molecules and walls), averaged over the capillary cross-section. The long-range force field curves the linear trajectory of molecules thereby changing their mobility in the direction of the capillary axis.

The results of calculations of $D_{\|}$ for microcapillaries of various radii are presented in Fig. 6.4. The $D_{\|}$ values pass through a maximum at a capillary radius r close to the molecular diameter. This capillary size corresponds to the deepest potential minimum (curve *3*, Fig. 6.3). Localization of molecules in a potential well in the center of a channel provides their high tangential mobility, corresponding to the transition from three- to one-dimensional motion. Further narrowing of the capillary (curves *5* and *6*, Fig. 6.3) reduces the molecular mobility due to the steric factor, i. e. increasing "friction" against the walls.

"Switching off" the field of surface forces (for instance by temperature elevation, when $\varepsilon << kT$) changes values toward the Knudsen coefficient of diffusion, if the calculations are performed by the same method. To compare with eq. (VI.1), the latter may be transformed into that of a diffusion type:

$$q = \mu D_k \nabla p / RT = K_1 \nabla p, \qquad \text{(VI.6)}$$

where D_k is the coefficient of the Knudsen diffusion, equal to

$$D_k = (8r/3)\,(kT/2\pi m)^{1/2}. \qquad \text{(VI.7)}$$

Equation (VI.5) gives the same dependence $D_k(r)$ at $\varepsilon << kT$. In this case the trajectories are disturbed only in a very thin (much lower than r) near-to-wall layer. As a result, the flow obeys eqs. (VI.1) and (VI.6).

The calculations showed that for microcapillaries the temperature dependence of the diffusion coefficient also changes. As seen from eq. (VI.5), $D_{\|}$ depends linearly on temperature, whereas in wider capillaries D is proportional to $T^{1/2}$, as follows from eq. (VI.7). The transition from one dependence of $D_{\|}(T)$ to the other takes place across the range of T from 100 to 800 K.

Calculations performed for a slit pore by the methods of molecular dynamics gave dependences similar to those shown in Fig. 6.4, passing a maximum at $r \sim 2\delta$. Here a more realistic model of pore walls, accounting for a discrete molecular structure of the latter, was used [198].

Nicholson and Petropoulos [199, 200], having solved numerically the equation of motion, obtained a series of possible molecular trajectories in a field of surface forces for flat and cylindrical bores of various cross-sections and length. This enabled them to calculate molecular flow in capillaries and compare this behaviour with Knudsen flow. Figure 6.5 schematically shows

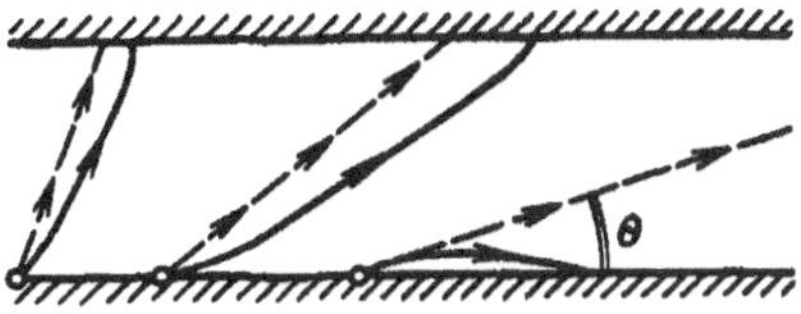

Figure 6.5 Change of molecular trajectories in thin channels with molecular attraction.

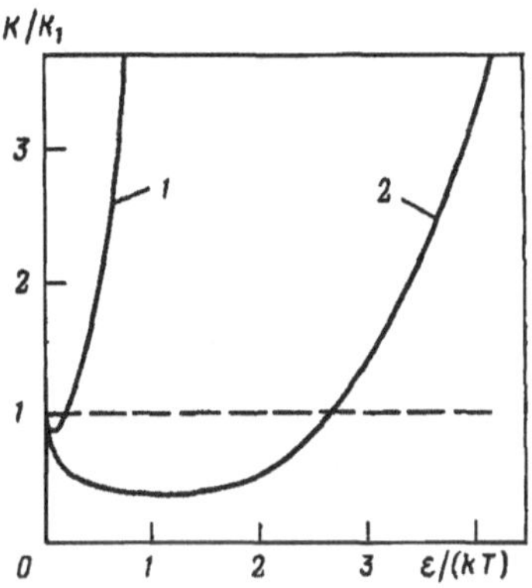

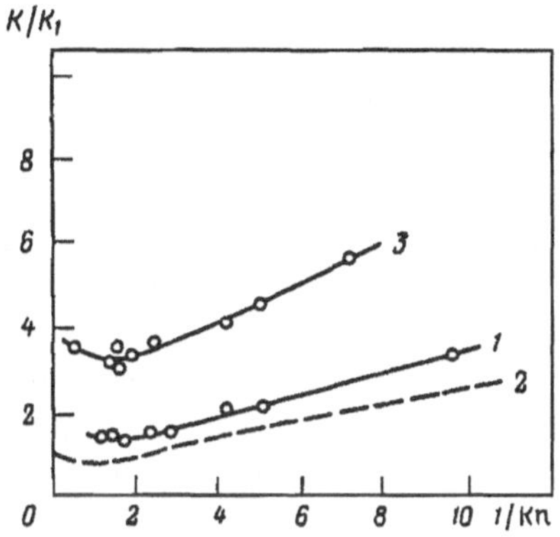

Fig. 6.6 Fig. 6.7

Figure 6.6 Results of calculations of the ratio of permeability coefficients, K/K_1, as a function of ε/kT for microcapillaries of radius $r = 1.84$ (*1*) and 11.08 nm (2) [199].

Figure 6.7 Dependences of K/K_1 on the reverse Knudsen number obtained in the course of water evaporation from quartz capillaries into undersaturated vapor at $p/p_s = 0.92$ (*1*) and 0.99 (*3*). The theoretical curve is shown by the dashed line (2).

various trajectories that correspond to different molecular take-off angles θ from the wall during diffuse reflection. Knudsen (at $u(\rho) = 0$) trajectories are shown by dashed lines and those curved due to forces of molecular attraction by the walls are shown by solid lines. At very low θ, the motion in the volume of a bore changes for the motion along the surface. With a reducing capillary radius, the contribution of surface diffusion increases as:

$$K/K_1 = 1 + (A/r^2), \qquad \text{(VI.8)}$$

where the constant A depends on the intensity of surface force field [200].

As seen from eq. (VI.8), surface diffusion magnifies the flux in comparison with Knudsen behaviour. On the other hand, longer trajectories than in the case of Knudsen flow (Fig. 6.5) reduce the value of K/K_1. Figure 6.6 shows K/K_1 values calculated as a function of ε/kT, the dimensionless potential in the Lennard–Jones equation, for two microcapillaries with different radii [199]. When $\varepsilon \to 0$, the values of K approach the Knudsen coefficient K_1. First the increasing energy of interaction between molecules and walls causes the values of K/K_1 to decrease, due to the curving of the trajectories, but then makes them grow due to the increasing contribution of surface diffusion. The increase in surface diffusion is associated with a growing number of molecules in the vicinity of the capillary surface due to the rising energy of interaction ε/kT. The narrower the capillary (curve *1*, Fig. 6.6), the earlier the effect of surface diffusion appears. For these systems we may still regard this as surface diffusion because the transition to the central motion in the capillary takes place at $r \sim \delta$.

The influence of surface diffusion of adsorbed molecules on vapor flux in microporous bodies has been investigated by many authors. The significance of this effect is reliably established. However, the complicated geometry of porous space hinders the transfer of the results obtained for one type of adsorbents in application to other porous bodies. Numerical calculations for various model structures of a porous body provide good agreement with the experimental results only at low coverage by adsorbed layers [201].

In this connection, we consider here the results obtained for single fine cylindrical capillaries, with simple geometry of porous space, enabling a comparison with the theory. Figure 6.7 shows the results of measurements of K/K_1 for water vapor, when the adsorption of water molecules on capillary walls does not exceed a monolayer (curve *1*). As seen from the comparison with the dashed curve *2*, corresponding to the vapor flow without adsorption, the values of K/K_1 are slightly greater than theoretical expectation. The difference between the curves *2* and *1* is associated with the surface flow in the adsorption film. At higher relative vapor pressure $p/p_s \sim 0.99$ (when the thickness of α-films is about 3–4 nm, Fig. 2.6, *b*), the values of K/K_1 are several times greater than the predictions by the theory (curve *3*). However, this case refers rather to the flow of wetting films considered below in Chapter VII than to surface diffusion.

§2 Vapor Diffusion Through Gas

When the pores free of liquid are filled both with vapor and a gas, for example air, the conditions of vapor transfer change. If the total pressure of the vapor–gas mixture in pores is equal to the pressure P_0 in the surrounding media, the isothermal vapor transfer at a partial pressure $p << P_0$ is governed by equations of diffusion through an immobile gas. Under these conditions the stationary mass flux is equal to:

$$q = -D\nabla C = -(\mu D/RT)\nabla p, \tag{VI.9}$$

where D is the coefficient of vapor diffusion through the gas and $C = \mu p/RT$ is the vapor concentration, g/cm^3.

A resolution of non-stationary situation is based on the law of mass conservation. For the one-dimensional case the law is written as follows:

$$\partial C/\partial t = -\partial q/\partial x. \tag{VI.10}$$

The change in vapor flow ∂q in unit length of a porous body ∂x is equal to the change in vapor concentration ∂C for this length within the time ∂t. Substitution of the expression for the derivative of the flux with respect to the coordinate $\partial q/\partial x$ from eq. (VI.9) into eq. (VI.10) gives the well-known equation for non-stationary diffusion:

$$\partial C/\partial t = \partial/\partial x[D(C)\partial C/\partial x]. \tag{VI.11}$$

Solution of this equation gives the distribution of vapor concentration in space and time, $C(x, t)$, that enables local values of the flux $q(x, t)$ to be determined using eq. (VI.9). In a number of cases we have to account for the dependence $D(C)$ of the diffusion coefficient on vapor concentration.

Since $D \approx (1/3)\lambda\langle v\rangle$, where λ is the free path and $\langle v\rangle$ is the average thermal velocity, in a general case the values of D reduce with increasing pressure and decreasing temperature. So, for water vapor in air we can use the empirical eq. (V.23) for the diffusion coefficient [202].

If a vapor diffuses into a dry porous body, the process is accompanied by vapor adsorption on a pore surface. In this case, a term that accounts for the changes in quantity of the substance adsorbed must be added into eqs. (VI.9) and (VI.11):

$$\partial a/\partial t + \partial C/\partial t = \partial/\partial x[D(C)\partial C/\partial x], \tag{VI.12}$$

where a is the mass of substance adsorbed in a unit pore volume, g/cm^3. The relation between a and C must be specified by the adsorption isotherm $a = a(C)$. For the capillary model, when $a = 2\Gamma/r$ (where Γ is adsorption, g/cm^2, and r is the capillary radius) eq. (VI.12) is replaced with the following expression:

$$(2RT/r\mu)\partial\Gamma/\partial t + \partial p/\partial t = \partial/\partial x[D(p)\partial p/\partial x], \tag{VI.13}$$

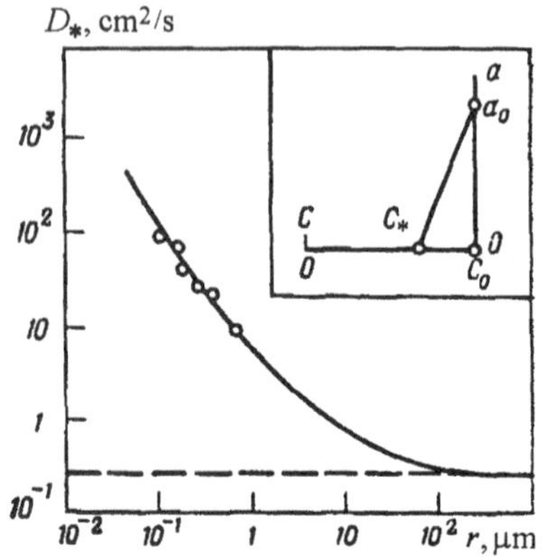

Figure 6.8 Dependence of the effective diffusion coefficient, D_*, of water vapor on the capillary radius r at $T = 299$ K and $C_0 = 2.44 \cdot 10^{-5}$ g/cm³.

where the concentration C is replaced by the partial vapor pressure $p = CRT/\mu$. As an example, let us consider the kinetics of vapor penetration into an initially empty infinite capillary from a medium, where constant vapor concentration $C_0 = \text{const}$ is maintained [203]. For an arbitrary isotherm $a(C)$ and an arbitrary relation $D(C)$, the solution of eq. (VI.12) under the initial and boundary conditions, $C(x, 0) = 0$ and $C(\infty, t) = 0$, is self-modeling. Under these conditions C and a depend only on the variable $\xi = (x/2)(D_0 t)^{1/2}$, where $D_0 = D(C_0)$ and eq. (VI.12) transforms into the ordinary differential equation:

$$\mathrm{d}/\mathrm{d}\xi\,[(D(C)/D_0)\mathrm{d}C/\mathrm{d}\xi] + 2\xi[(r/2)\mathrm{d}\Gamma/\mathrm{d}\xi + \mathrm{d}C/\mathrm{d}\xi] = 0, \qquad \text{(VI.14)}$$

which can be solved for any particular case by numerical methods.

For linear isotherms $a(C)$ and $D = \text{const}$ the analytical solution of eq. (VI.14) can be approached. The equation for the mass flux in a capillary has the same form as in the case of vapor diffusion without adsorption, but with some effective diffusion coefficient $D_* > D$ [203]:

$$q = -D_0 \mathrm{d}C/\mathrm{d}x|_{x=0} = C_0(D_*/\pi t)^{1/2}; \quad D_* = \pi(\rho A / 2C_0)^2, \qquad \text{(VI.15)}$$

where the constant A depends on the parameters of the adsorption isotherm.

Figure 6.8 shows the results of processing the experimental data for water vapor diffusion into fine quartz capillaries accompanied by polymolecular vapor adsorption [204]. Points on the graph, plotted in logarithmic coordinates, represent the experimentally obtained dependence of D_* on capillary radius r. The values of D_* were determined from eq. (VI.15) on the basis of the measured mass flux q as a function of time t. The solid curve is plotted using the solution of eq. (VI.14) for the adsorption isotherm consisting of two linear sections (see Fig. 6.8). The values of $a_0 = h_0\rho r/2$, where $h_0 = 4$ nm and $C_*/C_0 = 0.97$, correspond to the experimental isotherm of water α-films (Fig. 2.6, *b*). The horizontal dashed line shows the values of the coefficient of vapor diffusion through air according to eq. (V. 23): $D = 0.26$ cm^2/s.

As seen from Fig. 6.8, the adsorption sharply increases the mass flux q so that the values of D_* increase with decreasing r, and are two orders greater than D at $r = 0.1$ μm. In quartz capillaries of such radii, the polymolecular adsorption is a dominant mechanism governing the kinetics of mass exchange with the surrounding media. Vapor adsorption on capillary walls maintains high concentration gradients at the inlet of the capillary. At the same time, the adsorption hinders deep vapor penetration. The boundary of polymolecular adsorption, where $C = C_*$, in a capillary of radius $r = 0.22$ μm advances after 100 hours to 3.7 cm, while in much thinner capillary, $r = 0.07$ μm, to only 2.2 cm. In the absence of adsorption, the point where $C = C_*$ would travel to greater distances. In capillaries of radii greater than 100 μm, the influence of adsorption practically vanishes, and here $D = D_*$.

In the general case the influence of surface diffusion may be accounted for by introducing the effective diffusion coefficient [205] into eq. (VI.12):

$$D_* = D + \beta D_s, \qquad \text{(VI.16)}$$

where $\beta = \partial a/\partial C$ is the parameter of adsorption isotherm and D_s is the coefficient of surface diffusion. For example, for water adsorbed on silica D_s is about 8×10^{-7} cm^2/s for a surface coverage of $\theta = 0.15$; $D_s \cong 5\times10^{-6}$ cm^2/s for a complete monolayer ($\theta = 1$) and $D_s \cong 7\times10^{-6}$ cm^2/s for $\theta = 2$.

The equations of diffusion (VI.9)–(VI.11) are also applicable in weakly non-isothermal conditions, when it may be assumed that the different temperatures of menisci only specify different values of the partial vapor pressure. In this case, the pressure gradient is defined as $\partial p/\partial x = (\partial p_m/\partial T)(\partial T/\partial x)$, where p_m is the saturated vapor pressure over the menisci. For capillaries of $r \geq 0.1$ μm, we can enter no corrections for a decrease of pressure over concave menisci and assume that $p_m = p_s$. Using the Clausius–Clapeyron equation, $\partial p_s/\partial T = \lambda_0 p_s(T)/RT^2$, instead of eq. (VI.9), we obtain:

$$q_T = -[\mu D\lambda_0 p_s(T)/R^2T^3]\mathrm{d}T/\mathrm{d}x = [\mu D\lambda_0 p_s(T)/(RT)^2](\mathrm{d}\ln T/\mathrm{d}x), \qquad \text{(VI.17)}$$

where λ_0 is the heat of evaporation and T is the mean temperature in the region of corresponding measurements.

More exact calculations require adjustments for thermodiffusion of vapor and heat slippage. Generally, the influence of thermodiffusion, i.e. redistribution of components of a vapor–gas mixture under the temperature gradient, is insignificant. The concentrations of lighter components of the mixture, for example water vapor in air, increases in a hot region. For a temperature difference $\Delta T = 10°$, the change in concentration does not exceed 0.4%.

Heat slippage, which in nature is similar to the above-mentioned isothermal gas slippage (§1), is of greater importance. If a cylindrical capillary of radius r connects two volumes of gas, then the heat slippage will cause gas transfer to the hot side which results in a steady state with a constant thermal pressure difference Δp_T. Here the total gas flow is equal to zero: the amount of gas transferred due to heat slippage is equal to that transferred by viscous counter-flow driven by Δp_T. The Maxwell equation for the thermal slip velocity, refined by Derjaguin and Bakanov [206], gives the following expression for the thermo-molecular effect:

$$\Delta p_T / \Delta T = 12.2\, R\eta^2 / \mu p r^2, \qquad \text{(VI.18)}$$

where η is gas viscosity, μ is the mass of the mole and ΔT is the temperature difference.

Equating the slip velocity v_0 to the average rate of the Poiseuille flow, we obtain from eq. (VI.18):

$$v_0 = 1.52(\eta R/8p\mu)\nabla T \equiv (D/3T)\nabla T. \qquad \text{(VI.19)}$$

For example the thermal slip velocity of water vapor in air is about 0.003 cm/s at $\nabla T = 10$ degrees/cm under normal conditions. Under the same temperature gradients, the diffusion vapor flux calculated by eq. (VI.17) is more than 50 times as great as the vapor flow associated with heat slippage.

Equation (VI.9) is also valid when the solution concentration under the menisci separated by porous space is different, giving rise to a gradient of vapor pressure $\partial p/\partial x = (\partial p/\partial C)(\partial C/\partial x)$. As known, the saturated vapor pressure over solutions decreases with increasing concentration obeying the Raoult law: $p = p_s\exp(-\gamma C_p/\rho)$, where ρ is the density of liquid, C_p is the solution concentration, g/cm^3, and γ is the activity coefficient. As a result, we find that:

$$q = (v_m D p_s \gamma/RT)\exp(-C_p\gamma/\rho)\mathrm{d}C/\mathrm{d}x. \qquad \text{(VI.20)}$$

Here the vapor flux depends not only on the concentration gradient, but on solution concentration, C_p, itself being in the exponent and determining the extent of vapor pressure decrease, as compared with the saturated vapor pressure, p_s, over the pure solvent.

§3 Vapor Diffusion in a Non-uniform Electric Field

If a porous body or a single capillary is exposed to a non-uniform electric field (NEF), a force equal to $f = p_e \nabla E$, where p_e is the molecular dipole moment and E is the electric field strength, act on the dipole of the molecules of the vapor. Under the action of this force, the dipole molecules acquire a velocity component directed towards increasing values of E and equal to $u = Df/kT$, cm/s. In this case the vapor diffusion is accompanied by the convective flow $q_k = uC$, where C is the vapor concentration, g/cm^3. The total vapor flow is given by [207]:

$$q = -D\nabla C + uC = -D\nabla C + (Dp_eC/kT)\nabla E = -D_*\nabla C, \qquad \text{(VI.21)}$$

where D_* is the effective coefficient of diffusion. The NEF accelerates or impedes vapor transfer depending on the sign of ∇E ($D_* > D$, at $\nabla E > 0$). As seen from eq. (VI.21), the influence of the NEF is most pronounced at low gradients of vapor concentration and for polar liquids with high values of dipole moment, such as water, glycerine and alcohols. As a consequence of the NEF, the vapor transfer can be generated even in the absence of a gradient of vapor pressure.

Similar effects occur in non-uniform magnetic fields. Here the force exerted on a vapor molecule is equal to $f = b_m \nabla B$, where B is the strength of the magnetic field and b_m is the molecular magnetic moment. Paramagnetic molecules are drawn into the region of higher B values, and diamagnetic ones are repelled. The expression for the vapor flux has the same form as eq. (VI.21), except $p_e \nabla E$ is replaced with $b_m \nabla B$.

As an example of application of eq. (VI.21), consider a solution of the problem of vapor diffusion through a capillary exposed to NEF. For simplicity, assume that D = const and neglect the influence of vapor adsorption. The equation of non-stationary mass transfer is written as follows:

$$\partial C/\partial t = D\partial^2 C/\partial x^2 - u\partial C/\partial x, \qquad \text{(VI.22)}$$

where $u = Df/k$T.

For the same initial and boundary conditions as employed above (VI, §2): $C(0, t) = C_0$; $C(x, 0) = 0$; $C(\infty, t) = 0$; we obtain for the vapor flux into a capillary [207]:

$$q = q_n + q_c = -D(dC/dx)|_{x=0} + uC_0 = C_0(D/\pi t)^{1/2}\exp(-u^2t/4D) +$$
$$+ uC_0[1 - (1/2)\text{erfc}(ut^{1/2}/2D^{1/2})]. \qquad \text{(VI.23)}$$

When $u = 0$ this equation transforms into eq. (VI.15) for vapor diffusion. In the initial stage ($t \to 0$) the flux is driven chiefly by vapor diffusion ($q_n >> q_c$). In contrast, later when $t \to \infty$ the flux is driven only by NEF, and $q_d \to 0$. For $f > 0$ the vapor inflow into a capillary can be accelerated. This is confirmed by the comparative investigations of kinetics of vapor adsorption on the silica gel KCC-4 with average pore size $r = 2.2$ nm [207]. When $\nabla E = 8.5 \times 10^8$ V/m^2, the time it takes for adsorption equilibrium to be established is reduced by 20%. At the same time, such fields did not change the equilibrium adsorption isotherm and influence only the kinetics. This is because the applied external electric field was much weaker than the fields of the intermolecular forces responsible for adsorption.

The NEF can both impede or accelerate adsorption. When $f < 0$ ($\nabla E < 0$) a steady state can be approached at which the vapor cannot penetrate a porous body. In this case, in each part of a capillary:

$$q = -D(\mathrm{d}C/\mathrm{d}x) - uC = 0,$$

that leads, at u = const, to the time-independent distribution of vapor concentration inside the capillary:

$$C = C_0 \exp(-ux/D). \tag{VI.24}$$

Although a constant concentration C_0 is kept in the surrounding medium, the vapor fails to fill the capillary and penetrates only the region near to the inlet due to the repulsion of vapor molecules by NEF.

When $f > 0$, the steady state corresponds to an exponential increase in concentration $C = C_0 \exp(ux/D)$, that first causes, at vapor pressure approaching p_s, polymolecular adsorption and subsequently capillary condensation. After elimination of the external NEF, the condensate will evaporate and the initial constant concentration C_0 is established.

A rigorous verification of the influence of NEF on vapor diffusion was obtained through observations of the rate of water evaporation from the capillaries of rather large radii $r = 10$–20 μm, for which surface diffusion can be neglected [208]. The capillaries were placed in a thermostatted glass chamber in which various relative pressures p_0/p_s were maintained by saturated salt solutions. Two electrodes, flat and needle, were inserted into the chamber. They generated a NEF accelerating water evaporation from the capillary (as shown in Fig. 5.4). The distribution of the electric field strength E was measured using the method of dielectric ball. The rate of evaporation was determined by a cathetometer measuring the distance L between the capillary inlet and the receding meniscus.

According to eq. (VI.9), the mass flux for each fixed value of L is equal to:

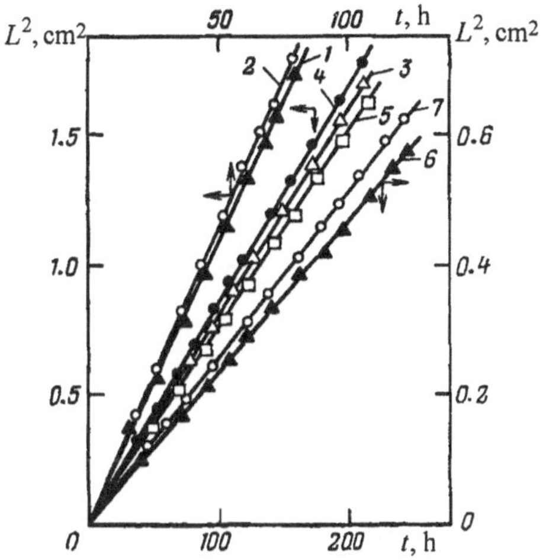

Figure 6.9 Rates of water evaporation from quartz capillaries (r = 10–13 μm) into air at atmospheric pressure and humidity $p_0/p_s = 0.324$ (*1*, 2); 0.78 (*3*, *4*, *5*) and 0.92 (*6*, *7*) without NEF (*1*, *3*, *6*); under the NEF of $E = 2.3 \cdot 10^4$ V/cm, $\nabla E = 7.8 \cdot 10^4$ V/cm^2 (2, *4*, *7*), $E = 2.1 \cdot 10^4$ V/cm, $\nabla E = -7.7 \cdot 10^4$ V/cm^2 (*5*).

$$q = (\mu D/RT)(p_s - p_0)/L,\ \text{g/cm}^2\text{s}, \qquad \text{(VI.25)}$$

where p_s is the vapor pressure over the meniscus and p_0 is the vapor pressure in the chamber. The rate of the meniscus displacement, $v = dL/dt$, is linked with q by simple relation: $v = q/\rho$, where ρ is the liquid density. The substitution of $q = \rho(dL/dt)$ into eq. (VI.25) and its integration gives the following expression for the quasi-stationary conditions of evaporation:

$$L^2 = 2\mu D(p_s - p_0)t/\rho RT. \qquad \text{(VI.26)}$$

The dependence $L(t)$ is linear in $L^2(t)$-coordinates enabling the coefficient of vapor diffusion to be determined using the values of p_s, p_0 and $v_m = \mu/\rho$, the molar volume of liquid.

The results obtained are shown in Fig. 6.9. The time of evaporation, t, is plotted on the x-axis and L^2 on the y-axis. According to eq. (VI.26), the experimental dependences are linear and pass through the origin. Curves *1*, *3* and *6* show the results of measurements in the absence of NEF. The diffusion coefficient $D = 0.255$ cm^2/s calculated by the slope of graphs agrees with that calculated by eq. (V. 23). Figure 6.9 shows (curves *2*, *4*, *7*) that the NEF increases the rate of evaporation. Corresponding values of the effective

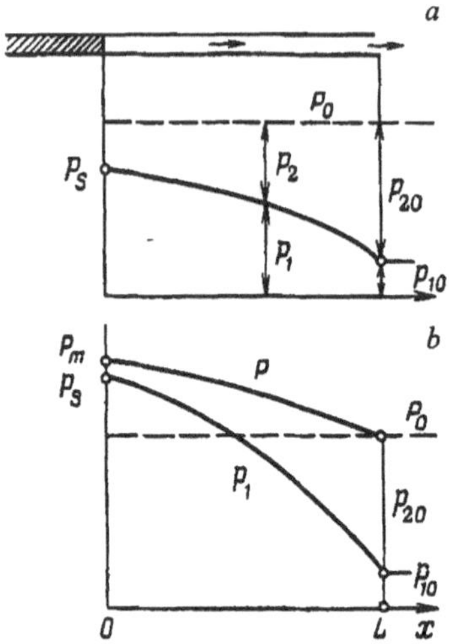

Figure 6.10 Distribution of partial vapour pressure, p, and the pressure, P, of vapor–gas mixture in the course of stationary evaporation of liquid from the capillary at temperature below (a) and above (b) the boiling point.

diffusion coefficient exceed D by 5–10%. The experimental data agree with values calculated by eq. (VI.21), where we assumed that $p_e = 1.84$ Debye for the water molecule. The agreement between the experiment and the theory was also satisfactory when the electrodes were changed so that the NEF impeded the evaporation (Fig. 6.9, curve 5).

The influence of NEF is more pronounced in the case of low evaporation velocities (at high values of p_0/p_s) also in agreement with the theory. Application of NEF in these circumstances can significantly intensify the evaporation from porous bodies.

§4 Flow of Vapor–Gas Mixture

In the case where we cannot consider the partial vapor pressure p to be small compared to the total pressure of a vapor–gas mixture P_0, the diffusion of vapor and gas becomes mutual. Application of a vapor concentration gradient generates a counter-diffusion of gas that will change the method of calculation of the mass transfer processes.

Let us consider the stationary vapor transfer in the capillary (Fig. 6.10) between the meniscus, $x = 0$, and the surrounding medium ($x > L$), where a constant pressure of vapor–gas mixture $P_0 = p_{20} + p_{10}$ is maintained. Here p_{20} and p_{10} are partial pressures of gas and vapor, respectively. The vapor pressure over the meniscus is assumed to be equal to that of saturated vapors $p_s(T) > p_{10}$.

Under the concentration gradient, $dC_1/dx = (dp_1/dx)/RT$, the vapor diffuses from the meniscus towards the capillary end. The gradient of gas concentration $dC_2/dx = -dC_1/dx = (dp_2/dx)/RT$ generates a counter-diffusion of gas from the surrounding medium to the meniscus. Under the assumption that the liquid is saturated with the given gas, its surface is impermeable to gas molecules. As a result, the gas concentration near to the meniscus and the pressure of vapor–gas mixture increase, generating a viscous flow of the vapor–gas mixture towards the open end of the capillary [209, 210]. Therefore, the equations of mass transfer for vapor q_1 and gas q_2 must include terms accounting for both diffusion and viscous flows:

$$G_1 = q_1/\mu_1 = -(D/RT)dp_1/dx - (r^2p_1/8\eta RT)dP/dx, \tag{VI.27}$$

$$G_2 = q_2/\mu_2 = (D/RT)dp_1/dx - (r^2p_2/8\eta RT)dP/dx. \tag{VI.28}$$

Here G_1 is the vapor flux, mol/cm^{2}s; G_2 is the gas flux; $P(x) = p_1(x) + p_2(x)$ is the pressure of the vapor–gas mixture in the capillary and η is its viscosity. The terms accounting for the viscous flow include partial pressures p_1 and p_2 but not the total, because G_1 and G_2 account for the fluxes of particular components. The sum of the viscous terms in eqs. (VI.27) and (VI.28) produces the above-given eq. (VI.3) for a viscous flow of a vapor–gas mixture. The diffusion term in eq. (VI.28) is expressed as $dp_1 = -dp_2$ for convenience.

Under stationary conditions (at L = const), the gas flux $q_2 = 0$ and q_1 = const. Adding eqs. (VI.27) and (VI.28) term by term, we get

$$q_1 = -(\mu_1r^2P/8\eta RT)dP/dx. \tag{VI.29}$$

Integration of this equation from $x = 0$ to $x = L$, where L is the length of the empty part of the capillary, gives:

$$q_1 = \mu_1r^2(P_m^2 - P_0^2)/16\eta RTL, \tag{VI.30}$$

where P_m is the still unknown value of vapor–gas mixture pressure near to the meniscus. To determine it, let us divide eq. (VI.27) by p_1 and eq. (VI.28) by p_2 and then subtract the second equation from the first. Since $G_2 = 0$, and second terms in these equations are equal, we obtain:

$$q_1 = -(D\mu_1/RT)(p_1/p_2 + 1)dp_1/dx.$$

Using the known inverse dependence (see eq. (V. 23)) of the diffusion coefficients on pressure, $DP = D_0 P_0$, where D_0 is the diffusion coefficient in the surrounding medium, we obtain after some transformation:

$$q_1 = -[D_0 P_0 \mu_1 / RT(1 - p_1/P)][\mathrm{d}(p_1/P)/\mathrm{d}x].$$

Integration of this equation gives:

$$q_1 = -(D_0 P_0 \mu_1 / RTL) \ln \{[1 - (p_0/P_0)]/[1 - (p_s/P_m)]\}. \qquad \text{(VI.31)}$$

Equating the right hand side of eqs. (VI.30) and (VI.31) gives the equation necessary to determine P_m, i. e. the pressure of vapor–gas mixture near to the meniscus (Fig. 6.10).

Calculations performed for water [210] have shown that when $T < T_b$ (where T_b is the boiling temperature) P_m values differ little from those of P_0 for capillaries with $r > 1$ μm. Under this condition, i. e. at $P_m = P_0$, eq. (VI.31) transforms into the Stefan equation:

$$q_1 = (D_0 P_0 \mu_1 / RTL) \ln [(P_0 - p_0)/(P_0 - p_s)]. \qquad \text{(VI.32)}$$

Thus, the system of eqs. (VI.31) and (VI.30) includes the Stefan equation as a particular case, when the capillaries are rather wide and the pressure gradient of vapor–gas mixture is equal to zero (Fig. 6.10, *a*). The constancy of the total pressure $P = P_0 = \text{const}$ along the capillary ceases to be obeyed in fine capillaries, where the viscous resistance to flow is high. This leads to a solution different from eq. (VI.32).

As seen from eq. (VI.30), at $P_m > P_0$ the vapor transfer is driven chiefly by viscous flow. The role of diffusion is reduced to the accumulation of gas and growth of vapor–gas mixture pressure in the vicinity of the meniscus.

In wide capillaries, a vanishingly small pressure drop can draw off the gas accumulated by diffusion. In fine capillaries a higher pressure drop is necessary that changes the rate of vapor transfer depending here on pressure $P_m > P_0$. The pressure P_m significantly exceeds P_0 when $T \geq T_b$ (Fig. 6.10, *b*). In wide capillaries (for water; $r \geq 3$ μm at 100°C; $r \geq 0.3$ μm at 120°C) the values of P_m practically coincide with those of the saturated vapor, p_s, at given temperature T. In fine capillaries pressure P_m exceeds p_s, increasing with diminution of the capillary radius. For example at 100–120°C the total pressure P_m is by 1.5–2 times greater than p_s. Such an increase in pressure, e.g. taking place at high-temperature drying, may cause explosive destruction of porous bodies.

The correction for diffusion slippage of a vapor–gas mixture results in insignificant changes in the quantitative estimated values [210].

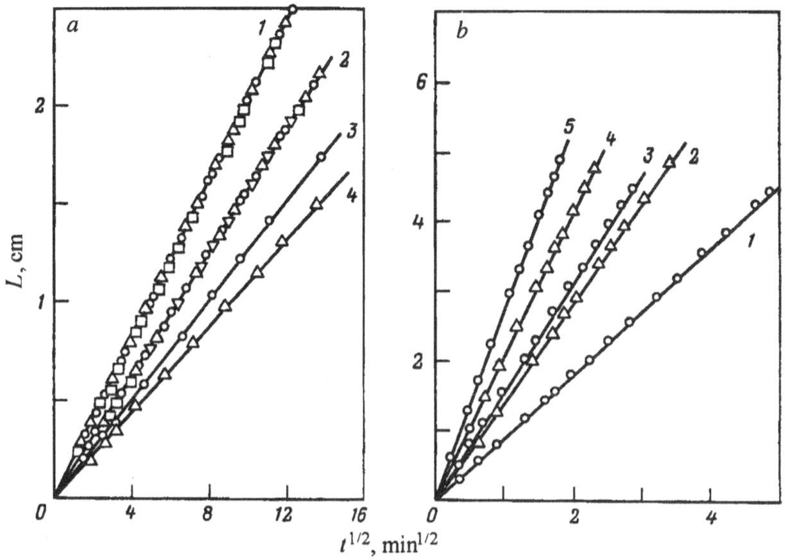

Figure 6.11 Meniscus displacement L in the course of water evaporation from quartz capillaries at temperature 365 K and 359 K (*a*) and 380 K (*b*). *a*) $T = 365$ K, $r = 10.6$; 8.6; 5.7; 4.3 and 14 μm (*1*); $T = 359$ K, $r = 11.5$; 6.1; 5.6 and 3.4 μm (2); $r = 0.31$ μm (*3*) and $r = 0.21$ μm (*4*); *b*) $T = 380$ K, $r =$ 1.65 (*1*); 3.7 (2); 4.6 (*3*); 6.3 (*4*) and 8.9 μm (*5*).

Experimental values of the rate of water evaporation from capillaries into air at atmospheric pressure [196] at temperatures somewhat below T_b (365 K and 359 K) are shown in Fig. 6.11, *a* and above T_b (380 K) in Fig. 6.11, *b*. The graphs are plotted in coordinates $L(t^{1/2})$, where L is the meniscus position in the capillary and t is the time. Since $q_l = \rho_l\ (dL/dt)$, in the quasi-stationary approximation eqs. (VI.30)–(VI.32) give after integration a linear relation between the meniscus shift L and $t^{1/2}$, as shown above (eq. (VI.26)). As seen from Fig. 6.11, this approximation works rather well. Using the slopes of the dependences obtained, the so-called evaporation constant $\underline{C}_e = q_l L/\rho_l$, independent of the meniscus position, can be calculated. A comparison of this expression for $\underline{C}_e$ with the equations presented for mass transfer gives for the Stefan regime:

$$\underline{C}_e = (D_0 P_0 v_m/RT)\ln(P_0 - p_0)/(P_0 - P_s) \qquad \text{(VI.33)}$$

and for the viscous flow regime:

$$\underline{C}_e = (r^2 v_m/16\eta RT)(P_m^2 - P_0^2), \qquad \text{(VI.34)}$$

where $v_m = \mu_l/\rho_l$ is the molar volume of liquid.

Comparing the experimental values of $\underline{C}_e$ with those calculated from eqs. (VI.33) and (VI.34), we have assumed that $P_0 = 0.1$ MPa, $v_m = 18$ cm^3/mol, $p_{0l} = 0$ and $D_0 = 0.364$ cm^2/s at 359 K and 0.377 at 365 K. The viscosity of vapor–gas mixture is averaged over the values of partial vapor pressure p_l and dry air p_2:

$$\eta = (\eta_1\langle p_1\rangle + \eta_2\langle p_2\rangle)/(\langle p_1\rangle + \langle p_2\rangle),$$

where values of p_1 and p_2 averaged along the capillary length and known viscosity for vapor η_1 and air η_2 [211] are used.

For capillaries of radii $r > 1.4$ μm at 365 K (Fig. 6.11, *a*, curve *1*) and $r >$ 3.4 μm at 359 K (Fig. 6.11, *a*, curve *2*) the slope of the graphs $L(t^{1/2})$ agree well with that calculated theoretically by eq. (VI.33). Deviations from the Stefan equation occur only in rather fine capillaries: $r = 0.31$ μm (Fig. 6.11, *a*, curve *3*) and $r = 0.21$ μm (Fig. 6.11, *a*, curve *4*). From the slope of these graphs and eq. (VI.31), the ratio P_m/p_s can be determined, being of 2.26 in the first case and 2.8 in the second. Thus, the pressure of vapor–gas mixture in fine capillaries near to the meniscus does considerably exceed the saturated vapor pressure, which is about 0.08 MPa.

The ratio P_m/p_s achieves even higher value at $T > T_b$ (Fig. 6.11, *b*). However, in this case the influence of meniscus cooling due to the effect of the heat of evaporation appears. This reduces the rate of evaporation, as compared to the theory, due to the decrease in vapor pressure over the meniscus. The effect of a decreased rate of evaporation, observed at $T > T_b$, is more pronounced at elevated T and wider capillary radii [212, 213].

An approach similar to that given above was used for the case of liquid evaporation from conical capillaries. Here the change in time of evaporation compared to cylindrical capillaries has been calculated [214]. One can use these results to calculate the drying time for porous bodies with anisotropic structure arising also as a result of non-uniform shrinkage. In the case of isotropic porous bodies with regular micro-widening and micro-narrowing of pores, one can use the model of a cylindrical capillary with some effective average radius.

The more complicated problem of evaporation of binary liquid mixtures consisting of components with different volatility has been investigated in [215–217]. In this case two vapor fluxes and a gas flow from the surrounding media must be taken into account. Moreover, due to the differences in volatility of the components, their concentrations in a liquid phase near to the meniscus are changed influencing the partial pressure of these components over the meniscus. It has been theoretically and experimentally confirmed that in this case the dependences $L(t^{1/2})$ become non-linear (Fig. 6.12). First the more volatile component evaporates at higher rate (*n*-hexane) and then the

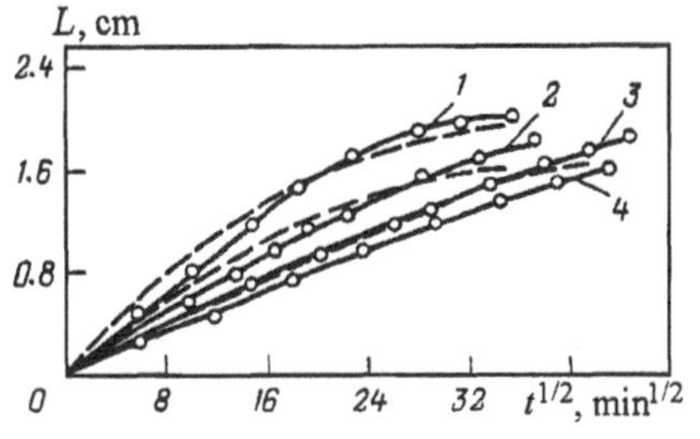

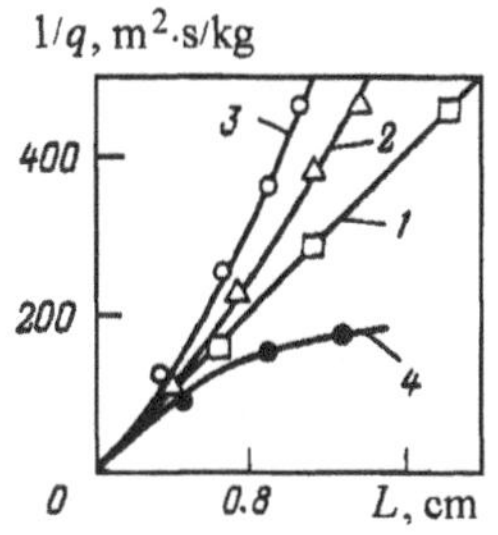

Fig. 6.12 Fig. 6.13

Figure 6.12 Meniscus displacement L in the course of evaporation of *n*-hexane/*n*-octane binary liquid mixture (T = 300.5 K) for the ratios of components in %%: 80/20 (*1*); 60/40 (2); 40/60 (*3*) and 20/80 (*4*) [215].

Figure 6.13 Dependences of the reverse rate of water evaporation $1/q$ from the capillary on L under temperature gradients different in sign and value, $\nabla T = 0$ (curve *1*); 3 deg/cm (curve 2); 6 deg/cm (curve *3*); −3 deg/cm (curve *4*) [218].

process is controlled by the transfer of the less volatile component (*n*-octane). The dashed curves in Fig. 6.12 were calculated by numerical methods on the basis of the theory developed. If the difference in volatility is small, the dependences $L(t^{1/2})$ approach linearity. The same effect takes place at elevated temperature, but here the reason is a dramatic reduction in the time necessary to remove the main portion of more volatile component. After a short non-linear region on the graph the linear dependence of the meniscus shift L on $t^{1/2}$ is established.

The influence of the temperature gradient on evaporation from capillaries has been investigated for one-component liquids [218]. Application of a temperature gradient also produces non-linear dependences $L(t^{1/2})$. The experimental results obtained for water are shown by points in Fig. 6.13. The graphs are plotted in coordinates $(1/q) - (L)$, where q is the mass flux, kg/m^2s, and L is the distance between meniscus and capillary end. According to eqs. (VI.30)–(VI.32), under isothermal conditions the dependences must be linear in these coordinates (curve *1*). The slope of the graph is proportional to the parameter $\underline{C}_e$ (eqs. (VI.33), (VI.34)). When the temperature increases in the direction of vapor flow ($\nabla T > 0$), the evaporation slows down (values of $1/q$ increase, as shown by curves *2* and *3*). With decreasing temperature ($\nabla T < 0$) the evaporation accelerates (curve *4*). In the latter case, at rather high temperature gradients the vapor condensation in the capillary may occur at some distance from the meniscus, where the temperature is below the temperature of vapor condensation.

Solid curves *1–4* in Fig. 6.13 have been calculated by numerical methods. They were obtained from the same equations for fluxes (VI.27), (VI.28), but taking into account the temperature-dependent changes in vapor density, diffusion coefficients and viscosity of a vapor–gas mixture along the capillary [218]. The calculated results were obtained under the condition of low and constant temperature gradients $(\nabla TL/T_m) \ll 1$, where T_m is the meniscus temperature. The slowing down of evaporation when $\nabla T > 0$ is caused by the growth of the coefficient of mutual diffusion D that leads to the reduction of the pressure gradient of vapor–gas mixture and relevant decrease in q values.

Kinetics of vapor transfer in capillaries accompanied by evaporation of a solid substance, that constitutes the walls and the bottom of the channels, are discussed in the monograph [189].

Chapter VII

Combined Transfer of Liquid and Vapor

Let us consider the two-phase flow when there is a combined transfer of material in both the vapor and liquid phases including capillary condensate and wetting films. Such combined transfer of a liquid and its vapor takes place in the case of porous bodies at a relative vapor pressure near to saturation. Equilibrium between the capillary condensate, wetting films and the vapor was considered in Chapter II.

The effect of surface diffusion of adsorption layers considered in Chapter VI influences the total mass transfer only in microporous bodies when the pore size is comparable with molecular dimensions. The description of the processes of mass transfer in porous bodies containing capillary condensate and wetting films is based on hydrodynamic equations (see Chapters IV and V) that include physical characteristics of liquids such as viscosity, density and surface tension.

First, evaporation of liquids from capillaries into surrounding air or an evacuated chamber will be considered. In this case, vapor diffusion is accompanied by liquid flow in wetting films being in local equilibrium with vapor. Most of the results were obtained for water evaporation from single capillaries of various radii and different surface hydrophilicity. The effect of relative vapor pressure in the ambient gas phase was investigated. In the last part of the chapter the processes of evaporation from model porous bodies consisting of communicating capillaries of different radii are analyzed.

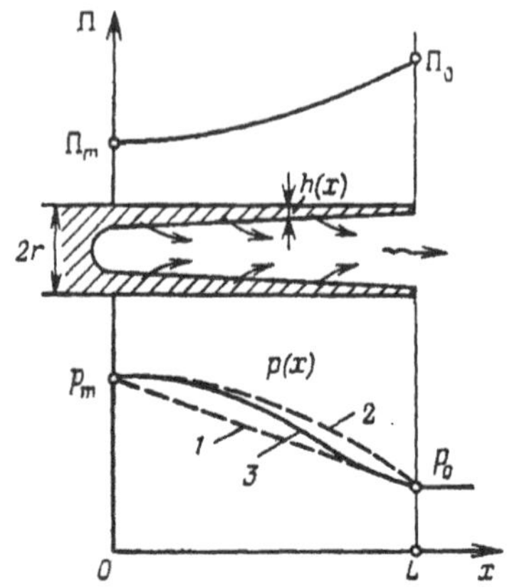

Figure 7. 1 Calculation of the combined transfer of vapor and wetting films during evaporation from a capillary.

§1 Film Flow and Vapor Diffusion

The problem of the combined transfer processes involving both the vapor and the liquid films applied to evaporation from single narrow capillaries, was investigated in the works [219, 220]. We consider as "narrow" the capillaries of radius r comparable with the thickness h of wetting films. For water and aqueous electrolyte solutions in capillaries with a highly charged surface the values of h achieve 0.1–0.3 μm for β-films (Fig. 2.6, *a*). Under these conditions the capillaries of radii up to 5 μm can be considered as "narrow".

When β-films become unstable and the capillary surface is covered with much thinner α-films (Fig. 2.6, *b*), the capillaries with radii smaller than 0.1–0.3 μm may be considered narrow in the above-mentioned sense. The same range of radii is characteristic of evaporation of low-molecular weight non-polar liquids. Polymer solutions can form much thicker stable adsorption films of several microns. In this case, capillaries with radii up to 100 μm may be considered as "narrow".

Thus, the contribution of film flow can reveal itself in capillaries of various radii depending on the thickness of wetting films that is determined by surface forces acting on them as summarised in the isotherm of disjoining pressure $\Pi(h)$ (see Chapter II).

Let us show how the problem of combined diffusion of vapor and film flow, applied to the isothermal evaporation of liquid from a capillary into the gas (for instance air), where a constant relative pressure of vapor is kept $p_0/p_s < 1$, can be solved. Consider, as usual, a stationary regime when the meniscus position, L = const, is maintained by continuous liquid inflow into the capillary from the bulk (Fig. 7.1).

We can write the fluxes of vapor Q_d and film liquid Q_f (in g/s) across an arbitrary section with the coordinate x on the basis of eqs. (VI.9) and (V.2):

$$Q = Q_d + Q_f = -\{\pi[r - h(p)]^2 \mu D/RT\}\, dp/dx + [2\pi r h^3(p)\rho/3\eta]\, d\Pi/dx = \text{const.} \quad \text{(VII.1)}$$

The constancy of Q values and their independence from x follows from the condition of continuous flow. Here the vapor pressure p and the disjoining pressure Π are linked by the condition of local thermodynamic equilibrium of the wetting film and its vapor expressed by eq. (II.1). This condition is met when $L >> r$ therefore limiting the application of the equation given below. At $L \sim r$, the vapor pressure depends not only on longitudinal but also on radial coordinate that requires consideration of the more complicated problem of two-dimensional mass exchange.

Expressing Π in terms of p with help of eq. (II.1) and integrating eq. (VII.1) from $x = 0$ to $x = L$, we obtain:

$$Q(L) = (\pi\mu D/RTL)\left[r^2(p_m - p_0) - 2r\int_{p_m}^{p_0} h(p)dp + \int_{p_m}^{p_0} h^2(p)dp\right]$$

$$-\,(2\pi r\rho RT/3\eta v_m L)\int_{p_m}^{p_0} [h^3(p)/p]dp = \pi r^2 \mu D^*(p_m - p_0)/RTL, \quad \text{(VII.2)}$$

where D^* is the effective coefficient of diffusion which includes the contribution from film flow.

To perform calculations by this equation, we should know the dependence $h(p)$ or the isotherm of disjoining pressure $\Pi(h)$. Integrating between the limits, for example from $x = 0$ (or $x = L$) to an arbitrary current value of x, one can find, if $h(p)$ is known, the distribution of the film thickness $h(x)$ or that of vapor pressure $p(x)$ along the capillary. In this case the p_m values, specified by the Kelvin equation, are used as a boundary condition at $x = 0$:

$$p_m = p_s \exp(-2\sigma v_m \cos\theta / rRT), \quad \text{(VII.3)}$$

where θ is the contact angle and σ is the surface tension of liquid. In capillaries of $r \geq 0.1$ μm, it can be assumed that $p_m \cong p_s$ without a significant error.

When the isotherms of disjoining pressure for flat films $\Pi(h)$ are used in eq. (VII.2), a correction should be made taking into account the curvature of the film interface inside the cylindrical capillary. Here the film thickness increases (due to the effect of capillary pressure equal to $\sigma/(r-h)$) compared with that of a film on a plane surface. Accordingly, the positive disjoining pressure of the film, Π_m, decreases approximately by half, from $2\sigma/(r-h)$ to $\sigma/(r-h)$.

In practice, a simplified solution of eq. (VIII.2), obtained with the assumption that $h << r$, can be used. For the isotherm of type $\Pi = A/6\pi h^3$, the ratio h/r is equal to 3×10^{-3} in capillaries of $r = 1$ µm; and it does not exceed 0.012 at $r = 0.1$ µm and 0.06 at $r = 0.01$ µm. Under this condition we obtain from eq. (VIII.2):

$$Q(L) = [\pi r^2 \mu D(p_m - p_0)/RTL] + (2\pi\rho r/3\eta L)\int_{\Pi_m}^{\Pi_0} h^3(\Pi)d\Pi$$

$$= \pi r^2 \mu D^*(p_m - p_0)/RTL. \qquad \text{(VII.4)}$$

Because the film thickness is usually determined from the isotherms of disjoining pressure, the vapor pressure in the second term of eq. (VII.2) is changed by the disjoining pressure using eq. (II.1). From the latter it follows that

$$dp = -(pv_m/RT)d\Pi. \qquad \text{(VII.5)}$$

Now we should specify analytically the isotherm $\Pi(h)$ to make eq. (VII.4) more concrete. For non-polar liquids, the isotherm of molecular forces $\Pi = A/6\pi h^3$ may be used*, that leads, at $r \geq 0.1$ µm and after integration, to the following dependence:

$$Q(L) = [\pi r^2 \mu D(p_s - p_0)/RTL] + (\rho r A/9\eta L)\ln \Pi_0/\Pi_m, \qquad \text{(VII.6)}$$

where A is the positive value of the Hamaker constant; $\Pi_0 = (RT/v_m)\ln(p_s/p_0) - (\sigma/r)$; $\Pi_m = \sigma/r$.

In many cases the approximation by the isotherm $\Pi = A_0/h^3$ is also possible for water films [219, 220]. The values of the empirical constant A_0 appropriate for rough estimations of the rate of water evaporation are about 5×10^{-21} J [221].

---

*Here and further positive values of the constant $A = -A_{132}$ (see eq. (1.1)) are used to simplify the writing of equations.

The rate of evaporation in eq. (VII.1) is expressed by vapor and liquid fluxes taken as being formally independent of each other. The first term on the right side of the equation can be determined from the diffusion law ignoring the film flow and the second one expresses the film flow in the absence of vapor diffusion. However, the fluxes must interact, otherwise the condition of local equilibrium (II.1) would not be obeyed. In the absence of film flow the vapor pressure changes linearly along the capillary (curve *1*, Fig. 7.1). In the case of pure film flow, the vapor pressure over the flowing film would change non-linearly (curve *2*, Fig. 7.1). The interaction of fluxes results in an intermediate distribution $p(x)$ (curve *3*, Fig. 7.1); the method of calculation of $p(x)$ has been discussed above.

Near to the meniscus, the value of dp/dx is lower than the average gradient equal to $(p_m - p_0)/L$. The contribution of the second term is greater here according to eq. (VII.1). The transfer is predominantly carried out by films because near the meniscus they possess the maximum thickness. As the capillary end is approached, the film thickness decreases (down to the thickness of a monolayer at $p/p_s \leq 0.9$–0.95), and vapor diffusion becomes the major contributor. The interaction of fluxes manifests itself in the fact that the liquid evaporates from the film, maintaining an equilibrium vapor pressure throughout the whole capillary. The evaporation in the presence of wetting films goes rather from the surface of films adjacent to the meniscus than from the meniscus itself. The films "suck off" the liquid under the meniscus displacing the latter. The evaporation region, L_0, is the greater, the higher the pressure p_0/p_s in the surrounding medium. Roughly, the relative length of the evaporation region is of about $L_0/L = 0.05/[1 - (p_0/p_s)]$, assuming that at $p_0/p_s \leq 0.95$ the film degenerates. At $p_0/p_s = 0.5$, the evaporation region constitutes a tenth part of L, and occupies half the length of the free capillary space when $p_0/p_s = 0.9$. When $p_0/p_s > 0.95$, when the capillary surface is completely covered with wetting films, the evaporation proceeds at all film surfaces, and the film flow is the major mechanism of transfer in this case. The effect of film flow on the rate of water evaporation from the capillaries into moist air is supported by direct measurements [222–227].

To perform these measurements, in [224] a thermostatted chamber, where the quartz capillaries were placed over the saturated solutions of salts, has been used. The system of thermo-regulation was supplied from a battery of accumulators providing continuous thermostatting during several months with an accuracy of ±0.05°C. The meniscus displacement was measured by a horizontal comparator with an accuracy of ±1 μm. The observations were carried out at dark-field light through a glass window on the plate of the chamber. During evaporation into air with relative vapor pressure, $p_0/p_s \leq 0.75$, the rate of water evaporation does not depend on the capillary radius and coincides with that calculated from the equation for pure vapor diffusion. The effect of film flow is not appreciable.

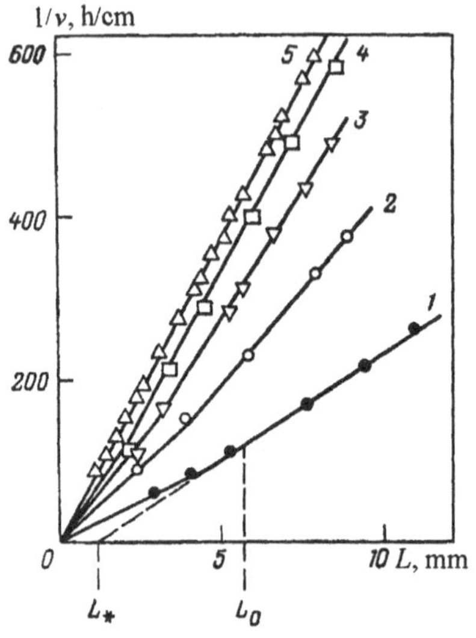

Figure 7. 2 Dependence on L of the inverse rate of water evaporation, $1/v$, from the capillaries of various radius $r = 0.75$ μm (*1*); 2.5 μm (*2*); 7.3 μm (*3*); 12.5 μm (*4*); 26 μm (*5*) at a relative humidity of the surrounding medium $\varphi_0 = p_0/p_s = 0.96$.

At higher air humidity, the film flow accelerates the water evaporation increasing with the diminution of the capillary radius. Figure 7.2 exemplifies the dependence of the inverse rate of evaporation, $1/v$, on the distance, L, between the meniscus and the capillary entrance. According to eq. (VII. 2), the velocity of the meniscus displacement, $v = dL/dt$, is:

$$v(L) = Q/\pi r^2 \rho = v_m D^*(p_m - p_0)/RTL. \qquad \text{(VII.7)}$$

It follows from this equation that the dependences of $1/v$ on L must be linear both with or without a film flow. In the latter case $D^* = D$, and the rate of evaporation should not depend on the capillary radius. When the film flow influences the evaporation, $D^* > D$ and, since the values of D^* depend on r (see eq. (VII.4)), the evaporation rate also depends on r. As seen from Fig. 7.2, the evaporation rate increases with decreasing r providing evidence of the growing contribution of the film transfer. However, these linear dependences do not pass through the origin. Their extensions intercept the L-axis at greater distances the smaller the capillary radius. This phenomenon, first revealed in [228, 229], is explained by specific form of the isotherms of disjoining pressure of water films on a silica surface (Fig. 2.6). Metastable β-films that are able to be formed behind the receding meniscus rupture as the

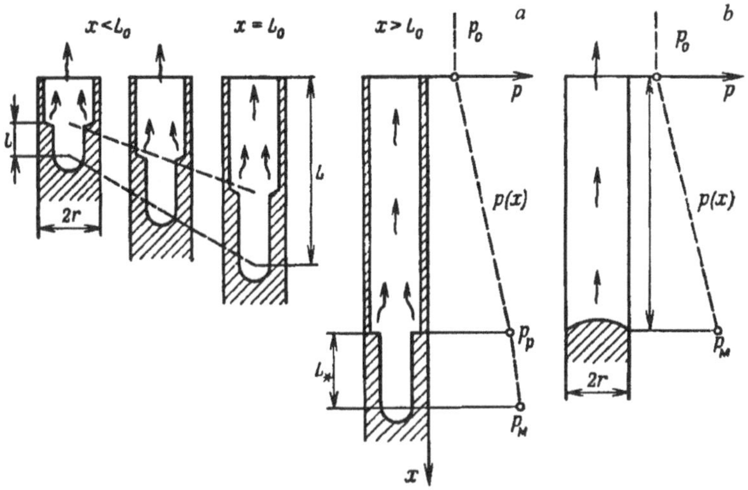

Figure 7. 3 Influence of the β-films (thicker, ~100–200 nm, than α-films, ~7 nm) on the rate of water evaporation from capillaries (see text).

ratio p/p_s goes down transforming into the thinner thermodynamically stable α-films at $\Pi > P_l$ (Fig. 2.6, *a*).

Figure 7.3 illustrates the influence of β-films on water evaporation from quartz capillaries. A thick metastable β-film ($h \sim 100–200$ nm) forms behind the receding meniscus. The film transforms into the more stable state of a thin ($h = 7–8$ nm) α-film as the pressure reaches a critical value, p_p/p_s. Since the point where the vapor pressure is equal to p_p/p_s shifts proportionally to L, the length of the β-film region also increases proportionally resulting (when $L < L_0$) in an initial linear dependence between $1/v$ and L (Fig. 7.2). However, as the β-film approaches its critical area, its length stops growing and then remains constant (at $L > L_0$) and equal to L^*. Due to the great thickness of β-films, mass transfer in the region L^* proceeds only in the liquid phase. This displaces the evaporation region by the distance L^* to the capillary end. Therefore, the segments intercepted on the L-axis are just equal to L^*. During evaporation from the same capillaries but after hydrophobization, when the wetting films do not form, the rate of evaporation is controlled only by vapor diffusion (Fig. 7.3, *b*).

Thus, instead of eq. (VII. 7), we can write for the linear part of the graphs when $L > L_0$:

$$v(L) = v_m D^*(p_p - p_0)/RT(L - L^*), \qquad \text{(VII.8)}$$

where p_p is the critical vapor pressure over a β-film. As shown in [230], in

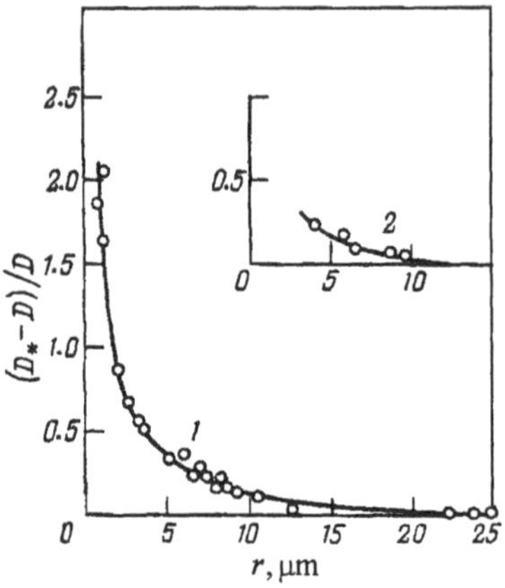

Fig. 7.4

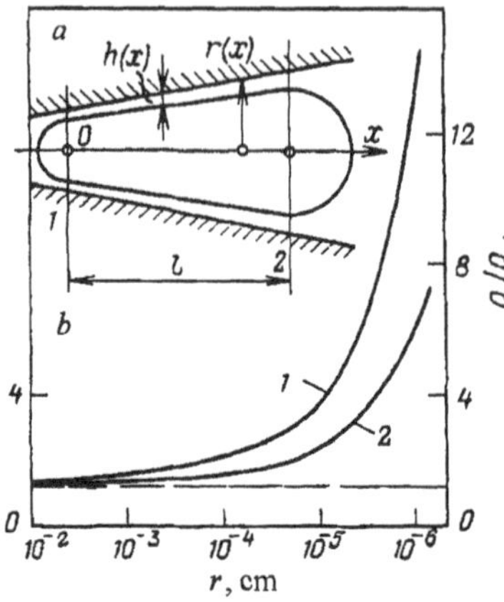

Fig. 7.5

Figure 7. 4 Effect of contribution of α-film flow into evaporation from capillaries of various radii at relative air humidity $\varphi_0 = p_0/p_s = 0.96$ (*1*) and 0.86 (*2*).

Figure 7.5 Scheme of calculation (*a*) and (*b*) the dependence of the ratio of total mass flux Q to the vapor flux Q_d on the radius of cylindrical capillaries r for decane (*1*) and hexane (*2*).

quartz capillaries $p_p \approx 0.996$. The values of D^* in eq. (VII. 8) are determined by the flow of α-films (Fig. 7.3, *a*) in the region $(L - L^*)$. Thus, a role of β-films is to push the region of the phase transition to the capillary end increasing the rate of evaporation approximately by a factor of $L/(L - L^*)$.

It should be mentioned that the co-existence of α- and β-films in a capillary is only kinetic. The metastable β-phase must transform, in a direct contact, into the thermodynamically stable state of α-films. In a narrow capillary, the necessity to remove excessive mass of liquid, as a result of the great difference in thickness of α- and β-films, hinders the transition $\beta \rightarrow \alpha$. As the capillary radius increases, the probability of the transition $\beta \rightarrow \alpha$ also increases which reduces the values of L^*; and at $r \geq 26$ μm β-films in the capillary completely disappear (Fig. 7.2). The experimental dependences L^* vs. r obtained in numerous experiments support this concept. When $r > 1$ μm, the values of L^* regularly decrease down to zero at $r \geq 15$–20 μm.

To estimate the contribution of mass transfer in α-films, let us use eq. (VII.8) and the expression for D^* from (VII.4). The experimental isotherm $h(\Pi)$ for α-films [231] has been used in the calculations. The dependence of the relative increase in the rate of evaporation vs. r, written as the ratio $(D^* - D)/D$, where D is the coefficient of vapor diffusion, is shown by the points in Fig. 7.4. The solid curves *1* and *2* show the results of calculations. The agreement of experimental and theoretical data was achieved with the assumption that the viscosity of α-films is approximately five times greater than that of bulk water. However, a significant increase in viscous resistance to the flow is rather caused by the discontinuity of α-films, i. e. by their local thinning generating rather high hydrodynamic resistance compared to that of a flat smooth film of an averaged thickness. The possibility of such a "discontinuous" state of α-films was discussed in [232, 233].

The theory and experiment show that the flow of α-films can significantly increase the rate of water evaporation from capillaries of $r < 5$ μm at a relative humidity of the surrounding gas $p_0/p_s > 0.85$. For example, at $r = 1$ μm and $p_0/p_s = 0.96$, the rate of evaporation increases by 3–4 times. A decrease in values of D^* with increasing r is associated with the fact that the vapor flux Q_d is proportional to r^2 but film flux Q_f depends on r linearly, as seen from eq. (VII.1).

A spontaneous $\beta \rightarrow \alpha$ transition has been observed in the course of bubble displacement in a conical capillary (Fig. 5.3). The mass transfer causes the air bubble to lengthen gradually due to its displacement to the narrower part of the channel. At some critical length of the bubble L^* (about several mm), the β-film ruptures forming drops on the surface covered by thin stable α-films. This transition is revealed by the changes in the rate of displacement. A gradual displacement is followed by jumps, when the meniscus approaches either regions of the α-film or the drops covering it [234].

Under non-isothermal conditions, the influence of thermocapillary film flow should be also taken into account. The combined transfer of liquid and vapor in non-isothermal conditions was considered for a region of a conical capillary (Fig. 7.5), when the film takes part both in a flow driven by the gradient of capillary pressure of the menisci and in a thermocapillary flow simultaneously [235]. In this case, the term Q_t enters into eq. (VII.1) that accounts for the thermocapillary flow of wetting films. Using the condition of local equilibrium (II.2), where $p = p(x, T)$, we obtain the following equations for the non-isothermal flux:

$$Q = Q_d + Q_f + Q_t = -\pi\mu D(r-h)^2\nabla p/RT - 2\pi r\rho h^3 RT/3\eta v_m$$

$$\times [\nabla p/p - \nabla p_s/p_s + (\nabla T/T)\ln p_s/p] + (\pi r\rho h^2/\eta)(\partial\sigma/\partial T)\nabla T, \quad \text{(VII.9)}$$

where $\nabla p = dp/dx$ and $\nabla T = dT/dx$ are the gradients of vapor pressure and temperature, respectively.

Boundary conditions on the menisci 1 ($x = 0$) and 2 ($x = L$, where L is the length of the bubble) are to be written as follows:

$$p(0) = p_1 = p_{s1}\exp(-2\sigma_1 v_m/r_1 RT_1)$$

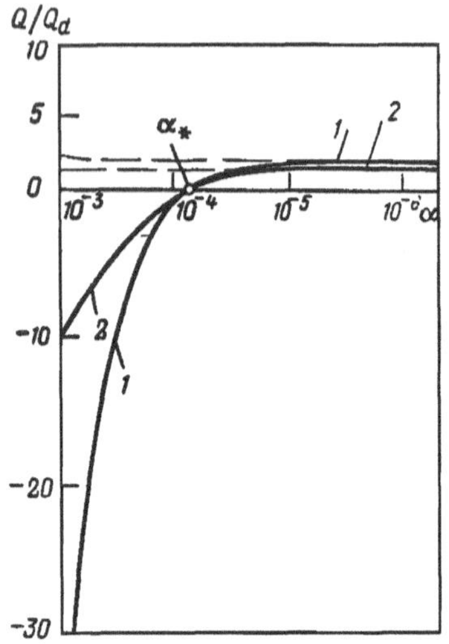

Figure 7.6 Dependences of Q/Q_d on the capillary conicity, α, for decane (*1*) and hexane (2). Average capillary radius in the region of the bubble location $r = 1$ μm, bubble length $L = 100$ μm, $\nabla T = -1$ deg/cm.

$$p(l) = p_2 = p_{s2}\exp(-2\sigma_2 v_m / r_2 R T_2),$$

where p_{s1} and p_{s2} are the pressures of saturated vapor at the temperature of the menisci T_1 and T_2, respectively.

The calculations using eq. (VII. 9) were performed by numerical methods for constant temperature gradients $\nabla T = \text{const}$ (from −1 to −100 degree/cm) and an isotherm of disjoining pressure of type $\Pi(h) = A_0/h^3$, under the assumption that $A_0 = 10^{-20}$ J and does not depend on temperature. The calculations were carried out at mean temperature 20°C for two liquids: a volatile fluid, like hexane ($p_s = 120$ mm Hg) and a less volatile liquid like decane ($p_s = 2.7$ mm Hg). The changes in capillary radius were specified by the equation $r = r_1 + \alpha x$, where α is the conicity of the capillary. The ratio L/r varied from 10 for large r to $L/r = 10^3$ for small and medium capillary radii r.

Figure 7.5, *b* shows the dependences obtained for the ratio of total flux Q to vapor flux Q_d for cylindrical capillaries ($\alpha = 0$) of various radii. The ratio Q/Q_d considerably exceeds unity when $r < 5$ μm. The contribution of the thermocapillary flow increases with decreasing capillary radius; and the effect is more pronounced for the less volatile liquid. Since all the fluxes depend linearly on ∇T and the mean temperature is constant, the values of Q/Q_d practically do not depend on the temperature gradient.

Figure 7.6 shows the ratio of the fluxes Q/Q_d calculated as a function of the capillary conicity α. If the conicity is small, the ratio Q/Q_d is practically constant and equals that of the cylindrical capillaries. An increase in conicity generates reverse film fluxes (Q_f) driven by an increasing difference in the capillary pressure of the menisci and retardation of vapor flow by the same reason. At $\alpha = \alpha^*$, the thermo-fluxes are completely balanced by reverse fluxes of liquid and vapor from the wider meniscus to the narrow. For the conditions chosen (mean radius $r = 1$ μm and $\nabla T = -1$ degree/cm), the value of α^* is close to 10^{-4}. Such circulation fluxes may take place in porous bodies incompletely saturated with liquid and are particularly generated, for example in thermal tubes [236].

When $\alpha > \alpha^*$, the fluxes change their direction, i.e. turn towards the hotter meniscus due to predominance of the film flow Q_f driven by the increasing difference in capillary pressure of the menisci. This effect is more pronounced in the case of the less volatile liquid.

In porous bodies with capillaries of variable radius, air bubbles moving under the temperature gradient towards the hotter end may stick in a narrowing of the pore at some local conicity $\alpha = \alpha^*$, where $Q = 0$. The value of α^* depends on the mean pore radius, the liquid properties and ∇T. The smaller the capillary radius, the more pronounced is the effect of conicity. At the same value of $\nabla T = -1$ degree/cm (as in Fig. 7.6), but at $r_1 = 0.01$ μm the values of α^* are reduced by approximately two orders. Even a small ($\alpha^* = 3\times10^{-8}$) capillary conicity can cause the retardation of thermo-fluxes.

In the case of mass transfer inside a porous body, the problem of combined transfer of vapor and wetting films can be simplified. Here a good model for the structure element may be a region of the capillary between the menisci covered with wetting films (Fig. 5.7). When the bubble is small, the pressure drop on the menisci is low thereby enabling the average values of film thickness to be used. Factoring out $h = \text{const}$ from the integral in eq. (VII.4) and taking into account that $(p_1 - p_2)/p_1 << 1$, we obtain instead of eq. (VII.4):

$$Q = [(\pi r^2 \mu D/RT) + (2\pi r\rho h^3 RT/3\eta v_m p)]dp/dx. \qquad \text{(VII.10)}$$

The combined transfer of vapor and wetting films under a temperature gradient has been discussed on the basis of this equation in Chapter V, §2.

The flux through the unit area of pore cross-section given by eq. (VII.10) is equal to:

$$q = [(\mu D/RT) + (2\rho h^3 RT/3\eta r v_m p)]dp/dx. \qquad \text{(VII.11)}$$

In Chapter VIII, §8 this equation will be used to calculate the coefficients of mass transfer in porous bodies as a function of mass content.

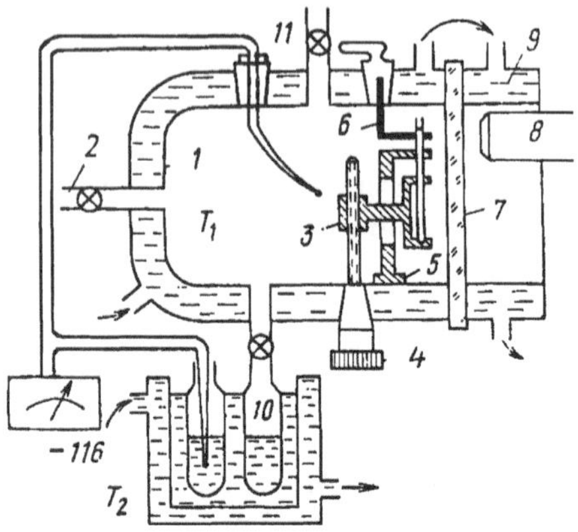

Figure 7.7 Apparatus for measuring the rates of evaporation from capillaries into vacuum and vapor atmosphere.

§2 Film Flow during Evaporation into Vapor Atmosphere

The expression for vapor flux, eq. (VII.1), must be corrected for the case of evaporation into a vapor atmosphere. Since various regimes of vapor flow can arise here (Chapter VI, §1), it is more convenient to use relative permeability coefficients K/K_1, where K_1 characterizes the free-molecular vapor flow expressed by eq. (VI.1)). Thus, for a combined flow of vapor and wetting films we get:

$$K/K_1 = K_d/K_1 + K_f/K_1 = K_d/K_1 -$$

$$\{[\rho RT(2\pi RT/\mu)^{1/2}]/[4\eta v_m r^2(p_m - p_0)]\} \int_{p_m}^{p_0} (h^3(p)/p)dp . \quad \text{(VII.12)}$$

The methods of calculation of the ratio K_d/K_1, taking into account various regimes of vapor flow, were presented in Chapter VI, §1. For water evaporation the ratio $K_d/K_1 \approx 1$ may be considered valid when $Kn > 0.5$.

For an isotherm of disjoining pressure such as $\Pi = A/6\pi h^3$, after integration and account of the curvature of the film surface, we have [233, 237]:

$$K/K_1 = K_d/K_1 + [A\rho(2RT/\pi\mu)^{1/2}/24\eta r^2(p_m - p_0)]$$

$$\times \ln[(v_m\sigma/rRT) - \ln(p_0/p_s)]/[(v_m\sigma/rRT) - \ln(p_m/p_s)]\,. \quad \text{(VII.13)}$$

Equations (VII.12) and (VII.13) were checked experimentally using quartz capillaries. Figure 7.7 shows the experimental apparatus. A glass chamber (*1*), thermostatted by water jackets, is evacuated through the tube (*2*) with a nitrogen trap to a pressure of 1.33 Pa. The capillaries are placed in a holder (*3*) movable in the vertical direction along a slit in the stand (*5*) with help of a screw rotating a ground cork (*4*). Rotating another cork, one can successively cut out upper portions of the capillaries by the knife (*6*). This enables several experiments to be performed in the same capillary. The chamber is covered with a plane-parallel glass (*7*), through which one can measure the meniscus displacement during evaporation by a cathetometer (at $r > 0.3$ μm) or a microscope (*8*) (at $r < 0.3$ μm). The glass is thermostatted by a water jacket (*9*).

A pre-selected vapor pressure p_0/p_s in the chamber is created by the difference ΔT of temperature in the chamber T_1 and in the ampoule (*10*) with the liquid under investigation thermostatted at $T_2 < T_1$. Deviations from constant ΔT values were registered by a differential thermocouple and did not exceed 0.05°C, this enabled the values of p_0/p_s to be maintained with an accuracy of ±0.003.

Experimental dependences $1/v = f(L)$ obtained during evaporation into the vapor atmosphere, are of the same form as for evaporation into moist air (Fig. 7.2). In the first stage, the evaporation follows a linear dependence, subsequently the slopes of the graphs change and their extensions intercept the L-axis the segments equal to L^*, the length of the metastable β-film. For wider capillaries, the values of L^* range from 0.5 to 1 mm. This effect is found not only for water but also in other polar liquids, for example ethanol. In the latter case the values of L^* are about 3 mm at $p_0/p_s = 0.94$ in a capillary of $r = 8.7$ μm. The length L^* of water β-film reduces gradually with capillary radius. Finally, β-films do not form when $r < 0.25$ μm (Fig. 7.8): here $L^* = 0$. In this case the evaporation follows linear dependence $L^2(t)$ passing through the origin in agreement with eq. (VI. 26). Here the vapor pressure is below the critical p_p/p_s value associated with the threshold of β-films existence (Fig. 2.6, *a*).

A comparison of these results with those given in Fig. 7.2 shows that the values of L^* as a function of capillary radius must pass through a maximum. At $r \to \infty$, the coexistence of metastable and stable phases (α and β) becomes impossible and at $r \to 0$ the metastable state does not appear at all. The maximum L^* value falls in the region of capillary radii from 1 to 10 μm, depending on external conditions.

Anomalies of evaporation (i.e. a break on the graphs of evaporation due to formation of β-films) disappear in hydrophobized capillaries where the vapor adsorption does not exceed a monolayer. In this case the process of

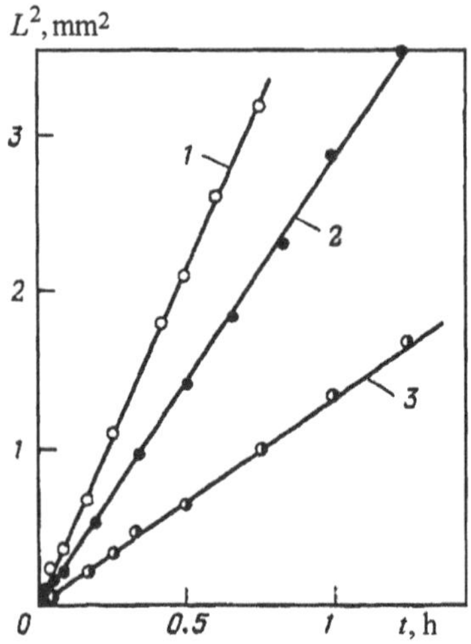

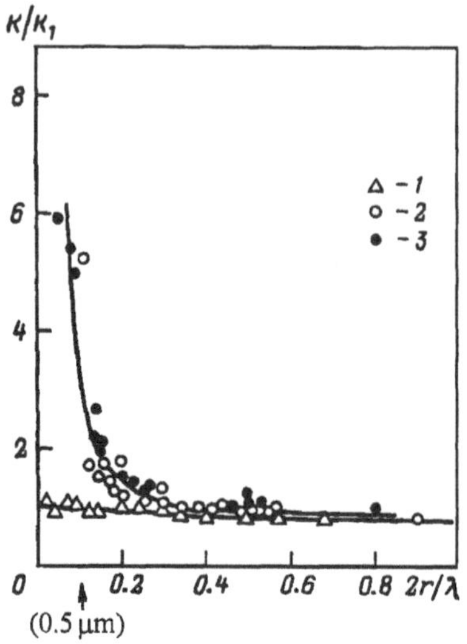

Fig. 7.8 Fig. 7.9

Figure 7.8 Dependence of square meniscus displacement in the course of water evaporation from narrow capillaries, L^2, on the time t: $r = 0.215$ μm (*1*); 0.21 μm (2); 0.112 μm (*3*); $p_0/p_s = 0.96$ and $T = 296$ K.

Figure 7.9 Dependences of the ratio, K/K_1, on the reverse Knudsen number, $1/Kn = 2r/\lambda$ at $p_0/p_s = 0.75$ (*1*); 0.97 (2) and 0.98 (*3*).

evaporation exactly follows the laws of vapor transfer (Fig. 7.3, *b*). On the other hand, the addition of species stabilizing the β-film tends to increase L^* values. For aqueous solutions of sodium oleate, the length of the region L^* in capillaries of r = 10–20 μm increases by 2–3 times compared with the evaporation of pure water.

Figure 7.9 shows a quantitative comparison of the results of calculations by the equation of two-phase transport (VII.12) with experimental data obtained for capillaries of different radii [233]. Reverse values of the Knudsen number $1/Kn = 2r/\lambda$ are plotted on the x-axis, whereas the values of K/K_1-ratio are given on the y-axis. Solid curves were calculated by eq. (VII.12) at $T = 296$ K and $p_0/p_s = 0.975$ using, as before (Fig. 7.4), the experimental isotherm $h(p/p_s)$ for water α-films on quartz obtained by Derjaguin and Zorin [231] and confirmed by Hall [238]. The agreement between the theory and experimental points *2* and *3*, corresponding to the experiments performed at p_0/p_s = 0.97 and 0.98, may be approached only under the assumption that the viscosity of α-films exceeds that of the bulk water by 4 times. The reasons for such an increase in hydrodynamic resistance of α-films were discussed above.

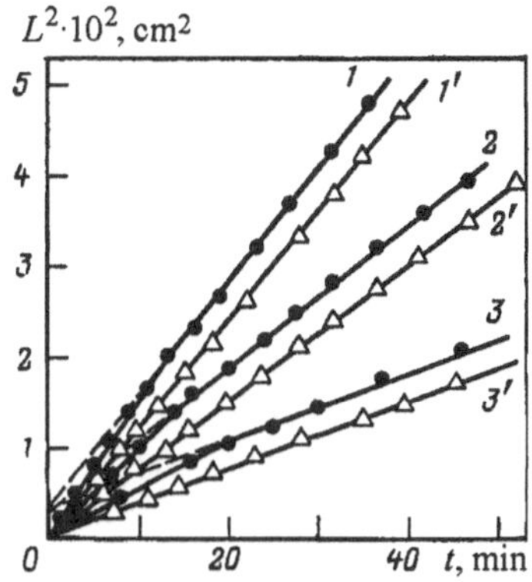

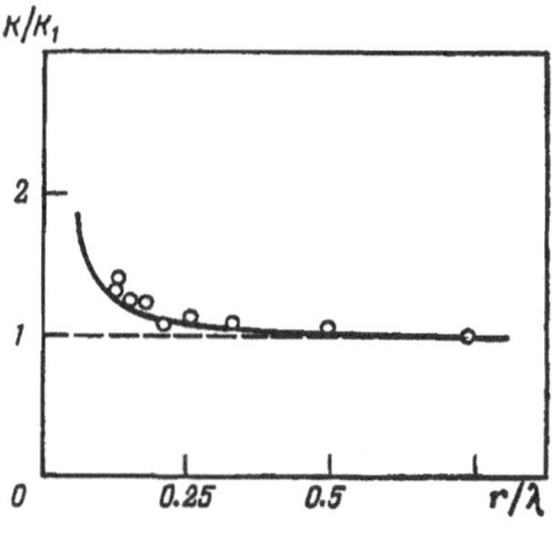

Fig. 7.10 Fig. 7.11

Figure 7.10 Influence of surface hydrophobization on the rate of water evaporation from the capillaries of $r = 2.2$ μm (*1*, *1'*); 1.35 μm (*2*, *2'*); 0.76 μm (*3*, *3'*) at $p_0/p_s = 0.95$. *1*–*3* – hydrophilic capillaries, *1'*–*3'* – hydrophobized capillaries.

Figure 7.11 Comparison of experimental data for decane (points) with theoretical dependence of K/K_1 on r/λ ($p_0/p_s = 0.955$, $T = 295$ K).

The curve passing through the points *1* in Fig. 7.9 shows theoretical values of K_d/K_1 calculated for the vapor flux. In the ranges of capillary radii and vapor pressures studied they are close to unity, that corresponds to a free-molecular vapor flow. When $2r/\lambda > 0.5$, all the experimental points do not deviate much from this curve, i. e. the effect of α-films is not pronounced. At $p/p_s = 0.97$–0.98 (points *2* and *3*) its contribution increases significantly with a reduction of r. In capillaries of radii 0.1–0.2 μm, the experimental values of K exceed the Knudsen coefficient K_1 by 5–6 times.

At a lower relative vapor pressure in the chamber, $p_0/p_s = 0.75$, (points *1*), the values of K/K_1 obtained coincide with those of K_d/K_1. As in the case of evaporation into moist air, when $p_0/p_s \leq 0.75$, film flow does not appear even in rather narrow capillaries. The reason for this is that the film thickness does not exceed a monolayer for the major part of the capillary surface. Polymolecular films form only within a short region close to the meniscus where the vapor pressure $p_0/p_s \geq 0.95$.

A comparison of the rate of water evaporation from hydrophilic and hydrophobic capillaries of exactly the same radii is presented in Fig. 7.10. In the hydrophobic capillaries the evaporation rate is determined only by the vapor diffusion and follows a linear dependence L^2 vs. t in agreement with eq. (VI.26). The rate of evaporation is higher in hydrophilic capillaries due to the flow of β-films forming behind the receding meniscus. The influence of α-films was not revealed because the capillary radius was large compared with the film thickness.

The same experimental apparatus was used to study the evaporation of non-polar liquid, decane, from capillaries when eq. (VII.13) with the constant $A = 1.57\times10^{-20}$ J may be used. The rate of decane evaporation follows a linear dependence between $L^2(t)$ passing through the origin, for the isotherm of disjoining pressure of decane films is monotonous and, in contrast with water, exhibits no metastable regions. Figure 7.11 shows the dependence K/K_1 vs. the reverse Knudsen number calculated from (VII.13). In calculations, the bulk values of viscosity and surface tension of decane were used. The experimental points, obtained for capillaries of radii from 1.5 to 0.175 μm at $p_0/p_s = 0.95-0.96$, agree with the theoretical curve. As seen from these data, the film flow of non-polar liquids can increase the rates of evaporation from capillaries of $r \leq 0.2$ μm by 30–40%. The thickness of decane films in these experiments (Fig. 7.11) was h = 3.3 nm close to meniscus and 1 nm near the capillary end, at a relative pressure of decane vapor in the chamber of $p_0/p_s = 0.94$ and r = 1 μm. In the case of narrower capillaries, $r \leq 0.05$ μm, the rate of evaporation can increase more than a factor of two due to input of the film flow.

For internal mass transfer in the region of the capillary between two menisci covered with a film, when the film thickness may be considered constant, we obtain instead of (VII.12):

$$K/K_1 = K_d/K_1 + [(\rho h^3 RT)/4\eta v_m r^2 p_m](2\pi RT/\mu)^{1/2}, \qquad \text{(VII.14)}$$

where p_m is the average vapor pressure in the capillary.

For capillaries of $r \geq 0.1$ μm, $p_m \approx p_s$. Then, approximating the isotherms of disjoining pressure by the dependence $\Pi = A_0/h^3$, eq. (VII.14) simplifies to:

$$K/K_1 = K_d/K_1 + [(\rho A_0 RT)/4\eta v_m \sigma r p_s](2\pi RT/\mu)^{1/2}, \qquad \text{(VII.15)}$$

where σ is the surface tension.

As seen from this equation, the input of mass conductance associated with film transfer increases with decreasing volatility of the liquid (characterized by p_s) and capillary radius and also with increasing constant A_0, that characterizes the field intensity of surface forces.

As an example, let us consider the results of some quantitative estimations by eq. (VII.15) for water at 300 K under the assumption that $A_0 = 5\times10^{-21}$ J. In capillaries of r = 1 μm, the values of K/K_1 increase by 5% due to additional film transfer. As the capillary radius reduces, the values of K/K_1 grow up to 40% at r = 0.1 μm and more than five times compared to vapor flux at r = 0.01 μm.

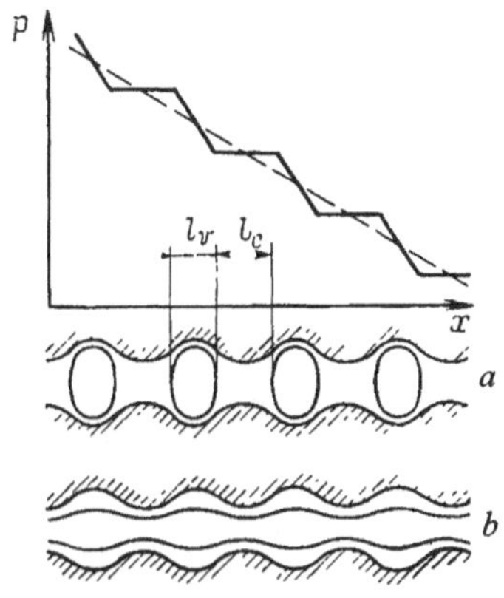

Figure 7.12 Scheme illustrating a combined transfer of vapor, wetting films and condensate in pores.

§3 Combined Flow of Vapor, Films and Condensate

Let us consider now the interaction between the liquid and vapor fluxes in the presence of a capillary condensate, filled narrow pore regions, and wetting films, where these cover the pore surface free from condensate. Figure 7.12 shows the most characteristic examples. The scheme *a* corresponds to the case when the vapor phase is discontinuous, separated into bubbles by the capillary condensate. The scheme *b* corresponds to the case of lower mass content, when the vapor phase is continuous and the regions of the capillary condensate, as in the first case, are connected by wetting bridges.

In both cases the rate of mass transfer is limited by vapor diffusion. The diffusive resistance is considerably higher than the viscous contribution from the liquid regions. The presence of the condensate tends, as shown at the top of Fig. 7.12, to increase the local gradients of vapor pressure in pore regions covered with films, compared to the average pressure gradient characterized by linear pressure decrease along the pore (dashed curve). Such a simplification enables the rate of mass exchange to be calculated by eqs. (VII.11) and (VII.14) as discussed in §2, but the value of pressure gradient must be N times greater, where $N = (l_v + l_c)/l_v$. Here l_v is the average length of regions free from condensate and l_c is the average length of the condensate regions. Roughly the values of N may be considered equal to that part of the pore space that is occupied by vapor.

In the case of the incomplete filling of pores, shown in Fig. 7.12, *b*, the applicability of the approach is based on the fact that a practically constant vapor pressure is established over the condensate regions due to the low resistance to viscous flow.

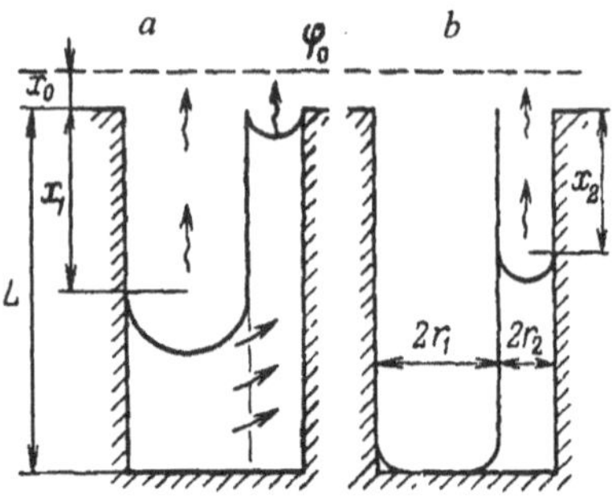

Figure 7.13 Two-capillary model of a porous body.

Let us find the ratio of the viscous resistance of the condensate $1/K_c$ to that due to diffusion $1/K_d$ at equal pressure gradients and in the absence of film transfer. Since, according to eq. (II.1), $\nabla P = (RT/pv_m)\nabla p$, the ratio of $1/K_c = \nabla p/q_c$ to $1/K_d = \nabla p/q_d$ is equal:

$$\omega = (1/K_c)(1/K_d) = 8\eta p v_m^2 D/(rRT)^2. \tag{VII.16}$$

For example, assuming for water at 300 K: $\eta = 10^{-3}$ N s/m^2, $p = p_s = 3.3\times10^4$ N/m^2, $v_m = 18$ cm^3/mol, at $r = 1$ μm we obtain $\omega = 3\times10^{-6}$. The value of ω remains low, about 10^{-2}, even when r reduces to 0.01 μm. Thus, if there is no film flow, practically in all the cases the rate of mass transfer is controlled by diffusive resistance to vapor flow in the regions where the condensate is separated by the vapor phase. As the liquid content of a porous body increases, the values of N also increase causing the mass conductance of the porous body to rise. When $N \to \infty$, the pore space is filled with condensate and the permeability becomes determined only by the viscous resistance to liquid flow (see Chapter IV).

Calculations of the combined transfer of vapor, films and condensate, based on this concept, were also performed to obtain the coefficients of vapor permeability of porous bodies [239–241].

§4 Evaporation from Communicating Capillaries

The influence of non-uniformity of pore space can be accounted for with the help of a simple model consisting of communicating capillaries with different radii. Such a two-capillary model (Fig. 7.13) was introduced in [242]. It allows a further increase in the number of capillaries of different radii.

Let us consider mass transfer in a system of communicating capillaries in a gravitational field, when the liquid evaporates into the medium, where a constant relative vapor pressure, p_0/p_s or relative air humidity $\varphi_0 = p_0/p_s < 1$ is maintained. The initial state corresponds to the complete filling of all capillaries. If φ_0 is lower than the relative vapor pressure φ_1 above the menisci of the ultimate curvature r_1 in wide capillaries, then evaporation will result in a drop in the level of the menisci. As long as the difference of capillary pressure of menisci in narrow and wide capillaries $\Delta P = P_2 - P_1 = 2\sigma(r_1 - r_2)/r_1 r_2$ is sufficient to draw up the amount of water capable of evaporating from narrow capillaries, the menisci remain near to the surface in these capillaries (Fig. 7.13, *a*). The amount of water evaporating from the surface of narrow capillaries is:

$$q_2 = Dp_s \mu F_2(\varphi_2 - \varphi_0)/RTx_0, \quad \text{(VII.17)}$$

where F_2 is the area and φ_2 is the relative vapor pressure in narrow capillaries; x_0 is the thickness of the diffusion layer over the surface of the porous body. The value of the flux is specified by external conditions (φ_0) and the size of the narrow capillaries (φ_2).

The flux of liquid flowing from wide into narrow capillaries is determined by the Poiseuille equation:

$$q = r_2^2 F_2(\Delta P - \rho g x_1)/8\eta x_1, \quad \text{(VII.18)}$$

where x_1 is the position of the menisci in wider capillaries; ρ is the density; g is the gravity acceleration; η is the viscosity.

The detachment of the menisci border in narrow capillaries will occur when $q_2 \geq q$. It is easy to determine the corresponding meniscus position in wide capillaries $x_1 = x_k$ from this condition:

$$x_k = [2\sigma(r_1 - r_2)/r_1 r_2][(8\eta q_2/r_2^2 F_2) + \rho g]^{-1}. \quad \text{(VII.19)}$$

We can neglect the effect of gravity in microporous bodies, so that the equation is simplified to:

$$x_k = \sigma F_2 r_2 (r_1 - r_2)/4\eta r_1 q_2. \quad \text{(VII.20)}$$

The rate of evaporation continues to be practically constant when $x < x_k$, because in narrow capillaries the menisci remain on the surface of the porous body. When $x > x_k$, the border of narrow menisci will detach from the surface leading to a sharp decrease in the rate of evaporation. The period of constant rate of evaporation will be the longer, as seen from eq. (VII.20), the larger the difference in capillary radii and the lower the rate of evaporation.

The menisci can separate from the surface earlier, when the characteristic length L of a porous body is shorter than x_k. When $x > x_k \geq L$, the menisci in wide and narrow capillaries recede simultaneously. Here the liquid is withdrawn from wide to narrow capillaries, from which the evaporation continues, but at a lower rate. The evaporation will proceed also from the menisci surface in wide capillaries. This flux is equal to:

$$q_1 = Dp_s\mu F_1(\varphi_1 - \varphi_0)/RT(x_1 + x_0), \quad \text{(VII.21)}$$

where F_1 is the area of wide capillaries.

When $F_1 \sim F_2$ this flux is much lower than q_2, because of the difference, x_k, between the positions of the menisci in wide and narrow capillaries.

When the wide capillaries become empty (Fig. 7.13, *b*), the evaporation proceeds as from a single capillary (§§1 and 2).

The two-capillary model was further refined in [243] by taking into account, in addition to the vapor diffusion and liquid outflow from wide capillaries into narrow capillaries, the film flow along the surfaces of all capillaries. It has been shown in this work that the model of two capillaries is well suited to calculating the rate of water evaporation from a biporous ceramic plate of pore sizes $r_2 = 1.5 \times 10^{-4}$ cm and $r_2 = 10^{-5}$ cm.

Chapter VIII

Non-stationary Mass Transfer. Evaporation, Drying

In previous chapters we have considered stationary processes to determine the mechanisms of mass transfer in porous bodies. In practice we more usually encounter non-stationary situations and the problem is to forecast and to calculate the kinetics of the mass transfer. Except the case of very fast, explosive processes, the theory of heat- and mass transfer usually uses a quasi-stationary state approach. This approach is applicable to porous bodies, because fine pores possess a high hydrodynamic resistance. Consequently, the mass transfer proceeds at a low rate as compared to that in tubes and open flows. This enables the rate of mass transfer at each moment to be calculated as in a stationary process with a given momentary distribution of mass content and temperature. This approach considerably simplifies the calculations and makes it possible to use the above mentioned stationary state computations for the processes of mass transfer. We also use this approach here to describe the processes of evaporation from porous bodies and their drying. However, it should be borne in mind that a quasi-stationary approximation is only valid for modest rates of mass exchange.

First, the peculiarities of evaporation of electrolyte, surfactant and polymer solutions will be considered. The difference as compared with pure water consists in the accumulation of non-volatile components near to evaporating menisci and in the influence of the dissolved species on the surface tension, vapor pressure and stability and thickness of wetting films. The rate of drying depends both on the surrounding environment regulating external mass transfer and on the processes within a porous body, that is, on internal mass transfer. The mechanism of internal mass transfer may be investigated using a radiotracer method. Some examples of application of this method are discussed. A particular case of evacuation of a non-volatile liquid from a porous body using a pseudofluidized dispersion of dry thin-porous particles is also considered.

Analysis of the rates of drying occupies the main part of the chapter. For this purpose various models of porous bodies are used and different mechanisms of liquid and vapor flow are taken into consideration. The calculations of the rate of drying involve coefficients of internal mass transfer determined on the basis of the pore size distribution that characterizes the structure of the porous body. Two special examples, deep drying and non-isothermal internal mass transport, are discussed at the end.

In practice, there are a great variety of methods of drying. The content of this chapter is limited to consideration of the basic physical mechanisms controlling the rates of drying.

§1 Evaporation of Electrolyte Solutions

The calculations of water evaporation from porous bodies discussed in previous chapters has been based on the physical characteristics of pure water: its vapor pressure, density and viscosity. In technical and natural porous bodies the aqueous phase must be regarded as a solution of complex composition. During evaporation such solutions become more concentrated, leading sometimes to crystallization influencing the kinetics.

For the case of stationary evaporation of a solution completely filling a porous body, one may use the equations of convective diffusion to calculate the distribution of solution concentration in a semi-infinite porous layer:

$$\partial C/\partial t = v\partial C/\partial x + D\,\partial^2 C/\partial x^2, \qquad \text{(VIII.1)}$$

where D is the coefficient of diffusion of the solute in the porous body. The stipulation of a constant rate of evaporation $v = \text{const}$ does not mean in this case that the process is stationary, because the concentration distribution continuously changes with time t.

The solution of eq. (VIII.1) for the concentration, C_*, in a surface layer with the boundary conditions $C(x, 0) = C_0$; $C(\infty, t) = C_0$ and under the condition of mass conservation of the non-volatile component in the porous body takes the form [244]:

$$C_*/C_0 = 2 + (v^2 t/D). \qquad \text{(VIII.2)}$$

This equation is valid across the whole interval of t values except the initial period t_0 during which the values of C_* formally grow to $2C_0$. Thereafter the concentration linearly increases with time as long as C_* values exceed the concentration of the saturated solution and crystallization takes place.

Concentration distribution with depth x and time (at $t > t_0$) is given by the following equation:

$$C(x, t)/C_0 = 1 + [1 + (v^2t/D) - (vx/D)]\exp(-vx/D). \tag{VIII.3}$$

At $x = 0$ this equation transforms into eq. (VIII.2); at $x \to \infty$ $C \to C_0$. Solution concentration in pores and at surfaces increases with a rise in the evaporation rate and a diminution in the diffusion coefficient.

The solutions obtained have been tested in experiments with water-saturated quartz sand (0.2–0.5 mm) and the electrolyte radiotracer $Na_2S^{35}O_4$ [244]. Due to the rather low tracer concentration and a low rate of evaporation ($v < 2\times10^{-5}$ cm/s) the process of solution concentration on the sand surface may be observed for a long period of time. The value of $D = 2\times10^{-5}$ cm^2/s close to the tabulated coefficient of molecular diffusion of this electrolyte has been determined from the slope of the linear graph C_*/C_0 (t).

When the evaporation takes place from a porous body incompletely filled with water, the values of D decrease due to the transition from diffusion of the tracer in filled pores to that in thin films on the surfaces of the sand particles separating the capillary condensate. At sand saturation less than 8% the D values decreased by one order of magnitude.

The problem of non-stationary evaporation of solutions from capillaries under conditions of the continuous displacement of the evaporating surface have been solved by P. Zolotarev [245]. Applied to electrolyte solutions (a non-volatile component) and low rates of evaporation, that is the case in fine capillaries, the solution concentration at the receding meniscus, C_m, according to the theory, should stay constant and equal to:

$$C_m/C_0 = 1 + (\pi\alpha/D)^{1/2}, \tag{VIII.4}$$

where α is the coefficient of evaporation relating the position of the meniscus in the capillary L to the time t from the beginning of evaporation: $\alpha = L^2/4t$. The value of α does not depend on time and is determined by external conditions only:

$$\alpha = v_m p_s D_0(\varphi_m - \varphi_0)/2RT, \tag{VIII.5}$$

where D_0 is the coefficient of vapor diffusion through gas, $\varphi_m = p_m/p_s$ is the vapor pressure over the meniscus and $\varphi_0 = p_0/p_s$ the relative vapor pressure in the surrounding environment.

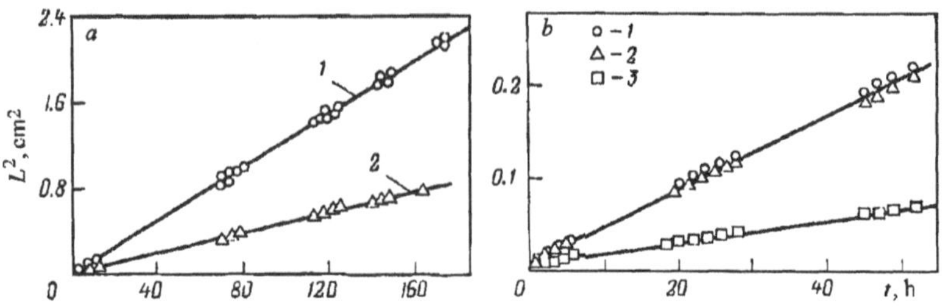

Figure 8.1 Comparison of the rates of evaporation of water and aqueous KCl solutions from capillaries of radius $r = 20$ μm at $T = 300.6$ K. *a*) $\varphi_0 = 0.75$; water (*1*) and solution $C_0 = 3.2$ mol/l (2); *b*) $\varphi_0 = 0.94$; water (points *1*), solution $C_0 = 10^{-2}$ mol/l (points 2) and $C_0 = 1$ mol/l (points *3*).

In the course of evaporation from the capillary the rate of evaporation $v = \alpha/2L$ gradually decreases due to the growing distance L between the meniscus and the capillary end. As shown in [245], a dynamic equilibrium is established between the decrease in the rate of the solute accumulation and its diffusion inside the capillary that leads to the constant concentration near to the meniscus.

This theoretical conclusion was checked by experiments [246] with the capillaries of radii ranging from 1 to 20 μm, from which KCl solutions of various concentration were evaporated: from $C_0 = 10^{-2}$ mol/l to $C_0 = 3.2$ mol/l which is close to the concentration of the saturated solution $C_H = 3.57$ mol/l. The measurements were performed at constant φ_0 (0.75; 0.85 or 0.94) in a chamber with prolonged thermostatting by a system of thermo-regulation supplied from accumulators [161]. This enabled continuous measurements of the rates of evaporation to be carried out over several days. The displacement $L(t)$ of menisci in horizontally positioned capillaries was measured by a comparator with a precision of 1 μm.

Some of the dependences obtained for $L^2(t)$ which should be linear under the constant conditions of evaporation according to theoretical relation (VI.26), $L^2 = 4\alpha t$, are shown in Fig. 8.1, *a*. As seen from the graphs, they are linear both for water (curve *1*) and also for the concentrated solution (curve 2). The values of α for water found from the slopes of the graphs $L^2(t)$ exactly coincide with the theoretical magnitudes calculated by eq. (VIII.5) using the known values of $v_m = 18$ cm^3/mol, $D = 0.257$ cm^2/s and $\varphi_m = 1$ for water. The rate of evaporation of the solution with low concentration, $C_0 = 10^{-2}$ mol/l, practically did not differ from that of water (Fig. 8.1, *b*).

In the case of more concentrated solutions the situation changes. Although the dependences $L^2(t)$ remain linear, the rate of evaporation falls due to the decrease of vapor pressure over the meniscus. The values of φ_m calculated by eq. (VIII.5) decrease to 0.84–0.96 depending on the rate of evaporation and the concentration of the initial solution C_0. The linearity of the dependences $L^2(t)$ confirms the constancy of φ_m values and the validity of eq. (VIII.4). At $\varphi_0 = 0.85$ and 0.94, when the rates of evaporation were not high, the difference between the concentration C_m calculated using tabulated data on the decrease in vapor pressure over the KCl solution and the experimentally found value of C_m did not exceed 15%. Higher deviations occur at higher rates of evaporation ($\varphi_0 = 0.75$) when the approximation (VIII.4) becomes unsatisfactory and more complex expressions for C_m should be used [245].

Therefore, the theory of evaporation of electrolyte solutions is well supported experimentally. The evaporation of binary liquid mixtures containing volatile components has been considered in Chapter VI, §4.

When concentrated KOH solutions evaporated from capillaries into the atmosphere of undersaturated vapor, the salt crept out and crystallized on the external surface of the capillary. This phenomenon is connected with the flow of wetting films along the capillary surface that transports the solute to the end of the capillary. Such films, as the measurements of their thickness on thin metal threads have shown [247], appear to be rather stable. Their equilibrium thickness is of 50–70 nm on the surface of copper threads and 30–40 nm on nickel at the disjoining pressure $\Pi = 1.6\times10^3$ N/m^2 and $C_0 =$ 10–15 mol/l. The greater thicknesses intensifying the transport of the concentrated solution in capillaries are probably connected with structural inhomogeneity of the concentrated solutions. Fluctuations of solution compositions grow when the saturation concentration approaches C_H. The scale of fluctuations becomes comparable with the film thickness, h, when $C_0 \to C_H$. Because of this, the thinning of the film is accompanied by the overlapping and restructuring of the structurally non-uniform regions. This makes the free energy of films and, consequently, the disjoining pressure (Chapter 1) dependent on h. Similar effects were found in thin (about 10 nm) interlayers of concentrated solutions (3 M $MgCl_2$, 5 M $CaCl_2$) between solid surfaces [248].

Crystal deposition at the end of the capillary during evaporation of water solutions of potassium bichromate has been visually detected in [249]. These deposited crystals narrowed the capillary bore, thereby reducing the rate of evaporation. However, in this case the salt concentration C_m should decrease near to the meniscus in accordance with eq. (VIII.4). The value of C_m may become less than that of C_H resulting in partial dissolution of a formed crystal. Such a decrease in crystal volume will again accelerate the evaporation making C_m grow, and the crystal volume will grow again when C_m approaches C_H. Therefore, the transition to a steady state proceeds through a regime of oscillations of crystal volume. The theory of this phenomenon was developed in [250]. The solutions obtained provided a method of controlling the kinetics of solution crystallization during their evaporation from porous bodies. To suppress the crystallization, it is necessary to reduce the rate of evaporation or the capillary length facilitating the diffusion of the salt from the meniscus zone inside the capillary. The danger of crystal deposition also decreases at low concentration of pore solution. The condition of crystallization may be roughly expressed as:

$$v_0 \leq (D/L) \ln (C_H/C_0). \qquad \text{(VIII.6)}$$

This equation determines the rate of evaporation v_0 which, when exceeded, may cause crystallization in the capillary entrance.

§2 Evaporation of Polymer Solutions

The theory of solution evaporation from capillaries discussed in the previous section does not take into account the effect of film flow. Evaporation from films may significantly change the distribution of solutes during evaporation and the concentration C_m near to the meniscus. This effect has not been detected in experiments with electrolyte solutions because in the case of moderately concentrated solutions the film thickness is small. In the case of low concentrations of solutions this practically does not affect evaporation (Fig. 8.1) and, therefore, fails to influence noticeably the redistribution of solution concentration. The exception is the experiments with a radio tracer [244]. Here the transport of the radio tracer with the film flow to the surface of sand particles of low moisture content is clearly defined in the course of solution evaporation. In the absence of film transfer, if only the vapor transfer took place, the tracer should be concentrated under the front of receding menisci, i.e. in the depth of the sample. Actually all the non-volatile tracer accumulates in the course of evaporation on the surface and, therefore, is carried here with the liquid flow. The application of radio tracers for the analysis of the mechanism of processes of internal mass transfer at evaporation and drying will be discussed in §5.

The influence of the film flow on the evaporation of solutions should be the more pronounced the thicker are the films. Stable wetting films about 1 μm thick form in polymer solutions. These films are thick because polymer adsorption layers form on both film surfaces, accompanied by the effect of steric repulsion.

The evaporation of an ethanol solution of polydimethylsiloxane (PDMS) with the molecular mass of 500–5000 from glass capillaries of radius ranging from 2.5 to 120 μm has been experimentally studied [251]. The technique was described in §1. When the rate of evaporation was controlled only by diffusion of ethanol vapors (PDMS is non-volatile), then the dependence $L^2(t)$ should follow the linear equation:

$$\Delta L^2 = L^2 - L_0^2 = [2\mu D_c (p_m - p_s)t]/[\rho(1 - C)RT] = 4\alpha t, \qquad \text{(VIII.7)}$$

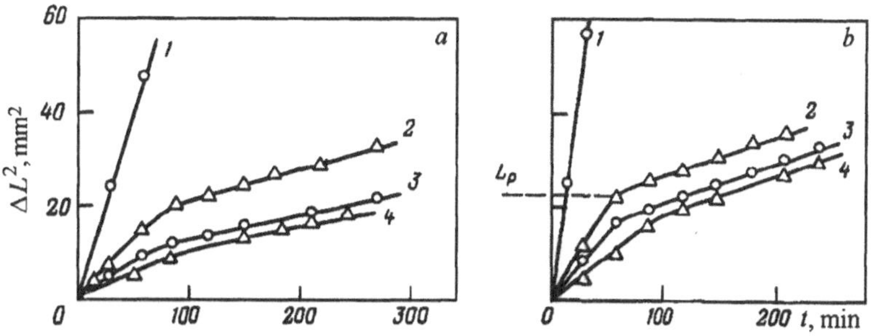

Figure 8.2 Kinetics of evaporation of ethanol (*1*) and PDMS solutions ($C_0 = 0.47$) from capillaries of various radius. *a*) 45°C; $r = 2.5$ (*1*, *2*), $r = 13$ (*3*) and $r = 120$ (*4*) μm; *b*) 60°C; $r = 2.5$ (*1*, *2*), $r = 10$ (*3*) and 120 (*4*) μm.

where $\mu = 46$ g/mol is the molar mass of the solvent and ρ is its density; L_0 is the initial position of the meniscus in the capillary corresponding to $t = 0$. Because the displacement of the meniscus is determined only by ethanol evaporation, the dimensionless mass concentration of the solvent, $(1 - C)$, g/g, stands in the denominator, where C is the mass fraction of the polymer. This correction may be neglected in solutions of low concentration.

The results obtained for ethanol and PDMS solutions evaporating into dry air at atmospheric pressure are shown in Fig. 8.2. For ethanol (curves *1*), the dependences $\Delta L^2(t)$ remain linear throughout the evaporation and intersect the origin. The influence of ethanol film flow does not appear because $p_0/p_s = 0$. The values of diffusion coefficients of ethanol through air, calculated according to eq. (VIII.7) at $C = 0$ and $p_m = p_s$, practically coincide with those tabulated.

The rate of solution evaporation (curves *2–4*) is much lower than that of the solvent. The graphs $\Delta L^2(t)$ consist of two linear regions. The transition to the low rate of evaporation occurs at some value $L = L_p$.

Before we turn to the discussion of the experimental results with PDMS solutions, let us consider the results of measurement of thick solid polymer films remaining on the capillary surface after the completion of evaporation. This experimental fact itself tells that a rather thick film of the solution remains behind the receding meniscus. The solid film thickness was measured at several points along the capillary length using a microscope MBI-6. First the capillary was viewed from the side in transmitted light, and the uniformity of the film coating (along the capillary) was estimated. Then the capillary was broken and its diameter and the film thickness were measured by an eyepiece-micrometer at several places (in normal direction to the capillary axis) to provide greater precision of the measurements. The measurements showed that the thickness h of the solid polymer film is practically constant along the capillary and depends only on its radius. The

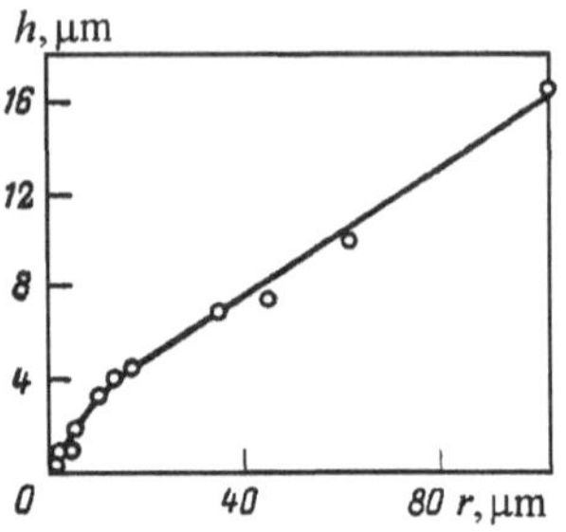

Figure 8.3 Dependence of the thickness of the dry polymer film, h, covering surfaces of glass capillaries on their radius, r, after evaporation.

experimental dependence of h on the capillary diameter is presented in Fig. 8.3. As seen from the graph, the film thickness, h, grows with decreasing capillary pressure of the meniscus $P_c = 2\sigma/r$ (σ is the surface tension of the solution).

The fact that the film thickness is constant along the capillary shows that the values of h do not depend on the rate of the meniscus recession during evaporation. This is possible, when an equilibrium wetting film forms in contact with the meniscus and its thickness is determined by the disjoining pressure of the forming film: $\Pi(h_*) = \sigma/(r - h_*)$. Here Π is the disjoining pressure of the flat film and h_* is the thickness of the liquid film of the solution remaining behind the receding meniscus.

It should be noted that at much higher rate of the meniscus movement, when the solution was displaced from the capillary by gas, wavy non-equilibrium films emerge on the surface as it was observed in other experiments with viscous liquids [252]. The viscosity of the polymer solution studied at $C = 0.47$ was about 1.5×10^{-2} Pa s at 25°C. Similar wavy films sometimes form at high rates of evaporation corresponding to high temperature and low values of L, i.e. near to the capillary entrance.

Let us consider now the rates of evaporation of solutions (Fig. 8.2, curves *2–4*). Equation (VIII.7) fails here, because the evaporation occurs not only from the meniscus surface but also from the surface of the residual film. Because the isotherms $\Pi(h_*)$ of wetting films are not known (data in Fig. 8.3 refer to the "dry" film, and $h < h_*$), quantitative calculations of the film flow are not possible. We therefore restrict ourselves to a qualitative discussion of the results obtained.

Let us consider the two stages of solution evaporation which differ by the slopes of the graphs. During the first stage the values of α responsible for the intensity of evaporation are higher than for the second stage, and they grow with decreasing capillary radius. This may be explained by the influence of polymer adsorption on the capillary walls. The smaller the capillary radius, the more distinct the decrease in initial solution concentration C in the capillary caused by adsorption. The values of C_m should correspondingly decrease, as follows from eq. (VIII.4), and the vapor pressure over the meniscus p_m should grow, accelerating the evaporation. Moreover, the intensification of evaporation may be explained by the increasing effect of the film flow.

However, even in the finest capillaries the rate of evaporation of a pure solvent was considerably higher than that of the polymer solution. This means that solution concentration in the capillary still remained rather high after adsorption, and, as a result, the values of p_m were lower than those of p_s for ethanol. The second stage of evaporation of polymer solutions at $L > L_p$ (Fig. 8.2) is characterized (in contrast to the first stage) by lower α values practically independent of capillary radius.

Note that for solutions of low-molecular weight species, in agreement with the theory, no transition was observed to the second stage (Fig. 8.1). The graphs $L^2(t)$ did not change the slope in the course of evaporation of electrolyte solutions throughout the experiment. A change in slopes of the graphs $\Delta L^2(t)$ when $L > L_p$ for polymer solutions is likely to be associated with the effect of polymerization beginning when the near-to-meniscus solution concentration rises to some critical value C_{mp}. One might expect that the values of C_{mp} are close to 0.75–0.8. At this concentration the viscosity of the solution increases approximately by one order. Solution structuring near to the meniscus makes the coefficient of polymer diffusion through the solution D in eq. (VIII.4) decrease and hinders the back diffusion of the polymer accumulating close to the evaporating meniscus. In the course of evaporation of PDMS solutions of various concentration from wide capillaries ($r \sim 1-2$ mm) a sharp decrease in the rate of evaporation at $C_0 \geq$ 0.75–0.8 has also been revealed.

If the polymerization takes place, a higher but constant value of $C_{mp} > C_m$ establishes, corresponding to new conditions of mass balance at the meniscus. This value is not dictated by the capillary radius, but by the nature of intermolecular bonds and molecular-weight distribution. As seen from Fig. 8.2, the slope of the second region of the graph is practically the same for the capillaries of different radius.

In narrow capillaries (here the thickness of liquid film h_* is comparable with r) an initial solution concentration due to the adsorption is lower than the bulk concentration of the solution $C = 0.47$. Therefore, relatively more solvent has to be evaporated to achieve the solution transition into a structured state in the vicinity of the meniscus. As seen from Fig. 8.2, the values of L_p regularly decrease with increasing capillary radius and also depend on temperature (Fig. 8.2). A decrease in temperature accelerates the transition into a structured state in the vicinity of the meniscus both due to the decrease in D values causing deterioration of the diffusive outflow of the polymer and because of the growing probability of intermolecular bond formation.

The graphs $\Delta L^2(t)$ for wide capillaries (when $r >> h_*$), where adsorption ceases to contribute, are practically coincident. The transition to the second stage of evaporation here is not associated with the influence of polymer adsorption or the presence of films on the capillary surface.

Therefore, evaporation of polymer solutions is complicated by the effect of polymerization leading to a sharper decrease in rates of evaporation than in the case of solutions of low-molecular weight species. The thickness of a deposited polymer film depends on the capillary pressure of the meniscus and decreases with capillary radius. In porous bodies with varying pore size, the film thickness should be greater in pore widenings which may change the pore size distribution. The formation of polymer films by evaporation may be used to modify the surface of porous bodies and also to create firm composite structures by polymerization in the course of the solvent removal at the points where particles of the disperse material contact.

The development of a quantitative theory of evaporation of polymer solutions adsorbing on pore surface requires a study of the isotherms of the disjoining pressure of wetting films for solutions of different concentrations.

§3 Influence of Surfactants on Water Evaporation from Porous Bodies

Surfactants are often used to suppress or depress evaporation from a free surface [253]. Surfactant monolayers formed on a water surface provide a barrier which molecules must overcome to transfer to the vapor phase from liquid. This is a kinetic effect which results in a decrease in vapor pressure Δp over evaporating water surface coated with the surfactant adsorption layer. In equilibrium, the vapor pressure is equal to that of saturated vapors over the solution. At a low concentration of surfactant molecules in the bulk, this pressure is practically the same as in pure water p_s.

According to the Hertz–Knudsen theory, a drop of vapor pressure for the rate of evaporation v (cm/s) is equal to:

$$\Delta p = (v\rho/\alpha_0)(2\pi RT/\mu)^{1/2}, \tag{VIII.8}$$

where α_0 is the coefficient of condensation. Its value depends on the height of the potential barrier ε: $\alpha_0 \sim \exp(-\varepsilon/kT)$. For water under normal conditions $\alpha_0 = 0.04-0.05$ that corresponds to ε of about 3 kT. This is close to the energy of the intermolecular hydrogen bond in water. An increase in temperature makes α_0 values grow a little. For electrolyte solutions the value of α_0 is higher the stronger the ion action on water structure. For KOH at concentration of 40% by weight, the values of α_0 increase up to 0.2 and for

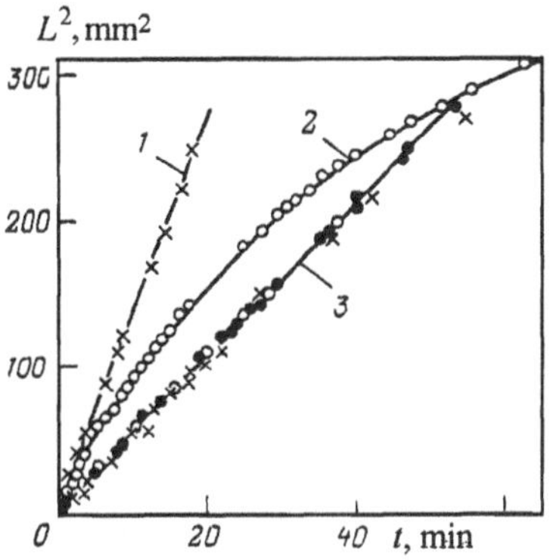

Figure 8.4 Kinetics of evaporation $L^2(t)$ of water and aqueous solutions of syntamide-5 (curve 2) from capillaries. *1*) water, $r = 8.2$ μm, $p_0/p_s = 0$; *2*) aqueous syntamide-5 solution $C = 0.25\%$, $r = 8.2$ μm, $p_0/p_s = 0$; *3*) aqueous syntamide-5 solutions $C = 0.05\%, 0.1\%, 0.25\%$, $r = 7.4$ μm; $p_0/p_s = 0.54$.

CsCl at concentration of 60% up to 0.11 [254].

For water at $v \leq 10^{-5}$ cm/s, the values of $\Delta p = p_s - p_0$ are small, and the rate of evaporation v is governed only by vapor diffusion into the surrounding environment. Here the vapor pressure over the liquid surface p_0 is equal to the that of saturated vapor p_s. Thus the influence of the coefficient of condensation on mass transfer in porous bodies, where v values are small, is often negligible. It may become significant for example during the initial stage of evaporation into vacuum. However, in this case a decrease in temperature of the evaporating surface may play a leading role.

The presence of surfactant adsorption layers on a surface sharply reduces, by several orders, the coefficient of condensation. Therefore, for the same conditions of evaporation as in pure water, the vapor pressure drop Δp becomes significant, and the evaporation is controlled by kinetic rather than by diffusion factors. The rate of evaporation v becomes dependent primarily on the coefficient α_0. For example $\alpha_0 = 3.5 \times 10^{-5}$ for a saturated monolayer of insoluble cetyl alcohol [255].

To study the effect of surfactants, experiments using quartz capillaries of radius of 10–20 μm were performed [256]. The results obtained for evaporation of pure water (curve *1*) and 0.25% solution of non-ionic surfactant, syntamide-5 (curve 2), from the capillaries of strictly equal radius $r = 8.2$ μm are shown in Fig. 8.4. The capillaries filled with water and with surfactant solution were placed in the evacuated chamber (Fig. 7.7) near to each other on the same support. Curves *1* and *2* refer to evaporation in vacuum ($p/p_s = 0$) at $\alpha = 5 \times 10^{-4}$ cm^2/s, where α is the coefficient (see eq. (VIII.5) characterizing the rate of evaporation. At first the curves *1* and *2*

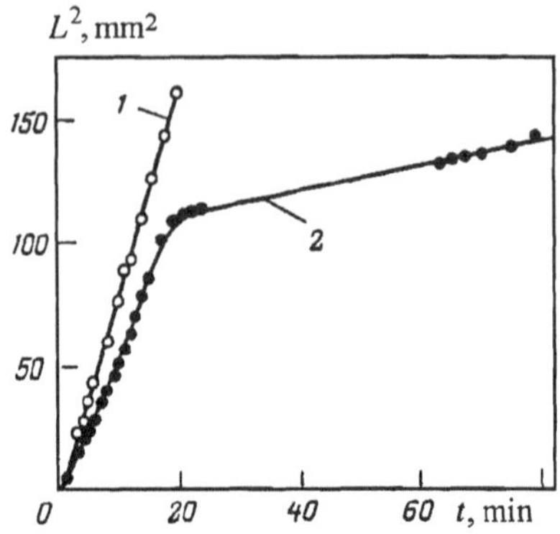

Figure 8.5 Kinetics of evaporation $L^2(t)$ of water (*1*) and water from the capillary with an adlayer of cetyl alcohol (*2*). $r = 7.6$ μm, $p_0/p_s = 0.48$.

practically coincide, but later evaporation in the capillary filled with the surfactant solution gradually slows down. This can be explained by surfactant concentrating in the vicinity of the meniscus surface.

At less than half the coefficient of evaporation ($\alpha = 2\times10^{-4}$ cm²/s), when a relative vapor pressure $p/p_s = 0.54$ was maintained in the chamber, there was no difference in the rate of evaporation of pure water and aqueous surfactant solutions (curve *3*). Probably, in this case surfactant molecules effectively escaped into the bulk of the capillary by diffusion. Similar results were also obtained for lower rates of evaporation at $p/p_s = 0.915$ ($\alpha = 0.37\times10^{-5}$ cm²/s) and $p/p_s = 0.95$ ($\alpha = 1.2\times10^{-5}$ cm²/s) and not only for syntamide-5 but also for cationic surfactant (CTAB) at concentrations of $C = 0.25$ and 0.5% by weight.

No effect of syntamide-5 ($C = 0.05$ and 0.25%) on the rates of evaporation from the capillaries into dry air, when α-values were much lower ($\alpha = 3.3\times10^{-6}$ cm²/s), has been revealed. This series of experiments was carried out in the thermostatted apparatus supplied from accumulators.

Consequently it must be concluded that the surfactants studied impede water evaporation only at a sufficiently high rate of evaporation $\alpha \geq 10^{-3}$ cm²/s. Such rates are rarely achieved in practice, thus the mechanism of action of surfactants used as depressors of evaporation from porous bodies lies not in slowing down the rate of water evaporation but in pore surface hydrophobization that has been supported in [257, 258].

For further experiments we have chosen cetyl alcohol (CA), a surfactant insoluble in water whose adsorption on water surface significantly decreases the coefficient of condensation [255]. However, the problem arose of how to apply the surfactant on the meniscus surface in a narrow capillary. First CA was applied to the whole capillary surface from its solution in ethanol. The

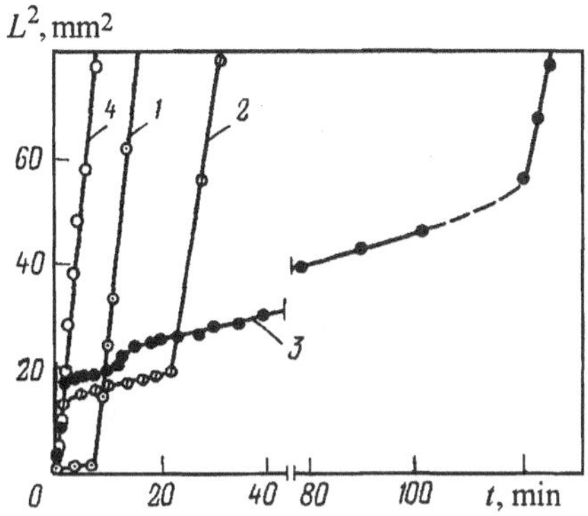

Figure 8.6 Kinetics of evaporation $L^2(t)$ of water from the capillary of r = 9 μm, p_0/p_s = 0.54: (*1–3*) capillaries with locally deposited cetyl alcohol on the distance z = 0.8 mm (*1*), 4.23 mm (2) and 6.14 mm (*3*) from the capillary end; (*4*) water in a clean capillary.

capillary was filled with 0.5–1% solution, which was then displaced, retaining on the capillary surface a thin wetting film, then the capillary was dried. It was filled then with water and sealed. The observation of evaporation was initiated after the capillary was broken in the region filled with water, so that at the beginning there was no CA on the meniscus.

The results of one such experiment are shown in Fig. 8.5. It is seen that at first the rates of evaporation from pure capillaries and those pre-treated with CA coincide. Then, as the meniscus becomes covered by CA molecules, due to their diffusion from the capillary surface, the rate of evaporation decreased. As shown in [255], a sharp decrease of the coefficient of condensation occurs after a complete monolayer is formed. This corresponds to a sharp decrease in the rate of evaporation after 20 min in the experiments presented.

In another experiment, CA was deposited on the capillary surface locally. For this purpose, only a part of the capillary was filled with CA solution in ethanol and the boundary was marked on the external capillary surface. Ethanol was then evaporated from the capillary, the empty side of which was preliminarily sealed. Thus a region covered with CA was created on the internal surface of the capillary. Next, the capillary was filled with pure water and evaporation was observed from the empty side of the capillary. As seen in Fig. 8.6, the rate of evaporation from the capillaries (curves *1–3*) was equal to that from the control capillary, not treated with CA (curve *4*), unless the meniscus approached the region covered with CA. At this moment the evaporation sharply slowed down due to the CA transition from the capillary surface onto the meniscus. It is seen that the evaporation actually slowed down when the distance from the entrance of the capillary to the meniscus was $L = z$.

In the first case (Fig. 8.5) CA concentration on the capillary surface was low. Therefore, a monolayer of CA was formed only when the meniscus had travelled a rather long distance. In the second case (Fig. 8.6) much more CA was deposited locally on the capillary surface. Thus the saturation of the monolayer was achieved quickly, practically at the meniscus contact with the CA layer deposited on the capillary surface.

It is interesting to note that after the meniscus has passed the region covered with CA the initial rate of evaporation re-establishes corresponding to the rate of evaporation of pure water (Fig. 8.6). This means that the meniscus displaces the CA due to reverse diffusion of its molecules to the clean capillary surface. A sharp transition to the initial rate of evaporation is evidence of a rapid disruption of the compact CA monolayer.

Effective inhibition of evaporation is associated with a sharp decrease of the coefficient of condensation, α_0, by the CA monolayer. According to eq. (VIII.8), the values of α_0 might be found using the values obtained for the rate of evaporation and $\Delta p = p_s - p_0$, where p_0 is the vapor pressure in the surrounding medium. The diffusion resistance to the vapor flow in the capillary between the meniscus and the entrance may be in this case neglected. For different regions of the graphs presented in Fig. 8.6, the values of α_0 ranged from 1.8×10^{-4} to 2.7×10^{-4}. In another experiment with a capillary of diameter 17.5 μm ($z = 4$ mm, $p/p_s = 0.54$) almost the same values of $\alpha_0 = 3\times10^{-4}$ were calculated.

The values of α_0 obtained from the observations of evaporation from the capillaries are lower by two orders than those for pure water (0.03–0.04), but approximately by one order higher than those obtained for CA earlier for evaporating droplets [255]. Probably, the conditions under which the surfactant layer forms on the meniscus contacting the quartz surface, also adsorbing CA, are different from the conditions in the case of evaporating drops.

It would be interesting to undertake further investigations of evaporation of surfactant solutions from narrow capillaries to follow the surfactant influence on the film flow along their surface. If the surfactant hydrophobizes the capillary surface, a decrease in the rate of evaporation should be expected due to the elimination of the film transfer. If the surfactant stabilizes the wetting films, improving wetting, then the rates of evaporation may increase. Therein lies the possibility of governing the processes of internal mass transfer by injection of surfactants.

In the technology of drying the effect of surfactants primarily manifests itself as changes in structure and rheological properties of porous bodies due to a change in the forces acting among the particles. The processes of structure formation will be discussed in more detail in Chapter IX.

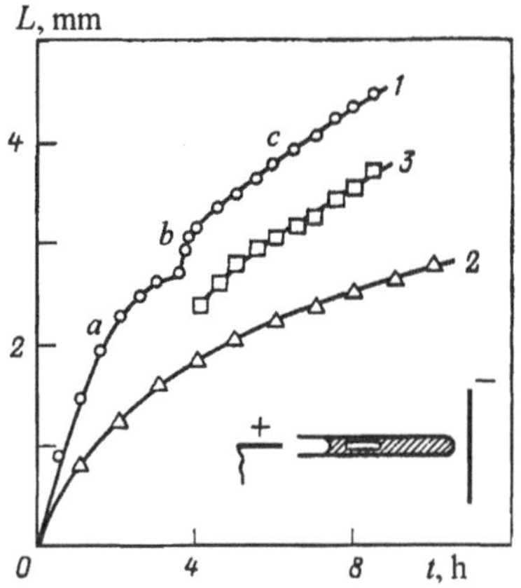

Figure 8. 7 Meniscus displacement of the main water column, L, with time, t, during evaporation into humid air ($\varphi_0 = 0.8$) in FCD (*1*) and without it (2). $r = 3.2$ μm; $T = 297.4$ K; $E = 2.7 \cdot 10^6$ V/m, $\nabla E = 1.2 \cdot 10^9$ V/m^2. Curve *3* shows the growing length of the condensate column.

§4 Evaporation in the Field of Corona Discharge

In Chapter VI, §2 and Chapter V, §1 we have discussed the effect of non-uniform electric field on vapor diffusion and film flow. When the intensity of the field increases further, a corona discharge is generated that changes the mechanism of evaporation from capillaries [259] and intensifies the drying of porous bodies [260]. The scheme of measurements is the same (Fig. 5.4) except the fact that higher voltage up to 5×10^6 V/m was applied to electrodes and the electrodes were oriented differently with respect to the capillary.

The action of the field of corona discharge (FCD) considerably accelerates the evaporation as compared to the non-uniform electric field (Fig. 6.9). This acceleration cannot be explained by a mere increase in the field strength E and the gradient ∇E that, according to eq. (VI.21), determines the vapor diffusion, because the rate of evaporation rises more rapidly than ∇E values. An increase in the rate of evaporation at a relative air humidity in the surrounding media of $\varphi_0 \leq 0.75$ (when the effect of film flow is weak) is associated with a convective action of FCD and a decrease in φ close to the capillary entrance. Due to this, the rate of evaporation increased by a factor of 1.2 at $\varphi_0 = 0.34$ and by 1.6 at $\varphi_0 = 0.75$. The values of φ near to the capillary entrance were 0.21 instead of $\varphi_0 = 0.34$ and $\varphi = 0.6$ instead of $\varphi_0 = 0.75$, respectively.

At higher air humidity ($\varphi_0 \geq 0.8$) the character of the effect changes qualitatively. The rate of water removal from the capillary changes non-monotonically with time. In the experiment, the results of which are presented in Fig. 8.7, the evaporation first proceeds (curve *1*, region *a*) according to theory accounting for vapor diffusion and film flow. For comparison, curve *2* shows the result of evaporation under the same conditions but in the absence of FCD. However, after 3.5 h the displacement of the meniscus in FCD abruptly accelerated (curve *1*, region *b*). Simultaneously, with the help of a microscope of a cathetometer used to measure the meniscus displacement $L(t)$, the formation of columns of condensate 0.1–0.3 mm long were observed (as shown below, Fig. 8.7) in front of the meniscus. When $L < 4$ mm, the condensate forms mainly in the vicinity of the capillary entrance. When the meniscus has receded to a greater distance L, the condensate forms then not further than 3–4 mm from the receding meniscus. The distance from the entrance to the evaporation region is reduced by the value of the length of the condensate column and a captured air bubble, which explains the effect of the sharp acceleration of evaporation. The rate of the displacement of the main meniscus, dL/dt, is kept approximately constant (curve *1*, region *c*), despite the extension of the distance between the meniscus and the capillary entrance. The reason for this is that the mass transfer between the main liquid and the condensate column proceeds through the wetting film rather than in the vapor phase. In so doing the length of the condensate column grows continuously (Fig. 8.7, curve *3*) and the air bubble is displaced into the depth of the capillary. Similar effects take place not only in water but in ethylene glycol and butyl alcohol. This effect is not, however, observed in non-polar benzene.

Condensate formation and a peculiar "fragmentation" of the liquid in the capillary bore are caused by the action of the force f on the wetting films (see eq. (V.8)) near to the receding meniscus. This force is proportional to $\nabla(E^2)$ according to eq. (V.8). The pulling of the film by the field leads to its gradual thickening and further coalescence and formation of the condensate column. Measurements performed with water solutions of 10^{-2} mol/l KCl confirm the fact that this effect is associated with the film flow. At this electrolyte concentration the film thickness is drastically reduced, preventing condensate formation. In this case, because of the reasons presented, FCD only accelerates the vapor diffusion.

The FCD action makes it possible to increase the rate of evaporation by 1.5–2 times at $\varphi_0 < 0.8$ through a decrease in φ near to the capillary entrance. When $\varphi_0 > 0.8$, the rate of water evaporation increases up to 3–5 times due to the intensification of the film flow and the formation of condensate columns.

The application of FCD enables the drying of linen thread to be accelerated considerably, so reducing the power expenses for dehydration [261].

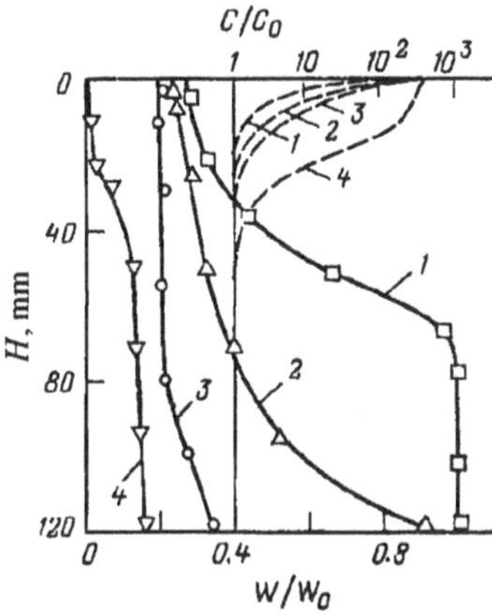

Figure 8.8 Distribution of moisture content W/W_0 (solid curves) and tracer concentration C/C_0 (dashed curves) over the height, H, of quartz sand samples: $W_c = 0.16$ g/g (*1*); 0.106 g/g (2); 0.05 g/g (*3*) and 0.025 g/g (*4*); $\varphi_0 = 0.9$.

§5 Study of the Mechanism of Internal Mass Transfer by Radiotracers

Evaporation leads to an increase in concentration of a non-volatile component of the solution in the region of evaporation. Thus by measuring the changes in solute concentration, it is possible to identify where and how intensely the evaporation occurs within the porous body. To prevent changes either in capillary pressure or in the thicknesses of the wetting films, the non-volatile tracers should be used at very low concentration. At the same time concentrations must be measured with a high precision. Radiotracers meet all these contradictory requirements [23]. The mass concentration of the tracer in the solution does not exceed 10^{-5} g/g and high specific activity of the solution (for $Na_2S^{35}O_4$ – 5–10 μCi/l provides the accuracy of the concentration measurement of several percent.

Figure 8.8 shows, as an example of the application of a radio ionic tracer ($S^{35}O_4^{2-}$), the changes of distribution with depth within a layer of fine sand (grain size less than 0.17 mm) in the course of moisture evaporation. First the sand in the test tube of a length of $H = 12$ cm and a diameter of 1.2 cm was completely filled with the solution and placed into a chamber, where a constant air humidity φ_0 and temperature of 26±0.1°C were held. The samples were regularly weighed and the specific activity of the surface was measured with help of a front counter MST-17 and a radiometer B-2. Each series comprised 3–4 similar samples. At different stages of the process, they were subject to layer by layer removal. The thickness of the layers was 2–5 mm in upper part of the sample and 10–20 mm in bottom. The moisture content and the tracer concentration were measured in each layer according to the technique described in [262].

The results of these measurements were used to build the graphs of the relative intensity of evaporation i/i_0 and the surface activity N/N_0 as a function of the mean moisture content W_c of the sample. The intensity of evaporation from the bulk water surface i_0 was measured in the chamber under the same conditions. The counting rate N_0 was measured before the beginning of drying. The distribution of local moisture content W/W_0 (W_0 is the moisture content corresponding to the complete filling of the sample) and tracer concentration C/C_0 (C_0 is the initial tracer concentration in pore moisture) along the sample length are presented in Fig. 8.8. The preparation of the samples was performed at various mean moisture contents W_c.

The analysis of the results obtained [263] shows that during a constant rate of evaporation the radiotracer concentrates in a relatively thin surface layer, but its concentration inside the drying sample remains equal to C_0. This proves the absence of moisture evaporation from the receding front of menisci, the position of which is seen from the graphs W/W_0. In the region between the front of the menisci and the surface of the sample, water is transferred not by vapor diffusion but in the form of the liquid. Because the moisture content of conducting layers, as follows from the isotherms of capillary pressure $P(W_c)$, corresponds to the state when the bridges of capillary condensate are separated by films (Fig. 2.15), the main mechanism of liquid transfer is film flow. As the experiments have shown, this mechanism prevails in grain systems such as quartz sand with mean particle size ranging from 1 mm to 50 μm at $\varphi_0 = 0.2–0.95$ and $t = 20–30°C$. This supports the theory of water evaporation from single capillaries and capillary-porous bodies, which takes into account the film flow of liquid along the pore walls (Chapter VII).

At $\varphi_0 = 0.5$, the distribution of tracer concentration in the sample may be described by eq. (VII.3). Therefore, the evaporation occurs here from the external surface of the sample. Its intensity is determined by the conditions of external mass transfer, the rate of which does not change unless the film flow provides enough moisture to the evaporating surface. Tracer concentration in the surface layer grows linearly in this stage, confirming the constancy of the film flow. This persists until the front of menisci descends to the end of the sample; after that the dehydration of the surface layers begins. The transporting ability of the films decreases due to their thinning and this leads to a dramatic decrease in the intensity of drying. So, the first critical point on the graphs of the drying rate corresponds to the transition from film transfer of moisture in surface layers to the vapor diffusion.

In the second stage of drying, the moisture evaporation takes place from the surface of the receding front of the capillary condensate and films. As long as the distance from the external surface is small, the vapor diffusion inside the

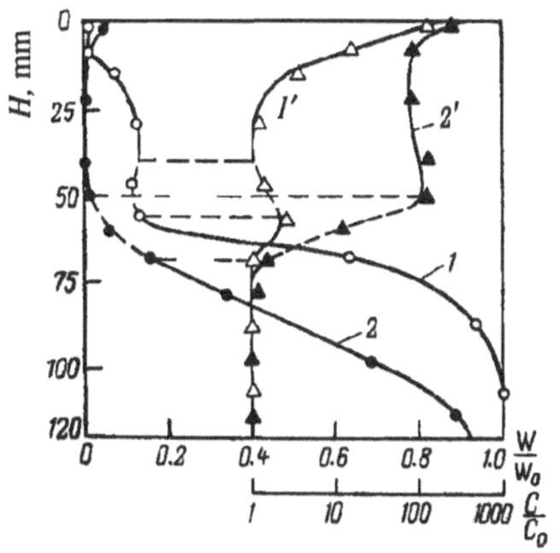

Figure 8.9 Distribution of moisture content W/W_0 (curves *1* and 2) and tracer concentration C/C_0 (curves *1'* and 2') over the height, *H*, of quartz sand samples (grains diameter 0.25–0.5 mm) at $W_c = 0.1$ g/g (*1* and *1'*) and 0.062 g/g (2 and 2'); $\varphi_0 = 0.5$; $T = 299$ K.

sample combines with the film flow, as demonstrated by the persisting transport of the tracer to external surface. Then the film flow of moisture to the surface stops and the tracer concentration does not change. The prevailing mechanism of transfer in the dried zone has become vapor diffusion. Below the front of the condensate and films the film flow continues, as far as the tracer concentration in these layers remains equal to C_0. Despite the low moisture content of sand ($W_c = 0.03$–0.04 g/g) and the availability of dry pores, the main mechanism of transfer is still the film flow. It should be noted that thin films transfer moisture along the particle surface in the regions between condensate bridges, so that the flow is not purely a film flow. In bodies with more complex geometry of pore space, where together with the wider pores there is also a network of narrow communicating channels, the outflow of the tracer may go through the micropores in preference to the films. This mechanism of moisture transfer corresponds to the two-capillary model described in Chapter VII, §4. In this case the tracer will accumulate near to the outlets of the fine capillaries.

At higher relative air humidity ($\varphi_0 = 0.9$), the process follows the same scheme as illustrated by Fig. 8.8. The difference from the evaporation at $\varphi_0 = 0.5$ is that the moisture evaporates not from the external surface but from the surface layer several mm thick. This follows from the fact that the tracer concentration on the external surface grows insignificantly, although the total quantity of the tracer transported to the surface layer is the same as at $\varphi_0 = 0.5$. An abrupt jump in tracer concentration at the transition to the second stage of drying is caused by the export of the concentrated tracer solution from the

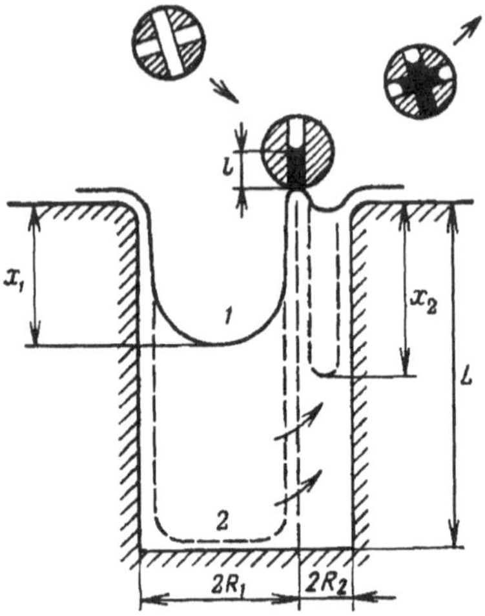

Figure 8.10 Evacuation of non-volatile liquid by suspended fine porous particles from a model porous body.

internal layers, where evaporation takes place, to the external surface of the specimen. The deepening of the evaporation zone at $\varphi_0 = 0.9$ was also observed for sand with particle size of 0.2–0.3 mm. For coarse-grained sand (0.5–1 mm) no change in the mechanism of moisture transfer at $\varphi_0 = 0.9$ was detected [264].

The results observed during transportation of water (containing the tracer) from a sample, in the middle part of which there was an interlayer of hydrophobized sand 13–15 mm thick, are presented in Fig. 8.9. The location of the hydrophobized interlayer corresponds to the dashed lines on the graphs. It is easily seen that the tracer concentration under the hydrophobic interlayer increases because the film transfer cannot occur here. The transfer through the hydrophobic interlayer proceeds chiefly by vapor diffusion that shifted the first critical point on the plots of drying towards the higher moisture content as compared with the control experiment. Correspondingly, the rate of sample dehydration is decreased. The film transport of the moisture from the bottom layers could not occur. As a result, the upper layers were fully dehydrated (curve 2), while the bottom ones remained over-moistened.

§6 Evacuation of Non-volatile Liquids from Porous Bodies

To remove non-volatile liquids from porous bodies (heavy oils, paraffins), the methods of sucking out through porous filters are usually used. However, this may be done more effectively with the help of suspended porous particles

[265]. This method was applied to remove a paraffin binder from technical ceramics. The products were placed into a pseudofluidized layer of particles with fine pores. When the particles touch the surface of a porous body, they had time to suck out a part of a binder from the pores of the ceramics. As a result of many collisions, a gradual evacuation of a low-volatile liquid from the porous body took place.

To describe the kinetics of the liquid evacuation, the two-capillary model discussed above (Fig. 7.13) was applied. Figure 8.10 shows how the mass transfer occurs between the surface of the porous body and the microporous particles colliding with it. The evacuation of liquid is possible unless the capillary pressure of the liquid in particle pores $P_p = 2\sigma\cos\theta/r$ is higher than the capillary (or disjoining) pressure of the liquid remaining in the porous body. Thus the pore size and the lyophilicity of particles of the dispersion determine the lower boundary of the moisture content in a porous body. However, at too small values of r the amount of liquid imbibed during the time of contact decreases due to the growth of viscous resistance. Therefore, in practice porous particles with pore size of about 10^{-6} cm are generally used.

The distribution of liquid in a porous body during the initial stage, when pores with smaller radius were still filled with liquid, is shown in Fig. 8.10 by the solid curve *1*. Impinging dry particles contact the surface of films and also menisci in narrow pores filled with liquid and imbibe it due to the difference of capillary pressures of particles P_p and the pressure P_0 on the surface of the porous body. The length of the pore l to which it imbibes the liquid is:

$$l = r\,[(P_p - P_0)t_k/8\eta]^{1/2}, \qquad \text{(VIII.9)}$$

where t_k is the time of contact and η is the viscosity.

The mass of liquid imbibed by one particle during the time of contact is:

$$M_1 = \pi r^2 l \rho N S \omega(P_0), \qquad \text{(VIII.10)}$$

where ρ is the density of the liquid, N is the number of pores per unit particle area; S is the area of contact between the particle and the porous body. The probability, ω, that dry regions of particle surface will contact the entrances of fine pores filled with liquid is:

$$\omega(P_0) = n_p n W_s(1 - W_p), \qquad \text{(VIII.11)}$$

where n_p is the particle porosity, n and W_s are the porosity and the surface moisture content of the porous body and W_p is the mean moisture content of impinging particles.

Then the flux of the liquid captured by the particles may be determined as:

$$q = M_1 N_p, \tag{VIII.12}$$

where N_p is the number of particles approaching in unit time per unit area of the porous body. On the other hand, in a quasi-stationary process the same flux should be brought to the surface of the porous body. In the case of the two capillary model and non-volatile liquid, the flux comprises the Poiseuille flux along fine capillaries q_2 and a film flow q_1 along pore walls in wider capillaries:

$$q = q_2 + q_1 = [\rho R_2^2 n_2(P_0 - 2\sigma/R_1)/8\eta(L - x_1)]$$

$$+ 2\rho A n_1 \ln(P_0 R_1/\sigma - 1)/3\eta R_1(L - x_1), \tag{VIII.13}$$

where x_1 is the position of menisci in wide pores; R_2 and R_1 are the radii of wide and narrow capillaries; L is the linear size of the porous body; n_2 and n_1 are the parts of the volume corresponding to wide and narrow pores respectively; A is the constant of the isotherm of disjoining pressure $\Pi = A/h^3$.

Equating eqs. (VIII.12) and (VIII.13) and turning from x_1 and L to the average moisture content of the porous body $W = n_2 + n_1(x_1/L)$, we get the following relation between the rate of the liquid evacuation $\partial W/\partial t$ and its content, W, in the porous body:

$$W_0 - W = [2An_1^2/(3\eta R_1 L^2 \partial W/\partial t)] \ln[(2R_1/r) -$$

$$-(\rho^2 R_1 L^2/\alpha^2 n_2^2 \sigma)(\partial W/\partial t)^2 - 1] +$$

$$+ [R_2^2 n_2 \sigma n_1/(8\eta R_1 L^2 \partial W/\partial t)][(2R_1/r) -$$

$$-(\rho^2 R_1 L^2/\alpha^2 n_2^2 \sigma)(\partial W/\partial t)^2 - 1], \tag{VIII.14}$$

where W_0 is the initial moisture content for the complete filling of capillaries with liquid and $\alpha = q/W_s(1 - W_s)(P_p - P_0)^{1/2}$.

It has been accepted that: $W_s = n_2$, $W_p = 0$, and the following expression for $P_p - P_0$ has been used:

$$P_p - P_0 = (n_1\rho/n_2\alpha)^2(dx_1/dt)^2,$$

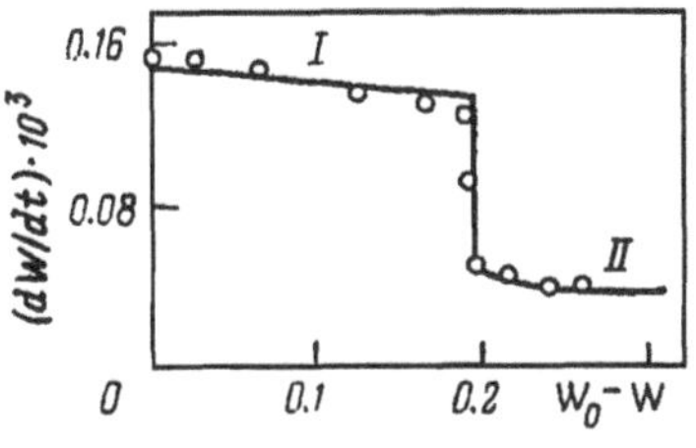

Figure 8.11 Comparison of calculations (solid line) and experimental data (points) on the rate of evacuation of paraffin from ceramics in the suspended layer of porous particles.

stemming from the determination of the mass flux i from the rate of the meniscus recession in the wide capillary:

$$-n_1\rho(dx_1/dt) = n_2\alpha(P_p - P_0)^{1/2} = i. \qquad \text{(VIII.15)}$$

When x_1 becomes equal to L and the wide pore becomes empty, the meniscus recession in the narrow capillary will begin. The position of the liquid front in the second stage of liquid evacuation is shown in Fig. 8.10 by the dashed line 2. In this case, we get instead of (VIII.13):

$$q = 2\rho A n_2 \ln(P_0 R_2/\sigma - 1)/3\eta R_2(L - x_2). \qquad \text{(VIII.16)}$$

This equation describes the film flow in narrow capillaries. The estimations show that the contribution of film flow in wide capillaries may be neglected.

Substituting the q values into eq. (VIII.14), we obtain the relevant expression for the rate of the liquid evacuation.

The results of calculation of the rate of liquid evacuation $\partial W/\partial t$ as a function of $\Delta W = W_0 - W$ for porous ceramics filled with an organic binder on the basis of paraffin are shown in Fig. 8.11.

Besides the two stages discussed above for the evacuation of liquid, an intermediate stage was taken into account, when the meniscus recession in narrow capillaries begins a short time prior to complete recession of the menisci in the wide capillaries. We assumed in the calculations that: $R_1 = 10^{-3}$ cm, $R_2 = 5\times10^{-6}$ cm, $r = 10^{-6}$ cm, $\sigma = 70$ mN/m, $\rho = 0.9$ g/cm^3, $A = 10^{-20}$ J, $T = 333$ K, $n = 0.5$, $n_1 = n_2 = 0.25$, $n_p = 0.4$, $\eta = 0.045$ Ns/m^2, $L = 0.7$ cm. The values of N and t_k were calculated proceeding from the parameters of pseudofluidization of a layer of porous particles 10^{-2} cm in diameter.

Points in Fig. 8.11 show the experimental results. As seen from the comparison of these points with the theoretical graph, the two-capillary model and application of equations for film flow provide satisfactory

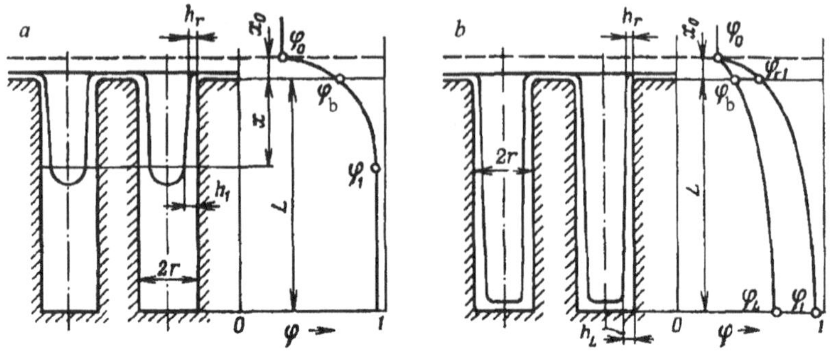

Figure 8.12 Model of a capillary-porous body and calculation schemes for the first (*a*) and the second (*b*) periods of drying.

agreement. This is a result of the bimodal distribution of pores by size in the ceramics and of the fact that the isotherm of disjoining pressure $\Pi(h)$ follows the equation like A/h^3. The values of the constant A were calculated on the basis of physical characteristics of the ceramics and paraffins.

After the low-temperature sucking out of the paraffin by porous particles down to a mass content of 2.5–3% by weight at ambient temperature, further removal of paraffin might be performed at an increased temperature when binder evaporation begins. In this case the products do not crack and shrink as could happen if attempts were undertaken to evaporate paraffin starting at a higher initial mass content. Such a change of regimes of liquid evacuation makes it possible to exploit more effectively liquid withdrawn and improve the quality of porous ceramics produced.

§7 Rate of Drying

Drying is a typical non-stationary process. Its rate, characterized by the derivative of mass content over time $\partial W/\partial t$, changes in the course of drying reflecting the change of the mechanisms of internal mass transfer. Usually the initial stage of drying of a porous body filled with liquid is characterized by an approximately constant (or slightly decreasing) rate. This later gives way to a period of decreasing rate of drying. The mass content corresponding to the transition to the second stage of drying is called the first critical point W_{k1}. In many cases additional characteristic periods of drying and critical points may be recognized [266].

The changes in the rate of drying are associated with peculiarities of the pore structures and sizes, lyophilicity of porous bodies. Shrinkage also considerably affects the rate of drying as will be discussed in the next chapter. Here we restrict ourselves to the analysis of the rates of drying of porous

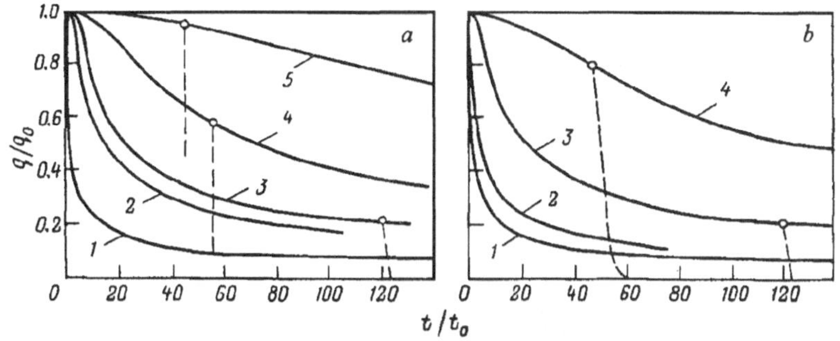

Figure 8.13 Dependence of the relative rate of drying, q/q_0, on dimensionless time of evaporation, t/t_0, for the first (solid lines) and the second (dashed lines) periods of drying. *a*) $r = 10^{-4}$ cm; only vapor diffusion (*1*); vapor diffusion and film flow: $\varphi_0 = 0.1$ (*2*); 0.5 (*3*); 0.9 (*4*) and 0.95 (*5*). *b*) $\varphi_0 = 0.5$; only vapor diffusion (*1*); vapor diffusion and film flow: $r = 10^{-3}$ cm (*2*); 10^{-4} cm (*3*); 10^{-5} cm (*4*).

bodies with a rigid skeleton held constant in the course of drying. It makes sense to calculate the rates of drying for various models of porous bodies and consider how the theoretical results agree with experimental observations.

The simplest model is a body with uniform pores represented by cylindrical capillaries of equal radius r. The difference from the case discussed for evaporation from single capillaries is that a diffusion zone of thickness x_0, depending on the conditions of convective gas mixing in the surrounding medium, forms over the porous body surface (Fig. 8.12).

In calculations of the rate of evaporation in the framework of this model [267] the films flowing along the pore walls and vapor diffusion were taken into account. The relevant equations for the combined transfer were presented in Chapter VII. A feature of these calculations was the role of the diffusion resistance of the layer x_0. Various values of air humidity, φ_0, in the surrounding media were specified and the boundary potentials φ_b (entering the equation for evaporation from capillaries) were calculated from the condition that the flux of liquid evaporating from the capillary, q, equals the flux into the surrounding medium;

$$q_0 = \mu p_s D(\varphi_b - \varphi_0)/x_0 RT. \qquad \text{(VIII.17)}$$

Figure 8.13 shows the results of numerical calculations of the rate of water evaporation into air at 300 K for the isotherms of disjoining pressure of wetting film $\Pi = A/h^3$ assuming $A = 5\times10^{-19}$ J, $x_0 = 0.2$ cm, $D = 0.26$ cm^2/s, and the porosity $n = 0.5$. The dimensionless time, t/t_0, is plotted on the x-axis where $t_0 = n^2 x_0/v_m q_0$, and the relative rate of drying, q/q_0, is plotted on the

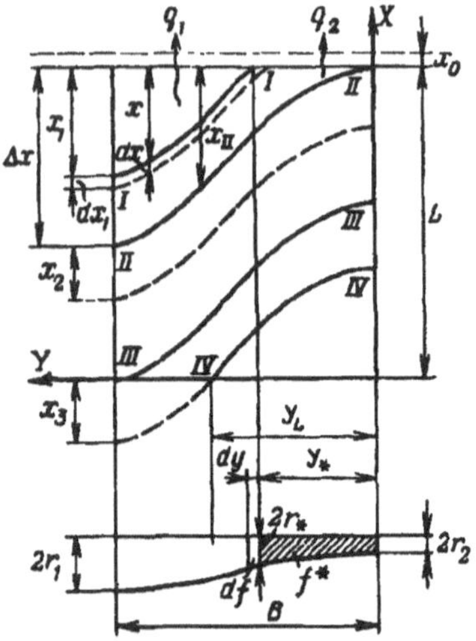

Figure 8.14 Slit model of a capillary-porous body.

y-axis, where q_0 is the rate of bulk water evaporation under the same conditions, determined from eq. (VIII.17) at $\varphi_b = 1$.

As seen from Fig. 8.13, the higher the relative air humidity, i.e. the greater the thickness of wetting films between the menisci and the capillary entrance, the longer the high relative rate of evaporation q/q_0 is maintained. Because the films actively supply the moisture to the capillary entrance, the evaporation proceeds as from an open water surface. The rate of evaporation falls more drastically with decreasing φ_0, receding in the limit to the curve *1* characterizing the evaporation from a hydrophobic capillary, where only vapor diffusion takes place. A decrease of capillary radius, increasing the input of the film flow, also promotes the maintenance of high q/q_0 values at a given value of φ_0 (Fig. 8.13, *b*).

When the menisci approach the bottom of the capillary (Fig. 8.12, *b*), a sharp decrease in the rate of evaporation takes place as shown in Fig. 8.13 by the dashed lines. The points marked on the graphs correspond to the first critical mass content. As seen from the above consideration, the transition from the first to the second regime of drying is associated with the change of the mechanism of internal mass transfer: the transition from the evaporation of capillary liquid (Fig. 8.12, *a*) to the evaporation of the retained film moisture (Fig. 8.12, *b*). The position of the first critical point depends on the linear size of the porous body L. At equal length L, the transition to the period of receding rate of drying occurs earlier, on the time scale t/t_0, at high φ_0 because in this case a higher relative rate of evaporation is maintained and the menisci approach the capillary bottom at an earlier time. The real time, when the first critical mass content is approached, is longer at high φ_0 due to the slow absolute rate of evaporation q_0 and, therefore, the characteristic time t_0 is longer.

The graphs of the rate of drying, presented in Fig. 8.13, are characteristic of many porous bodies, e.g. for sands and other single-granular materials. To see how a non-uniformity of pore size influences the rate of evaporation, consider two models. The first one represents the communicating pores of various size by a slit of variable cross-section [268]. Figure 8.14 shows the recession of the liquid front I–IV in the course of evaporation from a slit model characterized by minimal r_2 and maximal r_1 pore width. The distribution of pores by radii is selected to satisfy the exponential dependence of capillary pressure, P_c, on the mass content, known from experiments:

$$r = r_2 \exp[w \ln(r_1/r_2)], \tag{VIII.18}$$

where w is the volume fraction of liquid in a porous body corresponding to a given value of $P_c = \sigma/r$, and σ is the surface tension.

Initially the slit is assumed to be filled with liquid. If in the surrounding medium a relative air saturation $\varphi_0 < 1$ is maintained, then the evaporation begins leading to the recession of the liquid level in a wide part of the slit (curve *I*). Here the position of the meniscus front taking account of gravity may be expressed by the equation:

$$\rho g(x_1 - x) = (\sigma/r) - (\sigma/r_1), \tag{VIII.19}$$

where g is the acceleration due to gravity and ρ is the density of the liquid.

In further calculations we neglect, for simplicity, the effect of film flow and viscous losses in the liquid flow occurring in the filled part of the slit.

The position of the liquid level corresponding to the curve II is the limiting one. After this the front will shift down to the next position corresponding to the curve III. Thereafter the liquid remains only in the narrow part of the slit (curve IV) until complete evaporation has taken place. As seen, we can recognize here a series of critical points corresponding to changes in the evaporation regime. However, the transition from one regime to another goes smoothly according to the assumed smooth distribution of pores by size.

The rate of evaporation is calculated in the quasi-stationary approximation according to the equation of vapor diffusion written for local regions of the slit *df*. The vapor pressure over the menisci is set equal to p_s. Here the local rates of menisci displacement are:

$$v = p_s D(1 - \varphi_0)/\rho RT(x + x_0), \tag{VIII.20}$$

and the vapor flux from the slit is:

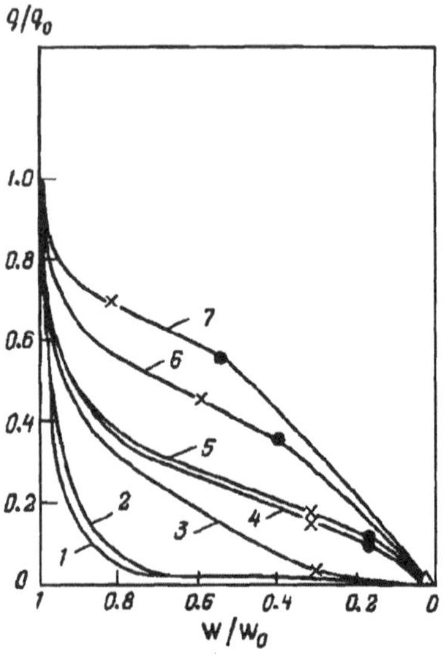

Figure 8.15 Dependences of q/q_0 on average mass content of the slit model of porous body, W/W_0, of various depth, L, and various parameters of distribution over pore size: $L = 16.9$ cm, $\alpha = 1$ (*1*); 2 (2); 10 (*3*); 100 (*4*); 1000 (*5*). $L = 15$ cm, $\alpha = 100$ (*6*); $L = 2$ cm, $\alpha = 100$ (*7*).

$$q = (\rho/F)\int_0^F v(x)df. \qquad \text{(VIII.21)}$$

Without going into detail of the rather complicated calculations performed for various front positions, let us consider the results obtained. They are presented in Fig. 8.15, where the values of relative rate of drying q/q_0 are plotted on the y-axis, where q_0 is determined by eq. (VIII.17); and the average values of mass content, related to the complete filling of the pore, W_0, are plotted on the x-axis. Curves *1*–*7* refer to different slit depth L and different pore-size distributions characterized by the average pore size and the ratio $\alpha = r_1/r_2$:

$$r_c = r_1 r_2 \ln(r_1/r_2)/(r_1 - r_2). \qquad \text{(VIII.22)}$$

The calculations were performed at $x_0 = 0.2$ cm and 293 K for $r_c = 10^{-2}$ cm and various α values characterizing the width of distribution. Black points on the curves denote the transition to the second period of drying characterized

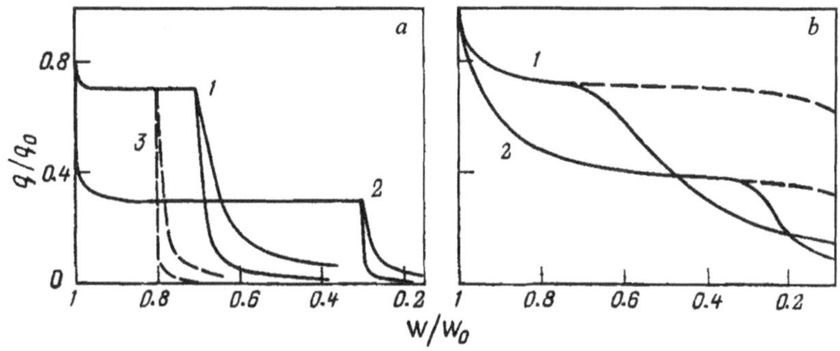

Figure 8.16 Dependences of q/q_0 on average mass content, W/W_0, for the two-capillary model of a porous body. *a)* L = 100 cm, fraction of narrow pores n_1 = 0.3 (*1*) and 0.7 (*2*); L = 200 cm, n_1 = 0.3 (*3*); *b)* L = 2 cm, n_1 = 0.3 (*1*) and 0.7 (*2*).

by the detachment of the menisci front from the surface of the sample (Fig. 8.14, curve II). These correspond to the first critical moisture content.

As seen from Fig. 8.15, an abrupt decrease in the rate of drying occurs for the slit of invariable width (curve *1*, α = 1) similar to vapor diffusion in cylindrical capillaries. The availability of communicating pores of different size ($\alpha > 1$) results in a higher rate of drying, because the front of liquid is kept closer to the sample surface in narrower pores. The higher the α values are, the higher the rate of drying.

The rate of evaporation is considerably greater in shorter samples due to the closer position of the evaporation front to the outer surface. Therefore, the wide distribution of pores by size allows a higher and less abruptly falling rate of evaporation to be maintained.

The second model of non-uniform porous medium has been already discussed in Chapter VII, §4. It is a two-capillary model, where the non-uniformity is exaggerated and reduced to the introduction of two sorts of capillaries of extremely different radii. As will be seen further, this type of porous structure enables an almost constant rate of evaporation in the first stage of drying to be maintained. The calculations of the rates of drying on the basis of this model were performed in the works [242, 243]. Both the evaporation from the narrow capillaries exposed to air and from the receding menisci front in wide capillaries were taken into account. Here the recession was caused by the liquid flow into narrow capillaries under the difference in the capillary pressure as well as by evaporation (to a less extent). Figure 8.16 shows the results of calculations of the dependences q/q_0 on W/W_0 for water evaporation, where we have assumed: $r_1 = 10^{-2}$ cm, $r_2 = 10^{-3}$ cm, $x_0 = 0.2$ cm and $T = 293$ K. At greater linear sample size L (Fig. 8.16, *a*) the rate of drying remains constant for a long time because the menisci front in narrow pores is retained at the sample surface for a long time. The recession of this front results in a dramatic decrease in the rate of evaporation. An increase in L shifts the first critical point towards lower values of W/W_0.

In the case of short samples (Fig. 8.16, *b*), the rates of evaporation change less dramatically. The account of the film flow (dashed curves) in all the cases sustains a higher rate of evaporation in the second stage of drying especially for small samples, where the distance to which the film flow extends is small.

The data considered in this section show that plots of the rate of drying, calculated for model porous bodies are close to that obtained in experiments with porous bodies not subjected to shrinkage. This leads to the conclusion that using one or another model of pore space, one can predict the behavior of porous bodies during isothermal drying. To calculate the rates of drying, no other characteristics are needed, apart from the physical properties of liquid, the isotherms of disjoining pressure of wetting films and the coefficients of vapor diffusion. The main difficulty arises in the selection of the structural model. Methods presented in Chapter 1, §3 may be used for this purpose.

The availability of fine communicating pores or wetting films, having thickness comparable with pore size, makes it possible to maintain a high rate of evaporation in the first stage of drying of the porous body. On the contrary, the hydrophobization of pore surface, although decreasing the binding energy of water, tends to impede evaporation. For example the method of hydrophobization of surface layers is widely used to retain soil moisture.

§8 Coefficients of Internal Mass Transfer

As shown in Chapters IV–VII, there is a variety of mechanisms of internal mass transfer. They comprise viscous flow and diffusion of vapor, viscous flow of capillary liquid and films. Since all these fluxes, under ordinary conditions of drying, depend linearly on the pressure gradients, they may be characterized, in general, by the unified coefficient of mass conductance K_m. Under isothermal conditions and in the absence of concentration gradients and electric field, we obtain from eq. (IV.1) for the one-dimensional case:

$$q = -K_m d\mu/dx, \text{ g/cm}^2\text{ s.} \tag{VIII.23}$$

In the non-stationary conditions, for a porous body incompletely filled with liquid, from the law of mass conservation of matter (see Chapter VI, §3) it follows:

$$-\gamma_0 \partial W(x)/\partial t = \partial q/\partial x, \tag{VIII.24}$$

where $W(x)$ is the local mass content of the porous body, in g/g, and γ_0 is the density of the dry porous body, in g/cm^3.

Having determined the values of derivative, $\partial q/\partial x$, from eq. (VIII.23), we come to the equation:

$$\gamma_0 \partial W/\partial t = \partial/\partial x(K_m \partial\mu/\partial x). \tag{VIII.25}$$

As shown in Chapter II, §4, the mass content of a porous body relates to the changes in chemical potential. Isotherms $W(\Delta\mu)$ have the form shown in Fig. 2.1. Here the lower branches of the isotherms should be used for the thermodynamic equilibrium states of a liquid in a porous body. Given this definition of the isotherm, the relation between the mass content and the chemical potential is quite unambiguous. The derivative $\partial W/\partial\mu$ characterizes, by the definition of A. Lykov [236, 266], the specific mass capacity of a porous body $C_m = \partial W/\partial\mu$, mol/J. This enables eq. (VIII.25) to be transformed as follows:

$$\partial W/\partial t = \partial/\partial x[K_m(\partial\mu/\partial W)(\partial W/\partial x)]/\gamma_0 = \partial/\partial x(a_m \partial W/\partial x), \tag{VIII.26}$$

where a_m is the generalized coefficient of mass transfer.

Equation (VIII.26) and eq. (VI.10) for the non-stationary diffusion are similar in form. However, in the former the mass content W is used instead of concentration of diffusing species and the driving force of the process is the gradient of mass content. Due to this, the coefficient a_m was called the coefficient of diffusion of liquid:

$$a_m = K_m/\gamma_0 C_m,\ \text{cm}^2/\text{s}. \tag{VIII.27}$$

Formally, it has the same dimensions as Fick's coefficients of diffusion. The advantage of the generalized approach, expressed by eq. (VIII.26), is that it enables the distribution of mass content across the depth of a porous layer at different times to be obtained; i.e. the function $W(x, t)$ may be found using different boundary conditions. This macroscopic approach enabled many important practical problems to be solved [236, 266, 269, 270]. However, all the complexity of the mechanism of internal mass transfer is embedded in the coefficient $a_m(W)$. The values of a_m in the framework of the macroscopic approach need to be experimentally determined. This, in a way, denudes the analytical approach of its capacity to predict, for $a_m(W)$ values may be obtained after the completing of drying, i.e. when you already know how it goes. Application of the function $a_m(W)$ obtained for other conditions of drying is not always possible, because the form of this function depends on temperature and external conditions of drying. For example the mechanism of vapor transfer changes in dependence of the Knudsen number for evaporation into a gas or into vacuum. Temperature variation changes the shape of the isotherms of disjoining pressure of wetting films $\Pi(h)_T$ and the viscosity of liquids in films and in narrow pores, where the viscosity differs from the bulk values.

The solution of eq. (VIII.26) with a variable coefficient of diffusion a_m may be obtained, for the general case, only by numerical methods. The analytical solution may be obtained only for the simplified dependences $a_m(W)$ not always approximating the real transfer properties of porous bodies.

In many cases, the values of $a_m(W)$ are considered to be independent of the mass content leading to a substantial simplification of eq. (VIII.26). However, this approximation may be only used when the change of mass content of the porous body is small during a given time interval. Often the constant a_m value is applied to particular separate layers of the porous body then joining the solutions obtained at the boundaries between layers by the condition of continuity of the mass flux.

Each approach, microscopic or macroscopic, has strong and weak features. The microscopic approach is focused on peculiarities of porous structure and uses the physically transparent approach, based on the equations of gas- and hydrodynamics, to describe the transfer processes. However, a solution for non-stationary problems is not always attainable. The macroscopic approach neglects the details of local mass transfer processes in a continuous medium characterizing all its peculiarities by the coefficient of diffusion $a_m(W)$. The macroscopic theory cannot specify the form of this function that is required to carry out independent experiments.

As will be shown below, there is a possibility of uniting the two approaches using their strong features. We can try to determine the coefficients $K_m(W)$, $C_m(W)$ and $a_m(W)$ for the model systems and then use the functions $a_m(W)$ obtained to find solutions of the diffusion equation of type (VIII.26). This allows the micro and macro approaches to be combined.

Let us consider, as an example, the simplest case of a porous body modelled by a set of cylindrical capillaries of different radii. Assume that the distribution of pores by radii is specified by some function $F(r)$. Let us neglect now, for simplicity, the effect of film flow. Then, depending on $\mu(x)$, some capillaries will be filled with liquid and others will be free for vapor diffusion in an elemental layer x of the porous body. The boundary value of capillary radius r_* can be easily determined from the Laplace equation for the capillary pressure P_c from eq. (II.1):

$$r_*(W) = -2\sigma v_m / \mu(W) = 2\sigma / P_c(W). \qquad \text{(VIII.28)}$$

When $r > r_*$, the pores are free of liquid, at $r < r_*$ they are filled with condensate. The mass content of liquid in a given layer, neglecting the vapor mass as compared to the liquid mass, may be found from the distribution function:

$$w = W/W_0 = \int_0^{r_*} F(r)dr, \tag{VIII.29}$$

where W_0 is the mass content at complete filling.

To determine K_m, let us use now eq. (VIII.23). The value of K_m is an additive function of the vapor flux, K_d, and that for the liquid, K_c. The vapor diffusion is determined by eq. (VI.9). Turning from the gradients of vapor pressure to those of chemical potential using eq. (II.1), we find:

$$K_d = [\rho v_m D/(RT)^2](1 - w)p(W), \text{ mol s/cm}^3, \tag{VIII.30}$$

where $(1 - w)$ characterizes the proportion of channels free from liquid and $p(W)$ is the vapor pressure that is determined from the isotherm $W(p/p_s)$.

The flux of condensate is described by eq. (IV.4) written for application to the set of capillaries $r < r_*$:

$$K_c = (\rho/8\eta v_m)\int_0^{r_*} r^2 F(r)dr, \text{ mol s/cm}^3, \tag{VIII.31}$$

where $F(r)dr$ characterizes the normalized fraction of area corresponding to the interval of radii dr.

It should be mentioned that all the information about the coefficients of mass conductance is embedded only in the distribution function of pores by radii $F(r)$. In addition, to calculate $K_m = K_d + K_c$, we require such physical characteristics as the viscosity, η, molar volume, v_m, density of liquid, ρ, and the coefficient of vapor diffusion D. In the framework of the adopted model for the porous body it is easy to obtain the isotherm $W(\mu)$ from the function $F(r)$. The values of w are determined by eq. (VIII.29) and the corresponding values of the chemical potential, μ, or P_c follow from eq. (VIII.28). This enables us to find, with a knowledge of $w(\mu)$, where $w = W/W_0$, the relevant dependence for $C_m = \partial W/\partial\mu$. From these there is no problem in calculating the required function $a_m = (W)$ by eq. (VIII.27).

Qualitatively, the form of the dependences of K_m, C_m and a_m on the pore filling w and on the isotherm $w(\mu)$ for the distribution function $F(r)$ with a single maximum (Fig. 2.13) is presented in Fig. 8.17. The values of C_m pass through a maximum in the region of mass content, where $F(r)$ goes through a

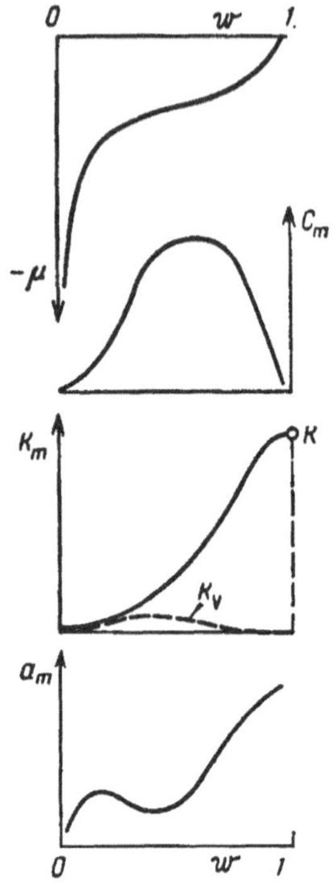

Figure 8.17 Qualitative dependences of transfer coefficients on mass content ω of a model porous body with structure specified by the distribution function $F(r)$.

maxima. In the same region the values of a_m are minimal. This is because at this mass content, high gradients of w correspond to low rates of change of μ, which control the intensity of mass transfer processes. To reconcile this physical fact with the formal description of mass transfer as a diffusion process under a gradient of mass content, the values of a_m should be decreased. In the region of low mass content, the values of a_m are the highest, because here, conversely, small gradients of w lead to high gradients of chemical potential.

Values of K_m only increase with w, because the values of K_c in eq. (VIII.31) grow as the condensate fills the pores. The values of K_d are usually comparable with those of K_c only in the region of low mass content and decrease with growing w, as follows from eq. (VIII.30). As shown by the dashed line, at some mass content, w, the values of K_d pass a maximum resulting from the competition of the growth of vapor pressure $p(W)$ and the decrease of $(1 - w)$ with growing mass content. The values of K_m at $w \to 1$ tend to the coefficient of filtration K in pores filled with liquid.

In experimental determination of the coefficients of mass conductance, stationary methods provide the most reliable results, when the measurements of K_m are performed at different mean mass contents of samples. For quartz sand with a grain size of 0.2–0.3 mm the measurements of K_m were carried out using a set-up allowing the moisture to be withdrawn from the sand to a particular mass content and then measure the flux through the sand applying modest diminution of capillary pressure [271]. Values of K_m in the range of mass content from saturation $W_0 = 23.2\%$ to $W = 3.25\%$ were obtained by this method. These values dramatically increased when $W_0 > 4.4\%$ due to the transition from a mainly film mechanism of transfer to the capillary process. However, even at $w = W/W_0 = 0.47$ the values of K_m comprised about 5% of the filtration coefficient.

The dependences in Fig. 8.17 qualitatively agree with those produced in experiments on various porous bodies. Using the functions of distribution of pores by size, found by the methods described in Chapter 1, §3, the relevant dependences of the diffusion coefficient a_m on mass content can be obtained. If the function $F(r)$ is analytically specified, the analytical dependences $a_m(W)$ can be obtained and then used to calculate the mass transfer on the basis of the diffusion equation (VIII.26).

The contribution of film flow along the surface of the capillaries $r > r_*$ presents no special problem. In this case one more component of the coefficient of mass transport should be only added:

$$K_f = (2\rho/3\eta v_m) \int_{r_*}^{\infty} [h^3(r)/r] F(r) dr, \tag{VIII.32}$$

where the film thickness h is determined from the isotherm $\Pi(h)$ using the value of the disjoining pressure $\Pi = \sigma/r$ (see Chapter II, §2).

Various models may be used to calculate the coefficients of mass transfer. For instance the expressions for the coefficients of transfer in the model of a porous body consisting of similar spherical particles were found in [272]. Another approach may be used for porous bodies with low mass content, when both vapor and adsorption layers control the mass transfer [205]. The essence of this approach is that there is a direct analogy between the known equation of adsorbing vapor diffusion (VI.12) and eq. (VIII.26). This allows the parameters usually used to solve the problems of adsorption kinetics [273] to be "translated" into the coefficients of the theory of mass transfer.

The mass content W can be expressed as the sum of the quantity that is adsorbed, a, and vapor concentration, C, in pores:

$$\gamma_0 W = a + C = f(C) + C, \tag{VIII.33}$$

where $a = f(C)$ is the adsorption isotherm. Having found from this the expression for $\partial W/\partial t$ and comparing eqs. (VIII.26) and (VIII.12) we obtain:

$$a_m = D_i/[1 + (\partial f(C)/\partial C)]. \tag{VIII.34}$$

For K_m and C_m we have, respectively, the following expressions:

$$K_m = D_i C/RT;\ C_m = (C/\gamma_0 RT)[1 + (\partial f(C)/\partial C)]. \tag{VIII.35}$$

However, in distinction from the usual expression for the coefficient of mass conductance caused only by vapor diffusion $K_d = D\mu p/(RT)^2$ (see eq. (VIII.30)), the coefficient D_i in eq. (VIII.35) includes the surface diffusion of the adsorbed species. Its value is equal to $D_i = D_* m$, where m is the porosity and the value of D_* is determined by eq. (VI.16). The values of K_m, C_m and a_m were calculated in [205] for three samples of silica gel (with mean pore radius r = 1.4 nm, 6 nm and 10 nm) at relative vapor pressures ranging from $p/p_s = 0.05$ to 0.912. For this purpose experimentally measured adsorption isotherms $a(C)$ and D_s(C) dependences were used. The variations of coefficients of mass transfer K_m and a_m with the mass content obtained are similar to those presented in Fig. 8.17. The values of coefficients increase with mean pore size being most pronounced in the region of low mass content. This takes place because, due to the small pore radii, the coefficients of the Knudsen diffusion were used in eq. (VI.16) which rise linearly with r, as seen from eq. (VI.7).

§9 Deep Drying

In drying of many porous polymer materials, medicines and food, the problem of intense (deep) drying arises, i.e. the necessity to remove liquids to a very low degree of residual mass content, less than 1% or even 0.1%.

The theory of deep drying may be built on the basis of eq. (VIII.12), because in this case a porous body contains only adsorption layers. An analytical solution for the rate of drying can be only obtained for the isotherms $a = f(C)$ approximated by two linear regions [274]. The equation of the isotherm can be written as follows:

$$a = a_1 = \beta_1 C;\ 0 \le C \le C_*$$

$$a = a_2 = \beta_1 C_* + \beta_2 (C - C_*);\ C \ge C_*. \tag{VIII.36}$$

Here β_1 and β_2 are the parameters of the isotherm: when $\beta_1 > \beta_2$ the isotherm is convex and when $\beta_1 < \beta_2$ it is concave.

The kinetics of desorption from porous half-infinite body for the isotherm of this type has been considered in [275]. When the coefficient of internal diffusion D_i = const the problem is expressed as follows:

$$(1 + \beta_1)\partial C_1/\partial t = D_i\partial^2 C_1/\partial x^2, \; 0 \le x \le l(t),$$

$$(1 + \beta_2)\partial C_2/\partial t = D_i\partial^2 C_1/\partial x^2, \; l(t) \le x \le \infty, \qquad \text{(VIII.37)}$$

$$C_1(0, t) = 0; \; C_2(x, 0) = C_0; \; a_2(x, 0) = a_0;$$

$$C_1(l, t) = C_2(l, t) = C_*; \; l(0) = 0;$$

$$(\partial C_1/\partial x)_{l(t)} = (\partial C_2/\partial x)_{l(t)}. \qquad \text{(VIII.38)}$$

Here the indices 1 and 2 refer to the concentrations C and a in the regions $0 \le C_1 \le C_*$ and $C_* \le C_2 \le C_0$, respectively; $l(t)$ is the movable boundary separating the regions 1 and 2, where $C = C_*$.

Using the results obtained in [275], we easily find the equation for the initial stage of deep drying for all types of isotherms and for an arbitrary form of porous bodies:

$$W(t)/W_0 = 1 - B\sqrt{t}. \qquad \text{(VIII.39)}$$

This corresponds to the rate of drying in the first stage equal to:

$$dW/dt = -W_0B/2\sqrt{t}. \qquad \text{(VIII.40)}$$

The constant B depends on D_i, parameters of the adsorption isotherm, and the particle sizes and shapes. For the isotherm (VIII.36) it appears to be equal to:

$$B = (2g_1/\sqrt{\pi})(C_*/a_0)[D_i(1 + \beta_1)]^{1/2}/\text{erf}\,[\xi_0\sqrt{1 + \beta_2}\,]. \qquad \text{(VIII.41)}$$

Here g_1 is the geometric factor, equal for spherical particles to $g_1 = 3/R_0$. The constant $\xi_0 = l(t)/2\sqrt{D_it}$ (where $l(t) \sim \sqrt{t}$) is determined from the transcendental equation:

$$\sqrt{(1 + \beta_2)}\,\text{erf}\,[\xi_0\sqrt{(1 + \beta_1)}\,][(C_0/C_*) - 1]\exp[-(1 + \beta_2)\xi_0^2]$$

$$= \sqrt{(1+\beta_1)}\,\mathrm{erfc}\,[\xi_0\sqrt{(1+\beta_2)}\,]\exp[-(1+\beta_1)\xi_0^2\,], \qquad \text{(VIII.42)}$$

which may be solved graphically or by the method of iterations.

The initial stage of drying described by eqs. (VIII.39)–(VIII.42) for concave and linear isotherms takes place at $W/W_0 \geq 0.5$. For convex isotherms the limits of applicability of the equations should be determined for each particular case.

In many cases further simplification of the formulae for deep drying is possible, for example in the case of the linear ($n = 1$) isotherm $a = \beta C$. For spherical particles:

$$B = 6\sqrt{D_i(1+\beta)}\,/\beta R_0\sqrt{\pi}, \qquad \text{(VIII.43)}$$

where D_i is determined by eq. (VI.16).

For a linear isotherm, full solutions up to the finite mass content $W_f = 0$, (similar to those of thermal conductivity) are known [276]. For instance the times of drying of spherical grains from W_0 to $W_f = 0.1\,W_0$ and to $W_f = 0.05\,W_0$ are, respectively:

$$t_0 \approx 0.182\,R_0^2\,a_0/C_0D_i;\; t_0 \approx 0.253\,R_0^2\,a_0/C_0D_i. \qquad \text{(VIII.44)}$$

The time of drying may also be estimated for concave isotherms that are close to the extremely concave one ($a = 0$, $C < C_0$; $a = a_0$, $C = C_0$) for which the full solution of the desorption problem is known [275].

All the above-presented solutions refer to the case when the surrounding media do not contain vapors of the substance to be removed and when the limiting stage is the internal transfer of mass. However, for linear and extremely concave isotherms the relevant forms of the equations may be obtained even at some finite rate of external mass transfer [275].

Analysis of the formulae obtained shows that three parameters mainly affect the kinetics of deep drying, they are: i) size of porous particles being dried, R_0; ii) the strength of binding of adsorbed molecules characterized by β; iii) the coefficient of internal diffusion, D_i, that depends on porosity, pore size, temperature and the contribution of surface diffusion. The ways of intensifying deep drying are mainly centred upon an increase of D_i values, because there is usually a technological limit to particle size reduction. There is also the possibility of decreasing β by surface modification of a porous body by physical or chemical methods. Unfortunately, there are few experimental data capable of examining the theory developed. Deep drying covers the region of low initial mass content when only the adsorbed substance is present in the porous body. Moreover, investigations in which adsorption isotherms and data on pore size distribution are present together with the graphs of the rate of drying and surface mobility of adsorbed molecules, are rare. First of all the theory should be tried on porous bodies with rigid skeleton, i.e. with a structure that does not change in the course of drying, and with a rather narrow distribution of pore radii. Spherical granules of silica gels of different specific surface area represent a good model. Observations of moisture desorption from such systems combined with sorption–desorption isotherms were completed by V. Dushchenko *et al.* [277].

A comparison of the theory of deep drying discussed above with the results of experiments was undertaken as follows. Values of the constant β were determined from the data on the rate of adsorption or desorption. Then, in the case of desorption, the values of D_i were calculated by eq. (VIII.41). In the case of adsorption the expression for the values of D_i was obtained in [278]:

$$D_i = [\pi/36(1 + \beta_2)]\{BR_0 \mathrm{erf}\,[\xi_0\sqrt{(1+\beta_2)}\,]/[1 - (C_*/C_0)](C_0/a_0)\}^2. \quad \text{(VIII.45)}$$

For a linear isotherm this formula simplifies and takes the form:

$$D_i = \pi B^2 R_0^2\, \beta/36. \quad \text{(VIII.46)}$$

During desorption, the values of a_0 and C_0 refer to the initial state of the system and are calculated from the initial moisture content of the sample W_0 and from the relative vapor pressure p/p_s corresponding to this value of W_0 on the isotherm. Then assuming that the pores are small ($r \le 10^{-5}$ cm) and the mass transfer in the gas phase proceeds as Knudsen diffusion, the mean radius of cylindrical pores was determined from eq. (VI.7). Surface diffusion was neglected here; it follows from [205] that this contribution should be taken into account only in narrower pores.

Pore sizes obtained in this way were compared to the values of r_K from the Kelvin equation obtained independently for the same samples and to the values of the average hydraulic pore radius $r_0 = 2m/S$, where S is the specific surface area and m is the porosity.

In experiments carried out in [277], two samples of silica gel were used: KSK-2 with mean pore radius $r_K = 4.0$ nm ($r_0 = 5$–6 nm) and KSM-5 with $r_K = 1.3$ nm ($r_0 = 1.1$–1.2 nm). Data on the kinetics of vapor adsorption from a saturated atmosphere were processed and gave for KSK-2: $B = 4\times10^{-3}\,\mathrm{s}^{-1/2}$ at $T = 293$ K and $R_0 = 0.2$ cm. On the basis of the adsorption isotherm taken from the same work (approximation by two linear regions) we obtain $\beta_1 = 9.8\times10^3$; $\beta_2 = 5.6\times10^5$; $C_*/C_0 = 0.9$. The solution of eq. (VIII.42) gives $\mathrm{erf}^2\,[\xi_0\sqrt{(1+\beta_2)}) = 0.35$. Then we find from eq. (VIII.45) $D_i = 0.0145\ \mathrm{cm^2/s}$.

According to eq. (VI.7), this value of D_i corresponds to the mean pore radius $r = 3.7$ nm close to r_K and r_0 presented above for KSK-2. The results of these calculations are given in detail in [274].

We present here only the results of calculations from the data [277] on the kinetics of water adsorption and desorption at a step-wise gradual growth or decrease in p/p_s values in the medium surrounding the granule of silica gel KSK-2. As the kinetic curves were obtained in narrow regions of p/p_s variations, each region of the isotherm corresponding to this interval may be reasonably taken as linear. It enabled the use of the simpler eq. (VIII.46) to calculate D_i. The values of B at desorption were found from the equation:

$$B = (W_0 - W)/(W_0 - W_p)\sqrt{t}, \qquad \text{(VIII.47)}$$

and during adsorption from the equation:

$$B = (W - W_0)/(W_p - W_0)\sqrt{t}, \qquad \text{(VIII.48)}$$

where W_p is the equilibrium mass content corresponding to the given value of p/p_s in the surrounding media.

The values of B were determined from linear regions of the graphs of drying rates plotted in coordinates $W\sqrt{t}$. The values of D_i were determined by eq. (VIII.46) using as β the slope of the isotherm within this p/p_s range. The granule radius was $R_0 = 2$ mm. Then, setting $D_i = mD_k$, the pore radius was found from eq. (VI.7). The results of the calculations give values of r close to those of r_K and r_0 for KSK-2. The mean value of pore radius was 6.0 ± 1.7 nm close to $r_0 = 5\text{–}6$ nm and $r_K = 4$ nm.

Equation (VIII.39) allowed the process of deep drying of powders of polymer particles containing the residual solvent to be described qualitatively [279]. The kinetics of drying in vacuum or in a helium atmosphere were investigated by mass spectroscopy determining the amount of evaporated water and benzene by the peaks $m/e = 17$ for water and $m/e = 43$ for benzene. The calculations led to the conclusion that at mass content $W > 1\%$ Knudsen diffusion proceeds through a set of micropores. At lower mass contents, the mechanism of transfer changes. When $W < 0.5\text{–}1\%$, the rate of drying decreases being determined by the diffusion of solvent molecules through the solid polymer matrix. The corresponding D_i values are about $10^{-13}\text{–}10^{-12}$ m^2/s that is close to the values of the diffusion coefficient in solids.

The theory of deep drying of porous particles developed, using the isotherms of adsorption as a basis for calculations, agrees with the reported experimental data. The theory is recommended for use in calculating the rate of drying and to identify the mechanism of internal mass transfer in the course of deep drying of various materials with fine pores.

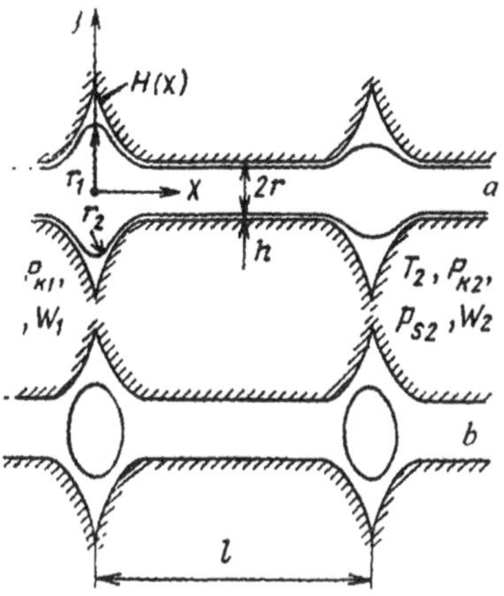

Figure 8.18 Mass transfer in a porous body with simultaneous effects of temperature and pressure gradients at low (*a*) and high (*b*) mass content.

§10 Thermo-transfer of Moisture

Mass transfer in unsaturated porous bodies driven by a temperature gradient is called the thermo-transfer of moisture. The vapor diffusion under the action of the temperature gradient is described by eq. (VI.17) and the flow of the capillary liquid and wetting films by eq. (V.12) and (V.13). Using these equations, let us find the thermo-gradient coefficient:

$$\delta_T = C_m\,(\partial\mu/\partial T)_W, \tag{VIII.49}$$

entering into the equation for the mass thermo-transfer:

$$q = -K_m\,(\partial\mu/\partial x - \delta_T \partial T/\partial x). \tag{VIII.50}$$

Let us use the model of an element of porous structure presented in Fig. 8.18. The thermal mobility of the liquid characterized by the coefficient $\delta_T = (\partial W/\partial\mu)_T(\partial\mu/\partial T)_W \approx (\partial W/\partial T)_{T,W}$ will be a maximum at some intermediate degree of saturation of pores with liquid (Fig. 8.18, *a*). At a very low mass content, the changes of W are small and $\delta \to 0$, that also follows from the fact that $C_m \to 0$ (Fig. 8.17). When pore filling is close to complete (Fig. 8.18, *b*), a small deformation of gas bubbles in pore widenings may occur without a significant change in mass content, as shown at the end of §1, Chapter VII. Due to this, also $\delta \to 0$. Therefore, at some mass content the values of δ_T, as those of C_m, should pass through a maximum.

Let us consider the mass fluxes in the case of incomplete filling of pores with liquid presented in Fig. 8.18, *a*. The shape of widening $H(x)$ may be selected to satisfy a real dependence between the mass content W and capillary pressure $P_c = \sigma(1/r_1 + 1/r_2)$, where r_1 and r_2 are the main radii of curvature of the meniscus surface. This will determine the dependence of the coefficient C_m on W which will be also specified by the profile $H(x)$. This approach enables the structures of different porous bodies to be modelled.

The capillary liquid is separated by film regions, and the film thickness is related to the capillary pressure by eq. (II.11). For simplicity the shape of the pore between the widenings is assumed to be cylindrical of radii $r >> h$.

When a constant gradient of temperature ∇T is imposed, vapor diffusion, the thermo-meniscus and thermo-capillary flows to the cold part of porous body will occur resulting in an local increase in mass content in this part. In an isolated system this will cause the reverse flux of vapor and film liquid and, when the fluxes compensate each other, a steady state will be established. This state may be described by eq. (VII.9) adopting the net flux $Q = 0$. Retaining only the main terms and turning from the gradients of vapor pressure, ∇p, to those of capillary pressure $\nabla P = -(RT/v_m p)\nabla p$, we get:

$$(\nabla\mu/\nabla T)_W = -v_m \nabla P/\nabla T = h^2(\partial\sigma/\partial T)/[(2h^3/3v_m) - (r\eta D p v_m/(R^2T^2)]. \quad \text{(VIII.51)}$$

The values of $\nabla\mu/\nabla T < 0$, for $\partial\sigma/\partial T < 0$. Therefore, the values of $\delta_T = C_m\nabla\mu/\nabla T = \nabla W/\nabla T$ are also negative following from the fact that an increase in mass content ($\nabla W > 0$) corresponds to a temperature decrease ($\nabla T < 0$).

Analysis of eq. (VIII.51) shows that the values of $\nabla\mu/\nabla T$ tend to 0 both at $h \to 0$ and at $h \to \infty$. Thus, the contribution of the film flow may increase the values of δ_T in the intermediate region of mass content. The growth of film thickness in this region of mass content increases the thermal mobility of moisture thereby explaining the increase in δ_T values.

The model shown in Fig. 8.18 may be also used to calculate the coefficients of specific mass capacity C_m that enables the theoretical dependences of thermo-gradient coefficient δ_T on mass content proceeding from eqs. (VIII.49) and (VIII.50) to be explained.

When $p/p_s < 0.9$, the effect of film transfer may be neglected. The main role here is vapor transfer. The stationary state corresponded in this case to the balance of vapor pressure over menisci $p_1 = p_2$, where:

$$p_1 = p_{s1}\exp(-v_m P_{c1}/RT),$$

$$p_2 = p_{s2}\exp(-v_m P_{c2}/RT). \quad \text{(VIII.52)}$$

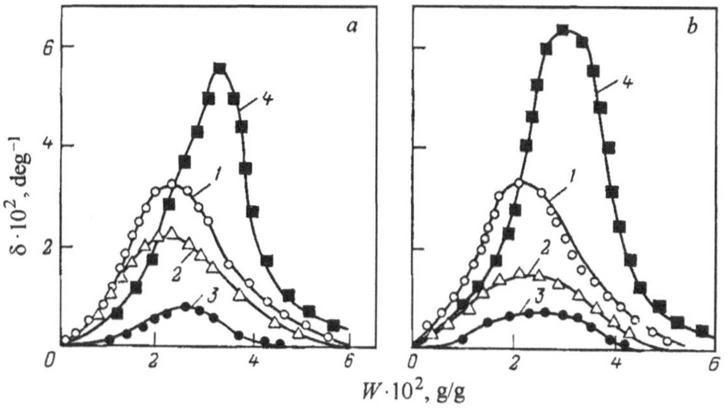

Figure 8.19 Thermo-gradient coefficient δ vs. mass content W of quartz sand and electrolyte concentration: *a*) NaCl, $C \leq 10^{-2}$ M (*1*); 10^{-1} M (2); 1 M (*3*); NaOH, C = 10^{-1} M (*4*); *b*) KCl, C ≤ 10^{-2} M (*1*); 10^{-1} M (2); 1 M (*3*); KOH, C = 10^{-1} M (*4*).

The increased pressure of saturated vapor $p_{s1} > p_{s2}$ when $T_1 > T_2$ is compensated by the corresponding change of menisci curvature $P_{c1} > P_{c2}$. From the condition $p_1 = p_2$ and the Clausius–Clapeyron equation we obtain:

$$\nabla P/\nabla T = \lambda / v_m T;\ \nabla \mu / \nabla T = -\lambda / T, \qquad \text{(VIII.53)}$$

where λ is the heat of evaporation of liquid, J/mol. As seen from eq. (VIII.53), the values of $\nabla\mu/\nabla T$ are practically constant. The form of the curve $\delta(W)$ in this case is mainly determined by the changes in C_m. As seen from Fig. 8.17, the values of C_m as a function of mass content pass through a maximum providing the same behavior of δ_T. The form of the function $\delta_T(W)$ is chiefly determined by the values of $C_m(W)$ depending on the peculiarities of the porous structure, the pore distribution by size, as has been shown in §§8 and 9. Only in the region where the contribution of film flow is substantial ($p/p_s > 0.9$) we may expect some increase in δ_T values compared to the pure vapor transfer.

Experiments with various porous bodies confirm that the thermo-gradient coefficients pass a maximum [266, 280–282]. The dependence $\delta_T(W)$ is obtained experimentally from determinations of the distribution of mass content along the porous body placed in a thermal and moisture isolated container by analyses of layer by layer separated samples (or by γ-rays). Here a constant temperature gradient ∇T is maintained at the ends of the sample. The sample is kept under these conditions during a time sufficient to establish a stationary distribution of mass content and temperature. From the dependences $W(x)$ and $T(x)$ the local values $\delta_T(W) = \partial W/\partial T$ are found by graphic differentiation.

The dependences $\delta_T(W)$ obtained for quartz sand with a particle size 0.1–0.25 mm [282], are shown as an example in Fig. 8.19. The sand was washed before the experiments with 0.1 N HCl and then with distilled water to neutral values of pH. The experiments were reproduced three times at $\nabla T =$ 10 deg/cm. The sand was moistened to the initial mass content $W_b = 2.5\times10^{-2}$ g/g, by spraying followed by mixing.

As seen from the comparison of curves *1–3*, an increase in electrolyte concentration results in a diminution of the thickness of wetting films which tends to decrease the values of thermo-gradient coefficient. This effect is partially associated with the fact that the vapor transfer gives rise to a gradient of solution concentration. As a result, the steady state establishes at a smaller mass content drop because the vapor pressure levels off not only due to the decrease in menisci curvature but to that of solution concentration in the cold part of the samples.

In the case of solutions of bases the thickness of wetting films increases due to the growth of electrical potential on the quartz surface. This results in noticeable growth of the values of δ_T (Fig. 8.19, curves *4*). In contrast, use of HCl solutions approaching the isoelectric state of quartz, reduces the maximum values of δ_T roughly by three times (at pH $\leq$ 3). Addition of anionic surfactants, increasing the charge of quartz and film surface that makes the values of Π_e rise, as well as polymers hydrophilizing the quartz surface, lead to a similar but stronger effect [282]. The increasing contribution of film flow shifts the maximum values of δ_T to higher mass contents.

In a general case, the maxima of C_m and $\nabla\mu/\nabla T$ values may diverge. The film flow may increase the values of δ_T only in the region of mass content corresponding to thick films.

In soils the contribution of film transfer is usually small. The velocity of thermal moisture transfer is, however, always higher than the values calculated according to the ordinary equation of vapor diffusion. This is associated with a combined vapor–liquid mechanism of transfer discussed in Chapter VII, §3 (Fig. 7.12). The contribution of film flow to the thermo-transfer in soils, according to the data reported in [281], increases from 5 to 20% with moisture content. Interaction of fluxes under the gradients of pressure and temperature plays a prime role in soils, where the temperature in upper layers changes in dependence on the time of the day and weather.

In more hydrophilic turf samples, the dominant mechanism of thermo-transfer at $W > 1.5$ g/g becomes the flow in the liquid phase [280]. Here the flow may occur not only in films but in thin pores filled with water perforating aggregates of particles. In the latter case the flow is caused by the change in capillary pressure of menisci under the gradient of surface tension:

$$\partial\sigma/\partial x = (\partial\sigma/\partial T)\nabla T. \qquad \text{(VIII.54)}$$

As seen from eq. (VIII.50), the rate of the total transfer depends on the magnitude and sign of thermo-gradient coefficient. The isothermal and non-isothermal terms of eq. (VIII.50) may have different signs. For example, during external heating of a porous body the thermo-flux is directed inside the sample and impedes the rate of drying. Because of this, the internal heating of samples is often used to accelerate the drying, for example by ultra-high frequency heating.

Chapter IX

Shrinkage of Non-rigid Porous Bodies During Drying

Deformable non-rigid porous bodies consisting of mobile particles interacting with each other through surface forces may restructure in the course of the processes of mass transfer. The restructuring shows up as changes in the relative dislocation of particles and in the number of contacts between them (aggregation, disaggregation). The mean distances between particles change with the thickness of separating liquid interlayers. The most common effects are swelling and shrinkage. These processes are accompanied by changes in pore sizes that makes the rate and the mechanism of the processes of mass transfer also dependent on the structural changes. The coefficients of transfer change correspondingly depending not only on mass content but on the changing structure of the porous body. All these variables considerably complicate the calculations. The theory of the processes of mass transfer must be supplemented with information about the simultaneously occurring restructuring.

One of the reasons for the restructuring are changes in capillary and disjoining pressure of thin liquid layers. It is necessary to take into account the displacement and repacking of particles of a disperse phase in order to perform the relevant calculations. The difficulties arising stem from the fact that usually the particles of a porous body do not occupy positions corresponding to the minimum of free energy of the disperse system. When a porous body is being formed, the defects of packing usually arise due to low mobility of particles, when only the condition of local mechanical equilibrium is obeyed. Closer packing, similar to that of liquids or crystals, would correspond to the thermodynamic equilibrium. Regular colloid structures and disperse bodies with regular packing may be obtained under the action of vibration and in the case of very slow formation of the structure [283]. For example in the course of processes extending to geological time scales, sedimentary rocks are formed with regular packing of plate particles and various minerals. However, many porous bodies met in practice have voids, i.e. vacancy defects of structure. When shear or compression deformation is applied to such a structure the voids may be partly occupied and the numbers of defects decrease. Rapid deformation may produce new defects.

As shown in Chapter I, the mean distances between particles are determined by potential curves of pair interaction (Fig. 1.4). These may have two minima, one of which, representing closer spacing, is the deeper and corresponds to the state of thermodynamic equilibrium. It may be so deep that the thermal energy of particles is insufficient to overcome the potential barrier and to transfer the particle to the far potential well. When the structure of a porous body forms too quickly, even in the absence of vacancy defects, some particles may reside in the outer potential well that gives rise to the looser and less strongly bound packing consisting of particle aggregates or regions with weakened bonds between particles. Deformation of such systems is accompanied by changes in the sizes and numbers of aggregates. Slow deformation should mainly cause transitions to the lower energy level of the system, because the imposed stresses allow the potential barrier dividing the regions of near and far potential wells to be overcome. This should result in compression of the system and the formation of a network of narrower pores. Rapid deformation may produce local fractures and make the structures loose and disordered [284, 285].

The greater the forces interacting between particles and the deeper the near potential minimum, the more difficult is to organize an equilibrium structure of a porous body. For example, when sediments of strongly interacting particles are forming, their structure is looser and more defective. When the attraction is weak and the repulsive forces prevail, gravity and the Brownian motion of particles provide a regular periodic structure of a sediment.

Microscopic investigations of ordered latex dispersions showed that macroscopically uniform dispersions are not uniform on a microscopic scale [286]. Local ordered structures were detected with local regions of higher density of packing being separated by regions with lower density. Depending on the volume fraction of the particles, φ, gaseous, liquid, crystal and glass-like states of disperse systems may be established [287–289].

§1 Shrinkage of a Periodic Colloid Structure

The calculations presented here are restricted to the case when shear stresses do not disturb the structure of the dispersion. This simplification may be valid

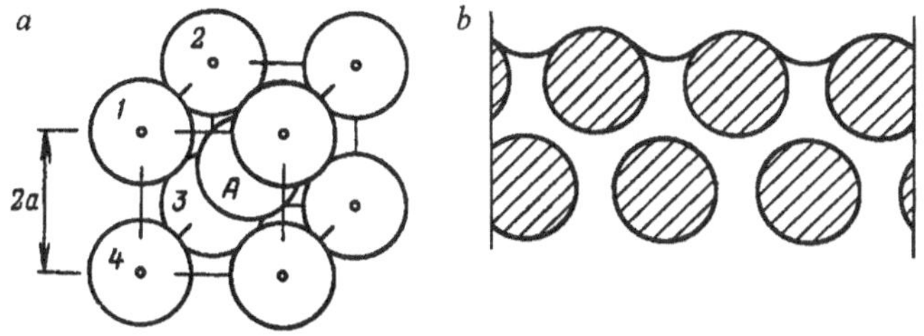

Figure 9.1 Model of a porous body consisting from spherical particles located in the sites of cubic lattice (*a*) and the surface structure of a wetted dispersion (*b*).

at relatively low rates of deformation. If the latter increase, a stratification of dispersion may occur accompanied by formation of a structurally disturbed weakened zone, where high rates of deformation develop [284, 285, 290, 291]. The shear flow may also cause changes in the shape and size of particle aggregates and changes in the conditions for coagulation of the dispersion [292, 293].

The case of regular structures spontaneously formed when a liquid is partly removed under the action of growing capillary pressure, can be easily theoretically analysed. To calculate the process of shrinkage, i.e. the change in volume occupied by a porous structure, it is enough to know how the forces of pair interaction between the particles depend on the distances between them.

Let us consider, for instance, a model of a porous body consisting of identical spherical particles of radius, r, forming a cubic lattice. The elemental cell of such a lattice is shown in Fig. 9.1, *a*, and a cross-section plane in Fig. 9.1, *b*. It is assumed that there are n particles per unit volume and the distances between the lattice sites are denoted as $2a$.

The force of pair interaction between particles is determined by the Derjaquin equation (I.18):

$$f(H) = \pi r G(h), \qquad \text{(IX.1)}$$

where H is the shortest distance between the particle surfaces. Let us calculate the energy of interaction $G(h)$ of plane surfaces having the same properties as the particles. We will restrict ourselves to consideration of the forces of electrostatic repulsion and molecular forces of attraction using for this purpose eqs. (I.3) and (I.11). Let us use eq. (I.2) to turn from $\Pi(h)$ to $G(h)$.

The calculated dependence $f(H)$ for quartz particles of radius $r = 0.7$ μm (curve *1*) and 0.6 μm (curve 2) in aqueous solution of a 1–1 electrolyte of concentration $C = 10^{-4}$ mol/l is shown in Fig. 9.2 [294]. The potential of the

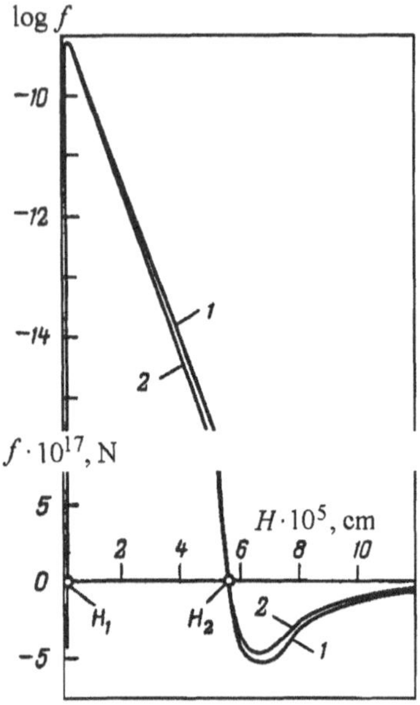

Figure 9.2 Dependence of interaction force $f(H)$ on the distance H between spherical particles of $r = 0.7$ μm (curve *1*) and 0.6 μm (curve *2*) in an aqueous electrolyte solution.

particle surface in this solution is taken to be $\psi = -150$ mV. The calculations of the energy of electrostatic interaction $G_e(h)$ used eq. (I.11) and those of molecular interaction $G_m(h)$ used eq. (I.3) assuming $A = -7\times10^{-21}$ J. Due to the high force barrier (about 10^{-9} N), the function $f(H)$ is plotted using two different scales: semi-logarithmic when $H < 5\times10^{-5}$ cm and linear when $H > 5\times10^{-5}$ cm for the region of the far potential well.

The distance $H_2 = 5.7\times10^{-5}$ cm corresponds to the location of particles in the far potential well. The depth of the well is small ($\sim10^{-2}$ kT) and the Brownian motion of particles in an infinite dispersion medium should destroy the lattice, transforming the dispersion system into a liquid or gaseous state. The high potential barrier hinders the transfer of particles to the deep near minimum corresponding to the distance $H_1 = 2.7\times10^{-7}$ cm. Naturally, this situation refers to the conditions selected: there is a high surface charge of particles and low solution concentration, when the electrostatic repulsive forces are great and long-range. However, the force–distance curves may be calculated and analysed for any particular case. For particle packing in the far potential well, shown in Fig. 9.1, *a*, when $H_2 = 5.7\times10^{-5}$ cm, the lattice spacing a_2 is equal to 2×10^{-4} cm at $r = 7\times10^{-5}$ cm, and $a_2 = 1.8\times10^{-4}$ cm at $r = 6\times10^{-5}$ cm. The displacement to each side from the central position of the particle A (Fig. 9.1, a) is associated with the work necessary to overcome the repulsive forces increasing with any displacement. This results in particle fixation in the lattice at $a < a_2$ and their participation only in vibrational thermal motion.

The pressure on the walls of a container bounding this ordered system is $P = F/S$, where $F = 4f/\sqrt{3}$ is the force of repulsion of the particle A from the particles 1–4, and $S = 4a^2$ is the area to which the force is related.

In non-rigid porous bodies the capillary pressure, P_c, conserves the stability of a system of repelling particles. It arises on the external surface of the sample provided by the concave menisci (Fig. 9.1, b). A simple relation may be approached between the volume of the system and its capillary pressure, assuming the process of restructuring is quasi-stationary, i.e. the condition $P = f/a^2\sqrt{3} = P_c$ is obeyed at each instant of time. This is possible at a low rate of drying when the particles have time to move to the position corresponding to a decreasing lattice spacing a, provided the mutual orientation of particles is conserved being specified by the same cubic lattice. The velocity of particle displacement is controlled by the viscous resistance of interlayers of the dispersion system. At a given lattice parameter a, the maximum capillary pressure is roughly equal to $P_m \cong 2\sigma/a$. When $P > P_m$, the dehydration of the surface layer will commence accompanied by the meniscus receding inside the porous body. This will take place if the rate of evacuation of the liquid mass due to evaporation exceeds the rate of its inflow due to the sample compression and the extrusion of liquid from the space between particles.

The rate of the linear shrinkage may be roughly estimated using the expression for the rate of particle approach along the lines of their centers v under the external force f_b [295]:

$$v = da/dt = 2H(f_b - f)/3\sqrt{3}\eta r^2, \qquad \text{(IX.2)}$$

where $H = (a\sqrt{3} - 2r)$ is the distance between the surfaces of particles and η is the viscosity of interlayers.

The force f_b should here exceed the force of particle repulsion f. At $f_b = f$, the system is in the state of mechanical equilibrium. When $f_b < f$, the liquid will be imbibed by the porous body that occurs on swelling. When $f_b > f$ shrinkage takes place: the lattice constant, a, decreases with velocity determined by eq. (IX.2).

To each value of a corresponds some equilibrium capillary pressure P_c that determines the mass content of the system $W = [(3a/\pi r)^2 - 1]\rho/\rho_s$, where ρ and ρ_s are the densities of the liquid and solid phase. The dependence $P_c(W)$ calculated for this particular case is shown by curve *1* in Fig. 9.3. In a system subject to shrinkage, the capillary pressure grows with decreasing mass

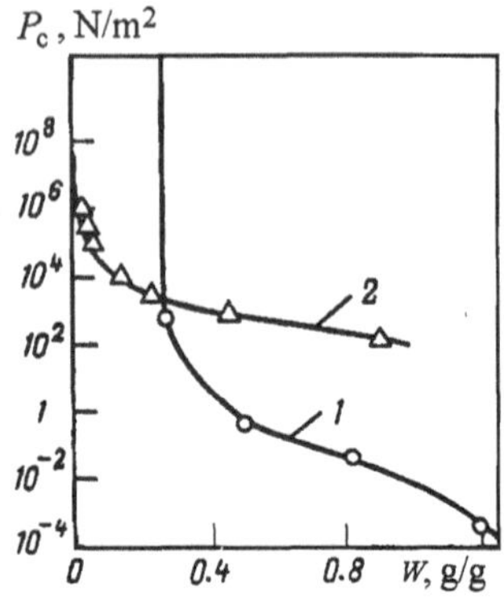

Figure 9.3 Theoretical dependences of the capillary pressure P_c on the relative mass content W of a deformable system.

content as in porous bodies with a rigid skeleton. However, contrary to the rigid porous structures, the mass constant changes here only due to the decreasing distances between particles. The pores, the spacings between particles, remain completely filled with liquid.

The data presented in Fig. 9.3 show that the equilibrium capillary pressure in the system under consideration is much less the maximal $P_m = 2\sigma/a$ corresponding to the limiting curvature of the menisci. Since the values of a vary insignificantly (from 1.08×10^{-4} at $W = 1.2$ g/g to 0.81×10^{-4} cm at particle contact), the values of P_m change correspondingly within narrow limits from 1.3×10^5 to 1.73×10^5 N/m². This means that the recession of the menisci front may start here only after the structure becomes formed from the particles being in close contact ($H = 0$). This structure is rigid and non-deformable because the particles are now fixed in the first deep potential minimum. The dependence P_c vs. W becomes sharper here because the capillary bridges begin to dehydrate as shown in Fig. 2.15.

The calculations show that even assuming the viscosity in eq. (IX.2) to be doubled, the viscous resistance to the particle approach is not the limiting stage controlling the system relaxation to equilibrium. If the value of f_b is of the same order as f, the velocity of the particle approach is between 10^{-2} and 10^{-1} cm/s. To provide that the system preserves the two-phase state in the course of shrinkage, the rates of evaporation should be kept low. The rate of dehydration should not exceed the velocity of the particle approach v.

During fast drying, when a limiting capillary pressure $P_c = P_m$ is established on the external surface, the rate of viscous deformation of the system also grows; the values of v are of the order of ≥ 1 cm/s. Here the controlling stage is not the process of particle approach but the viscous resistance to the liquid flow along the lattice of particles to the evaporating

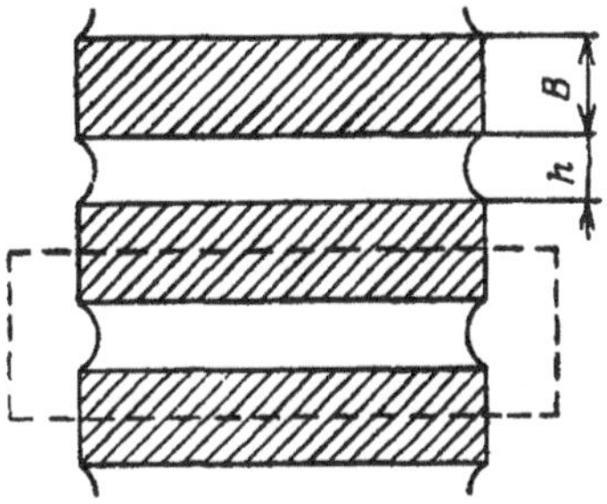

Figure 9.4 Model of deformable porous body formed from plane particles.

surface. Simultaneously a gradient of internal pressure develops leading to a non-uniform shrinkage in the direction of mass transfer. When the linear size of the porous body is great, this effect has been experimentally observed. The surface layers undergo a considerable, often ultimate, shrinkage and are no longer deformable.

In application to clay rocks built from thin plate particles, the calculations of swelling and compression, also in the course of drying, are performed directly in the framework of the model of plane interlayers [296]. This model allows the consideration of shrinkage in such porous bodies to be simplified and makes it more illustrative. An elemental layer of the porous body with the mass content $W = h\rho/B\rho_s$, where B is the thickness of the solid layer, is shown in Fig. 9.4 by the dashed line.

The disjoining pressure acting in a liquid interlayer of thickness h (gravity is neglected) is balanced by the capillary pressure of the menisci P_c. Taking into account, as a first approximation, only the electrostatic repulsion expressed by eq. (I.13) (when $B >> h$), we get the following relation between the mass content W and the capillary pressure:

$$P_c = (\pi\varepsilon_0/2)(\rho kT/ezWB\rho_s)^2. \quad \text{(IX.3)}$$

The results of calculations, in which it has been assumed that $\rho = 1$ g/cm^3; $\rho_s = 2.2$ g/cm^3; $B = 10^{-5}$ cm, $z = 1$, $\varepsilon_0 = 80$, are presented in Fig. 9.3 by curve *2*. Because the pores are filled with liquid, this dependence characterizes the equilibrium states of the porous body: changes in its volume with alternating capillary pressure P_c. In a mild regime of drying, when the evaporation goes from the outer surface of the sample, the quasi-equilibrium deformation of porous body takes place according to eq. (IX.3). During a strong regime of drying more liquid can be evaporated rather than extruded by capillary pressure. Here the front of the menisci will recede inside the sample, the

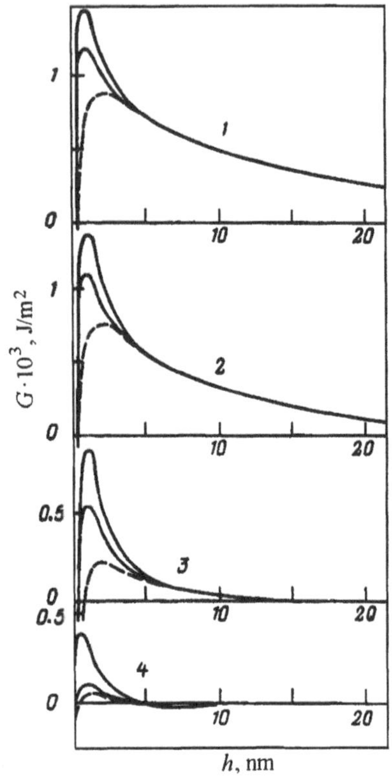

Figure 9.5 Potential curves $G(h)$ of interaction between plate particles in aqueous 1–1 electrolyte solutions of concentration: *1*) 10^{-4} mol/l, $\psi_1 = -150$ mV; 2) 10^{-3} mol/l, $\psi_1 = -100$ mV; 10^{-2} mol/l, $\psi_1 = -45$ mV; 10^{-1} mol/l, $\psi_1 = -25$ mV.

evaporation will slow down and the overdried skin surface will hamper further deformation. The internal stresses generated may result in cracks and crumbling. As will be shown further, such processes actually take place in porous bodies.

Figure 9.5 shows the potential curves $G(h)$ calculated for the case of plane interlayers between silica plate particles. The dashed curves are derived according to the DLVO theory taking account of two components of the disjoining pressure: electrostatic and molecular. Solid curves $G(h)$ are found taking additional account of structural forces: upper curves for $K_s = 2\times10^6$ N/m^2, lower at $K_s = 10^6$ N/m^2 (see eq. I.16) [9]. Analysis of the dependences built for various concentration of 1–1 electrolyte in the dispersion medium shows that the account of structural forces, which are most pronounced in the case of hydrophilic particles, tend to increase the potential barrier and correspondingly the force barrier. In hydrophilic porous bodies this leads to the fact that, even with a sharp growth of the capillary pressure, the force barrier cannot be overcome and a thin interlayer remains between the approaching particles. Direct force measurements of water interlayers between mica surfaces show that to extrude the last several water layers a pressure exceeding hundreds of atmospheres is required [10]. In hydrophilic systems the dependence P_c vs. W is smoother (Fig. 9.3, curve 2) and the shrinkage may be reversible because the "dry" contacts do not form.

An increase in electrolyte concentration, occurring for example in external layers of a porous body while the solution is being evaporated, decreases the height of the potential barrier and the structural repulsion. In this case the transition over the potential barrier is feasible accompanied by the formation of close and also irreversible contacts. Such behavior is characteristic of hydrophobic systems, when the particles come within atomic distances of each other under the action of the dominating forces of molecular attraction.

Therefore, the mechanism of the processes of two-phase shrinkage may be considered as proceeding from simple, primitive models of porous structures. The results can only provide a qualitative analysis of shrinkage and evaporation, because the structures of real porous bodies are much more complex and diversified. The application of the model porous structures and the theory of surface forces provides a basis for study of the structural changes in the course of mass transfer processes in non-rigid porous bodies. Hopefully improved methods of calculation and appropriate models of dispersions will make it possible to turn from qualitative to quantitative forecasts.

§2 Experimental Investigations of Shrinkage Mechanism During Drying

Measurements of linear and volumetric shrinkage resulting from drying have been performed for a variety of porous bodies: clays, turf, building materials, food and leather [266, 297–306].

The volumetric shrinkage is usually characterized by the coefficient β equal to the ratio of relative volume change, V/V_0, of the porous body with the changes in mass content W:

$$\beta = d(V/V_0)/dW. \qquad \text{(IX.4)}$$

where V_0 is the initial volume of the porous body.

Constant values of β in the course of drying demonstrate that the changes in moisture content are proportional to the changes in volume. For systems consisting of approximately equal particles like clays and some kinds of turf, the values of β are constant up to the first critical mass content. This shows that the evaporation takes place here mainly from the sample surface and the shrinkage occurs at the expense of the volume of the moisture removed. This mechanism agrees well with the models considered in §1.

When the porous body consists of deformable porous particles, higher capillary pressures are needed to cause deformation. Here the shrinkage does not develop immediately. First of all the wide pores between particles are emptied tending to decrease the rate of evaporation. Once the moisture removal from porous particles with narrower pores begins, the capillary pressure must increase sufficiently to initiate their deformation and cause shrinkage.

In porous bodies with two greatly differing pore sizes, the shrinkage arises due to the wide pores narrowing. After the formation of capillary menisci in narrower channels of porous particles, their deformation takes place affecting the rate of shrinkage. For a wide distribution of pores by size, a non-linear shrinkage may develop with a gradual change of the value of the coefficient β.

In porous bodies with rigid particles, shrinkage stops as the particles approach each other. If the "dry" contact is achieved, the system loses its reversibility because the particles consolidate. When liquid interlayers remain between the particles, the system may be returned to the initial state by moistening: the shrinkage is reversible. The degree of irreversibility depends on the lyophilicity of the system. In the case of lyophobic systems, agglomeration of the particles and formation of crystallization contacts may occur. Lyophilic systems at incomplete evacuation of liquid may remain reversible. Because the surfaces of real particles are heterogeneous, contacts of a different nature form, and in general the shrinkage is only partially reversible. A higher rate of shrinkage on completion of drying may be associated rather with spontaneous molecular attraction of particles than with capillary forces. The tightening of particles stops when they approach spacing of molecular size corresponding to the first deep potential well.

The degree of shrinkage in the last stage of drying depends on the distance between the position of the force barrier and that of the first potential well (Fig. 1.4). In lyophilic porous bodies, where structural repulsive forces act together with those of electrostatic repulsion, this distance is greater. Surfactant and polymer adsorption increase it further. The relation between the shrinkage and the form of potential curves of pair interaction is supported by the fact that the pH of the dispersion medium, its salt composition and surfactant additions affect the course of shrinkage [304, 306]. The addition of surfactant also changes the surface tension tending to decrease the capillary pressure and the shrinkage [305].

The first estimations of the stresses arising within a porous body during drying, were performed in [297, 301]. They were based on measurements of

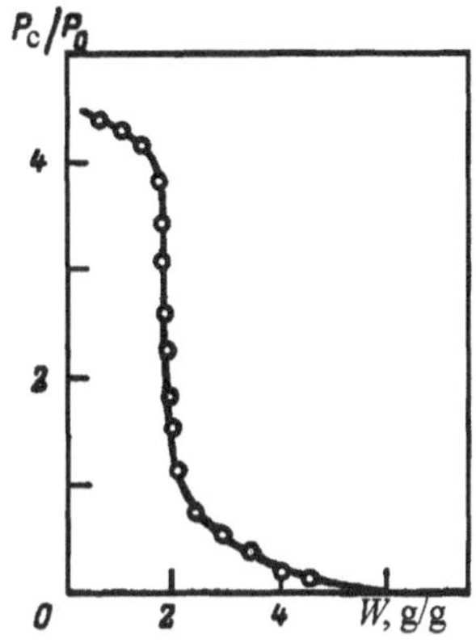

Figure 9.6 Dependence of capillary pressure P_c on the relative moisture content W in turf. P_0 is the atmospheric pressure.

the force necessary to prevent the shrinkage. However, the method of measurement influences the course of the shrinkage and this complicated the quantitative interpretation of the results.

The reason for shrinkage, at least in the first stage, is the capillary pressure. In [302, 303] has been developed an appropriate method of a "rigid" manometer to measure the stresses, P, arising in the course of shrinkage. A sensor with elastic walls was placed within the drying sample. The deformation of the walls, accepted by the liquid filled the sensor, was transmitted to the sensitive pressure sensor. Such a system does not noticeably influence the course of restructuring in the sample, because of an insignificant deformation of the walls and small sensor size (~5 mm in diameter). It has been shown by this method that in clays the internal pressure grows with the forces of capillary pressure in the course of drying. The coefficient of shrinkage remains roughly constant slightly decreasing (by 5–10%) with decreasing mass content. When $W < 0.1$ g/g, the values of P stabilize due to the growing rigidity of the developing structure.

Shrinkage measurements were also performed by another method with the help of a microtensiometer, consisting of narrow capillaries sealed at one end (0.1–0.5 μm in diameter). The capillaries were partly filled with water, so that an air column remains at the sealed end. As the capillary pressure rises in the course of drying, the liquid is partly sucked out from the capillary inserted into the drying sample. The value of the capillary pressure, P_c, is determined by the increase in the length of the air column. This method may be also used to establish the pressure distribution inside the porous body due to the small sensor size. The measurements show [303] that the capillary pressure is maximal in the surface layers of the sample and decrease towards the center. This is believed to be a consequence of the considerable viscous resistance to the liquid flow in pores being deformed. This causes non-uniform deformation in different layers of the sample, greater at the surface and far less in the center.

The dependence of the capillary pressure P_c on the mass content W of turf during drying [303] is presented in Fig. 9.6. The form of the curve qualitatively agrees with the theoretical dependence $P_c(W)$ shown in Fig. 9.3 (curve *1*). As the temperature increases, the capillary pressure P_c decreases due to decreasing surface tension. However, the thermal variation of the curves $P_c(W)$ depends not only on $\sigma(T)$, because it influences the forces of pair interaction of particles $F(H)$. In particular, an increase in temperature decreases the forces of structural repulsion due to the destruction of the specific structure of the boundary layers.

The reverse process, the swelling of a deformable porous body in contact with the bulk liquid, may be considered from the same point of view. For clay such calculations based on the model of plate particles (Fig. 9.4), have been performed in [307–311]. Here the dependence of the swelling pressure P_s on the mass content has been explained quantitatively on the basis of the isotherms of disjoining pressure $\Pi(h)$ of thin plane interlayers. When the plane clay particles are oriented parallel, which can be achieved by slow sedimentation, the swelling pressure is equal to the disjoining pressure. As shown by P. Low [308], the repulsion of hydrophilic clay particles is determined mainly by the structural forces. This was also supported by Yu. Tarasevich [310]. In many cases the dependences $P_s(W)$ may be explained on the basis of the theory of electrostatic repulsion [307, 311] or osmotic pressure caused by the difference between ion concentration inside thin pores and in bulk electrolyte solution [309].

The contribution of the different components of the disjoining pressure varies depending on the properties of clay and the chemical composition of its ionic exchange capacity. In any case, the theory of surface forces enables the mechanism of swelling and shrinkage to be explained. To describe these effects quantitatively, the electrical potentials of clay particles and the constants for the molecular and structural forces, which depend on solution composition, temperature and the degree of hydrophilicity of the surface, should be known.

Chapter X

Wetting and Capillary Impregnation

The phenomenon of wetting that results in the formation of a contact angle between the liquid and the solid surface demands a special consideration. The value of the equilibrium contact angle θ_0 is governed by the surface forces field, i.e. the energy of interaction between the liquid and the solid surface. Weak interaction produces non-wetting, and in contrast strong attraction leads to the spreading of the liquid over the surface, i.e. its complete wetting. We will consider in this Chapter the factors determining the contact angle, the phenomenon of wetting hysteresis and some mass transfer processes depending directly on the value of the contact angle and its variations during the course of flow.

Capillary rise and capillary imbibition will be considered as an example when the kinetics are influenced by contact angles, their dependence on flow rate and surfactant adsorption. Three different mechanisms of spontaneous and forced penetration of surfactant solutions into hydrophobic capillaries are discussed in dependence on solution concentration.

§1 Contact Angles

The development of a theory of surface forces has made it possible to determine quantitatively the contact angles from the dependence of the physical properties of the solid surface and the liquid interacting with it. In Chapter II, §3 eqs. (II.18) and (II.19) are presented enabling the values of the equilibrium contact angle θ_0 to be calculated characterizing the wetting on the basis of the isotherms of disjoining pressure $\Pi(h)$ or the adsorption isotherms $\Gamma(p/p_s)$. Equation (II.18) is applicable for small contact angles $\theta_0 \leq 40–45°$,

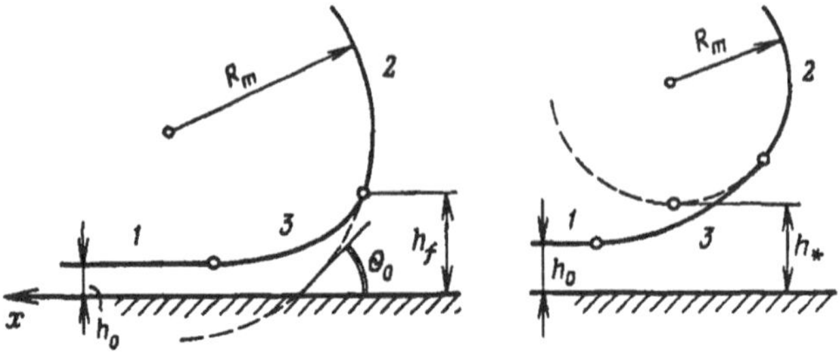

Figure 10.1 Various cases of equilibrium between menisci and films in a narrow plane slit.

when wetting films form on the surface. Equation (II.19) is applicable to higher values θ_0, when the surface is covered with thinner adsorption films.

First of all let us define the contact angle θ_0. At equilibrium, the contact angle should be determined at the point of intersection of the substrate and the extension of the drop or meniscus profile undisturbed by surface forces (Fig. 10.1). The meniscus contact here with the surface film with equilibrium thickness h_0 that is determined by the isotherms $\Pi(h)$ or $\Gamma(p/p_s)$. The values of h_0 correspond to the disjoining pressure $\Pi(h_0) = \Pi_0 = P_c$, where P_c is the capillary pressure of the meniscus, or to the relative vapor pressure $p_0/p_s = \exp(v_m P_c/RT)$, where v_m is the molar volume of liquid.

The Young equation relates the equilibrium value of the contact angle θ_0 with the interfacial tension on the boundaries solid–vapor σ_{SV} and solid–liquid σ_{SL} and with the surface tension of the liquid σ:

$$\sigma \cos\theta_0 = \sigma_{SV} - \sigma_{SL}. \tag{X.1}$$

However, because the values σ_{SV} and σ_{SL} cannot be measured directly, the value of the contact angle θ_0 cannot be determined from eq. (X.1). Thus the Young equation is used rather to solve the inverse problem, and namely to calculate the difference $\sigma_{SV} - \sigma_{SL}$ from the measured values of θ_0. Comparing eq. (X.1) and (II.18), one can notice that the latter equation just gives the definition of the difference $\sigma_{SV} - \sigma_{SL}$.

For a solid substrate covered with an adsorption film, this may be accepted only if we assume that the state of the solid surface does not undergo changes when the liquid contacting the surface is replaced by vapor. Then, according to the Gibbs equation of adsorption, the changes of σ_{SV} compared to the standard state may be found. (The state of a thick film with free energy equal to the sum of surface interfacial energies σ_{SL} and σ is taken as the standard state.)

The value of σ_{SV} is to be found from the Gibbs equation:

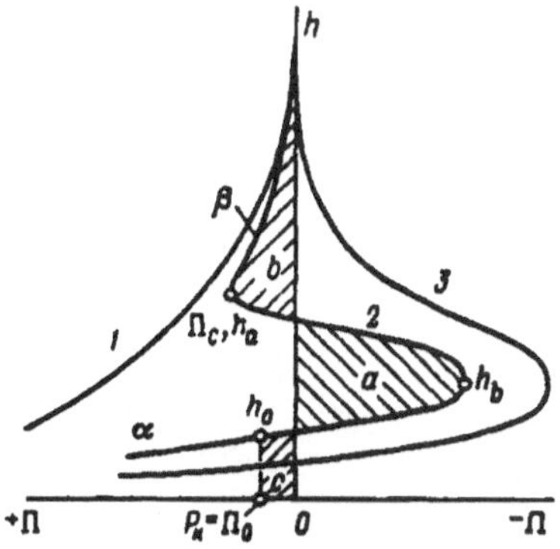

Figure 10.2 Various isotherms for disjoining pressure, $\Pi(h)$, of wetting films.

$$\Gamma = (d\sigma_{SV}/d\mu)_T, \tag{X.2}$$

where $\mu = -v_m\Pi = RT\ln(p/p_s)$ is the excess chemical potential of the liquid molecules. Integrating this equation:

$$\int_{\sigma_{SV}}^{\sigma_{SL}+\sigma} d\sigma_{SV} = -\int_{\mu}^{0} \Gamma d\mu\,,$$

and substituting the film thickness $h = \Gamma v_m$ for the adsorption Γ and expressing the chemical potential μ through the disjoining pressure, we obtain:

$$\sigma_{SL} + \sigma - \sigma_{SV} = \int_{\Pi}^{0} h d\Pi = \int_{h_0}^{\infty} \Pi(h)dh - \Pi_0 h_0\,.$$

Expressing the difference $(\sigma_{SL} - \sigma_{SV})$ by $\sigma\cos\theta_0$ using eq. (X.1), we obtain the formula for the equilibrium contact angle θ_0 relating it to the isotherm of disjoining pressure of wetting films:

$$\cos\theta_0 = 1 + (1/\sigma)\left[\int_{h}^{\infty} \Pi(h)dh + \Pi_0 h_0\right]. \tag{X.3}$$

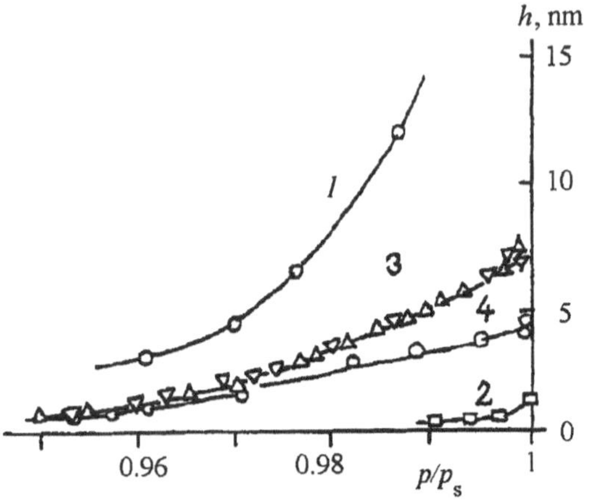

Figure 10.3 Isotherms $h(p/p_s)$ for water films on quartz surface with different degree of hydrophilicity: $\theta_0 = 0$ (*1*), $\theta_0 = 40°$ (*2*) [314], $\theta_0 = 5°$ (*3*), $\theta_0 = 10°$ (*4*).

This coincides with eq. (II.18) obtained by a different approach.

The most commonly occurring isotherms of different shape are shown in Fig. 10.2. The isotherm *1* corresponds to complete wetting, and isotherms *2* and *3* to the partial wetting when there is a finite contact angle $\theta_0 \geq 0$. According to eq. (X.3), the value of the contact angle is proportional to the difference between the areas bounded by the isotherm graph in the regions $\Pi < 0$ and $\Pi > 0$. In the case of the isotherm *2*, the contact angle is proportional to the difference of the shaded areas *a* and *b* equal to the value of the integral in eq. (X.3). The third term of this equation in the case of menisci or drops with curvature radius $r > 10^{-3}$ cm is much smaller than the first two terms and may be neglected. When the contact angle $\theta_0 = 0$, the areas *a* and *b* are equal. When $a < b$, there is complete wetting, and when $a > b$ wetting is partial. The isotherm shown by the curve *3* corresponds to poor wetting and its branch in the region $p/p_s < 1$, or $\Pi < 0$ may here correspond rather to adsorption films. As seen from the comparison of the curves *1*–*3*, equilibrium films of smaller thickness, h_0, at the same value of the capillary pressure P_c correspond to less wetting.

The isotherms in Fig. 10.2 have a variety of shapes due to the different contributions of the particular components of surface forces (see Chapter I). The isotherm of the type 1 is characteristic of non-polar liquids, where the forces of molecular repulsion of film surfaces mainly act, $\Pi_m > 0$. Isotherms of type 2 are characteristic of water films and aqueous solutions, where the forces of electrostatic repulsion $\Pi_e > 0$ predominating at high thickness are being replaced by the forces of electrostatic attraction with decreasing thickness, as takes place when the surface potentials of the film are different in value ($\psi_1 \neq \psi_2$). A rise in concentration of an electrolyte solution decreasing the forces of electrostatic repulsion shifts the isotherms from the curve *2* to the curve *3*. In a region of small thicknesses, structural forces also markedly contribute. Hydrophilization of the surface tends to raise the values of $\Pi_s > 0$ consequently increasing the film thickness h_0. Hydrophobization shifts the isotherms to the curve *3*, and the film thickness, h_0, may decrease down to a monolayer (Fig. 10.2) due to the growth of the hydrophobic attraction ($\Pi_s < 0$). High values of the contact angles ($\theta_0 \geq 90°$) are associated with this component of the surface forces [312, 313].

The role of the forces of hydrophilic repulsion ($\Pi_s > 0$) in the formation of the contact angle is illustrated in Fig. 10.3, where the experimentally obtained isotherms of disjoining pressure of water films on surfaces with different degrees of hydrophilicity are shown. The results, presented in Fig. 10.3, have been obtained by ellipsometric measurements of the thickness of films being formed as a result of polymolecular vapor adsorption. The isotherms *1* and *2* have been obtained at 100% and 10% coverage of a crystal quartz surface with hydroxyl groups [314], respectively. They correspond to complete wetting in the first case and to the value of the contact angle $\theta_0 = 40°$ in the second. The ellipsometric data used to plot the isotherms *3* and *4* were presented in the monograph [26]. The isotherms were obtained in the course of adsorption of water vapor on the polished silica surface. The hydrophilicity of the surface was characterized by the contact angles of $\theta_0 \approx 5°$ (curve *3*) and 10° (curve *4*). The isotherms *3* and *4* occupy an intermediate position between the isotherms *1* and *2*. As the molecular and electrostatic forces in water films on silica surface are nearly identical, the difference in the shapes of the isotherms may be ascribed to the forces of structural repulsion which depend on the hydrophilicity of the substrate. Weakening of these forces with lower degree of hydrophilicity of the substrate results in smaller thicknesses of the films formed.

It has been shown in [312, 313] that the form of the isotherm *1* can be quantitatively explained merely by considering the action of structural repulsive forces with parameters $K_s \approx 10^8$ N/m^2 and $\lambda = 1.5$ nm in eq. (I.16). The isotherms *3* and *4* may be quantitatively explained not only by the weakened structural forces, but also by the molecular and electrostatic forces acting inside the film. Finally, to describe the isotherm *2*, it is sufficient to consider only the electrostatic and molecular forces. At a surface hydrophilicity characterized by the contact angle 40°, structural forces do not appear.

The hydrophilic structural repulsion contributes markedly to the total isotherms $\Pi(h)$ only at $\theta_0 < 20°$, and the contribution increases dramatically as $\theta_0 \to 0°$. The action of the forces of hydrophobic attraction begins to appear when $\theta_0 > 50°$, and their contribution in the total isotherm $\Pi(h)$ rises with hydrophobicity of the substrate. The properties of the wetting films on

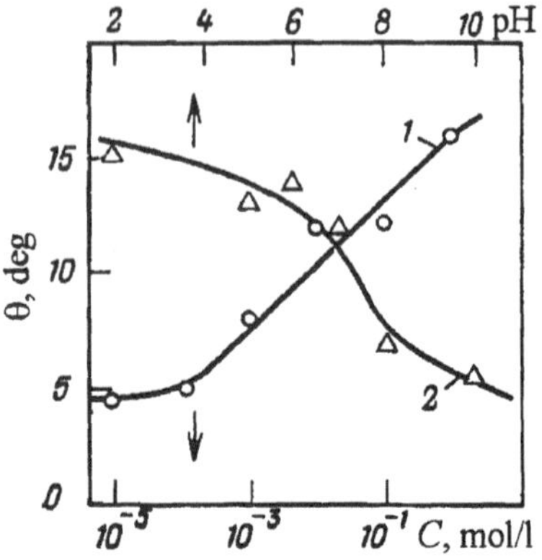

Figure 10.4 Contact angle θ_0 of aqueous solutions on quartz surface vs. KCl concentration (curve *1*) and pH (curve *2*) at a constant ionic strength 10^{-2} mol/l (HCl + KCl + KOH). Experimental data are shown by points, and the calculated results by solid curves.

surfaces characterized by contact angles in the range from 20° to 50° may be explained without resort to structural forces. In this case we may restrict ourselves to molecular and electrostatic forces being the basis of the DVLO theory [315, 316].

Figure 10.4 shows the results of calculations of contact angles by eq. (X.3) for aqueous solutions on silica surface [317]. Curve *1* corresponds to KCl solutions of different concentrations with account of the three components of surface forces: molecular, Π_m, calculated by eq. (I.3); electrostatic, Π_e, calculated by numerical methods, and structural, Π_s, determined from eq. (I.16). The parameters of the structural isotherm, $\Pi_s(h)$, were chosen to achieve the best fit of the curve *1* to the experimental data [318] depicted by points in Fig. 10.4. The values of K_s in eq. (I.16) equal to $K_s = 2\times10^6$ N/m^2 ($\lambda = 2$ nm) for the solution concentration $C = 10^{-5}$ mol/l sharply decrease with growing C, which reflects the effect of restructuring of boundary layers with increasing electrolyte concentration (Chapter II, §3). When $C = 10^{-2}$ mol/l, the value of $K_s = 0$ pointing to the absence of structural forces ($\Pi_s = 0$) at this concentration. Therefore, at an ionic strength of 10^{-2} mol/l and different pH (curve 2) we can get by with only two components: Π_m and Π_e in calculations of θ_0. As seen from Fig. 10.4, a good agreement is achieved here between the theoretical curve 2 and the experimental points.

Growth of contact angles at higher concentrations of KCl solution (curve *1*) is associated with the suppression of the forces of both electrostatic and structural repulsion. This leads to a shift of the isotherm to the region $\Pi < 0$ at constant Π_m. When $C = 10^{-5}$ mol/l, the isotherm $\Pi(h)$ has a form similar to the isotherm 2 in Fig. 10.2, and at $C \geq 0.1$ mol/l it approaches the isotherm

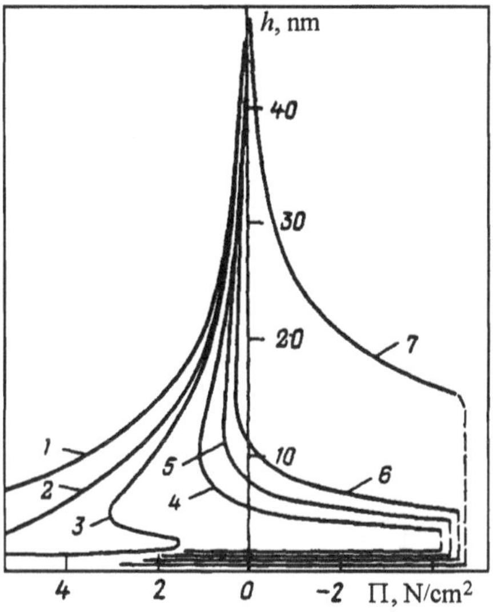

Figure 10.5 Theoretical isotherms $\Pi(h)$ illustrating the effect of ionic surfactants on calculated contact angles: complete wetting (curves *1* to *3*), 5° (curve *4*), 7° (curve *5*), 8° (curve *6*) and 28° (curve *7*).

3. The elevation of electrolyte concentration is used in flotation to hydrophobize the surface and to improve the adhesion of air bubbles to the particles.

A decrease in pH (curve 2, Fig. 10.4) brings the silica surface to the isoelectrical state. The forces of electrostatic repulsion decrease and the contact angles grow. At pH 2 very thin films remain on the surface which are stable only because of molecular forces. On the other hand, an increase in pH raising the concentration of potential-determining OH^- ions cause the negative surface potentials of silica to rise. The forces of electrostatic repulsion increase, and results in better wetting.

Therefore, the application of the theory of surface forces allows not only the factors governing the wetting to be identified, but in some cases values of contact angles that agree with the experiments to be determined.

To control wetting, the surfactants are often used. The isotherms $\Pi(h)$ calculated using the same methods for the systems, when the adsorption of ionic surfactants changes the film–gas electrical potential ψ_2 [319], are presented in Fig. 10.5. The potential of silica surface ψ_1 was taken as constant and equal to $\psi_1 = -100$ mV in these calculations. The calculations were performed for a background concentration of 10^{-3} mol/l KCl in solution. The same parameters of the isotherm of structural forces were used for KCl solutions, as before. The initial isotherm *6* (without surfactant) corresponds to the KCl electrolyte concentration of 10^{-3} mol/l. In this case the surface

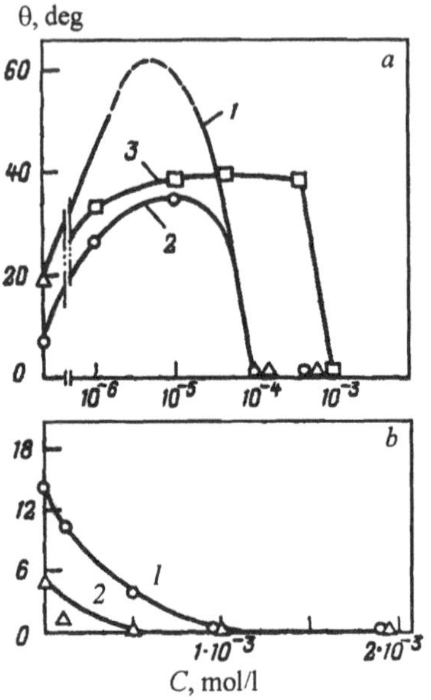

Figure 10.6 Effect of ionic surfactants on contact angles of aqueous solutions on quartz surface. *a*) CTAB, 10^{-1} N NaCl, θ_A (*1*) and θ_R (2); 10^{-6} N KCl, θ_A (*3*) [321]; *b*) SDS, 10^{-1} N NaCl, θ_A (*1*) and θ_R (2).

potentials were taken equal to $\psi_1 = -100$ mV and $\psi_2 = -25$ mV. The calculation gives the value of $\theta_0 = 8°$, as shown in Fig. 10.4.

An increase in negative potential ψ_2 due to adsorption of anionic surfactant on the film–gas interface shifts the isotherms to the region of positive values of disjoining pressure. When $\psi_2 = -35$ mV, the calculated value of the contact angle falls to 7° (curve *5*), and at $\psi_2 = -45$ mV to 5° (curve *4*). Further growth of negative ψ_2 values leads to complete wetting: the isotherms *3* ($\psi_2 = -65$ mV), *2* ($\psi_2 = -75$ mV) and *1* ($\psi_2 = -100$ mV) are entirely located in the region $\Pi > 0$.

When a cationic surfactant adsorbs on the film surface charging it positively, then, due to the electrostatic attraction between the negatively ($\psi_1 = -100$ mV) and positively ($\psi_2 = +100$ mV) charged surfaces ($\Pi_e < 0$), the whole isotherm shifts to the region $\Pi < 0$ that raises the contact angle to 28° (curve *7*).

These calculations agree well with the results of direct measurements of contact angles of aqueous surfactant solutions on silica surface [320]. Figure 10.6 shows the values of contact angles formed between wetting films of small radius and the bounding capillary meniscus, which were measured by the differential interference method. The dependence of advancing θ_A (curve

1) and receding θ_R (curve 2) contact angles for aqueous solutions of cetyltrimethyl ammonium bromide (CTAB) are presented in Fig. 10.6, *a*. Addition of this surfactant to 10^{-1} mol/l NaCl causes the contact angles to rise due to the prevailing surfactant adsorption at $C \leq 10^{-5}$ mol/l on the film surface. At higher concentration the adsorption of surfactants on silica surface becomes possible, that first neutralizes and then charges the silica surface to the opposite sign. In the latter case both film interfaces are charged identically (positively) that causes them to repel ($\Pi_e > 0$) and results in complete wetting. Similar results were obtained previously in [321] (curve *3*), where the background solution was 10^{-6} mol/l KCl. The dependences $\theta(C)$ passing a maximum were also obtained for CTAB solutions on silica in refs. [312, 322, 323].

Addition of anionic surfactant, sodium dodecyl sulfate (SDS), increasing the negative ψ_1 and ψ_2 potentials causes, in accordance with theory, only an improvement of wetting (Fig. 10.6) [319].

The notion of the contact angle is applicable unless the extension of the meniscus intersects the substrate. The ultimate case is $\theta_0 = 0$, when the extension of the meniscus profile only touches the substrate and the areas *a* and *b* in Fig. 10.2 are equal. For isotherm *1* in Fig. 10.2, the profile of the undisturbed meniscus no longer intersects and touches the substrate (Fig. 10.1, right). The degree of complete wetting is characterized here by the difference, h_*, between the slit width H (or capillary radius) and the radius of curvature of the meniscus: $h_* = H - R_m$. The larger the ratio $\gamma = h_*/h_0$, the better the wetting. When $h_* > 0$, eq. (X.3) is no longer applicable, because the integral becomes positive. In this case other equations of the theory should be used, namely eqs. (II.15) and (II.16), enabling the radius of curvature of the meniscus $R_m = \sigma/P_c$ to be calculated for a particular slit width H and then the value of h_* [319]:

$$h_* = h_0 + \left[\int_{h_0}^{\infty} \Pi(h) dh / P_c \right],$$

where $P_c = \sigma/H$ is the capillary pressure of the meniscus in the plane slit.

Therefore, both cases of complete and partial wetting may be quantitatively treated through the use of the isotherms of disjoining pressure of wetting films $\Pi(h)$. It should be mentioned that there is a variety of cases of complete wetting, because the values of h_* may range from 0 at $\theta = 0$ to $h_* > h_0$. When the films are stable due to the forces of molecular repulsion the value of h_* approaches $1.5h_0$. When the stability is provided by the forces of electrostatic repulsion the ultimate value of h_* achieves $2h_0$. Direct interference measurements of the thickness h_* for wetting films of electrolyte solutions on silica

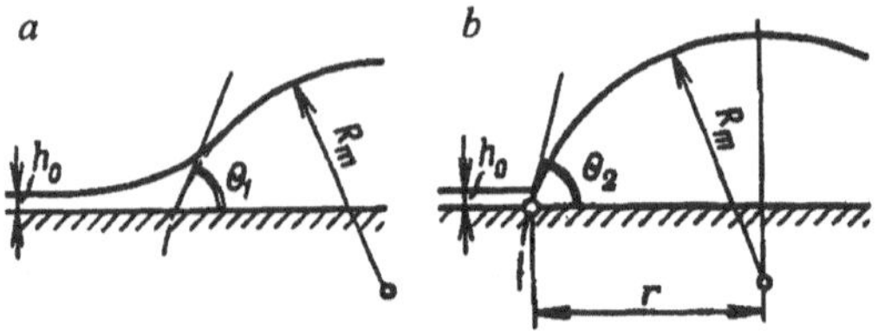

Figure 10.7 Determination of the line tension ϑ from the difference between two contact angles θ_1 and θ_2 calculated with (*a*) and without (*b*) the effect of a transition zone.

surface give the values of ratio h_*/h_0 ranging from 1.3 to 1.6 [324]. A rise of electrolyte concentration somewhat decreases the h_*/h_0 ratio [325].

As shown in Fig. 10.1, a transition zone (*3*) (where capillary and surface forces act simultaneously) forms between the meniscus (2) and the equilibrium wetting film h_0 (*1*). The profile of the transition zone may be determined from eq. (II.12) using the known isotherm of disjoining pressure $\Pi(h)$ for a given system substrate–liquid. The thickness of the transition zone depends on the radius of action of surface forces (10^{-6}–10^{-5} cm). The horizontal length of this zone is equal to $l \approx (h_0 H)^{1/2}$ [32]. For example, for pores of width $H = 10$ μm and $h_0 \approx 10^{-5}$ cm, l values are of the order of microns. Direct interference measurements give the l values of about 9 μm for water and decreasing to 3 μm values for 10^{-3} M KCl [324].

The existence of the transition zone results in the effect predicted in general by Gibbs, the effect of line tension [326]. By analogy with surface tension σ, when the transition zone between the liquid and its vapor is replaced with a plane of tension, the transition zone between the meniscus and the film may be replaced with the line of three-phase contact (Fig. 10.7) with radius r and line tension ϑ. Contrary to σ, line tension may be both positive or negative causing the circular perimeter of radius r to contract in the first case and to extend in the latter.

When the transition zone is replaced with the three-phase contact line, an additional term ϑ/r should enter into the Young equation:

$$\sigma \cos \theta_0 = \sigma_{SV} - \sigma_{SL} + \vartheta/r. \tag{X.4}$$

Positive values of ϑ tend to increase the values of θ_0, and negative values decrease them.

Experimental estimations of line tension for water and aqueous solutions show that values of ϑ are negative and of about 10^{-10}–10^{-11} N for free films and small contact angles [327]. Therefore, the effect of line tension ($\vartheta/r \geq 10^{-4}$ N/m) appears at $r \leq 10^{-4}$–10^{-5} cm, i.e. for drops of very small radius.

For example, the action of line tension is most pronounced in the initial stage of flotation contact when gas bubbles approach the particles and also during the water vapor condensation on solid substrates, during the stage of nuclei formation [328].

Theoretical estimation of the values of line tension were made comparing two equilibrium states of a droplet on a solid surface (Fig. 10.7). The first state takes into account the effect of transition zone (Fig. 10.7, *a*) that forms between a flat wetting film and bulk droplet. The profile of the transition zone was calculated on the basis of eq. (II.12) using a simplified form of the disjoining pressure isotherm [319, 329, 330]:

$$\Pi = 0 \text{ at } h > t_1, \qquad \Pi = a(t_0 - h) \text{ at } 0 < h < t_1,$$

where a, t_0 and t_1 are the parameters of the isotherm. It was assumed that the profile of the transition zone is gently sloping, and gravity is neglected. As a result the value of visible contact angle θ_1 was calculated using eq. (II.18).

In contrast, in Fig. 10.7, *b* the shape of a droplet is shown when the effect of the transition zone was neglected. The droplet has the same values of surface curvature, chemical potential and equilibrium vapor pressure as in the first case. The contact angle θ_2 that corresponds to this idealized state was calculated on the basis of eq. (X.3) using the same isotherm $\Pi(h)$.

The difference between the $\cos\theta_2$ and $\cos\theta_1$ values reflects the effect of transition zone, expressed in eq. (X.4) as the line tension term:

$$\sigma(\cos\theta_2 - \cos\theta_1) = \vartheta/r. \qquad \text{(X.5)}$$

Because $\theta_1 > \theta_2$, the line tension, in the framework adopted, is negative and depends on the radius r of contact line. When $r \to \infty$, values of ϑ tend to some limiting value $\vartheta_0 = -2\sigma t_1 \tan\theta_0$, where t_1 is the parameter of the isotherm which reflects the radius of action of surface forces. For a particular case of a common black foam film when $\theta_0 = 0.5°$ and $\sigma = 35$ mN/m, this equation gives $\vartheta_0 = -1.3\times10^{-11}$ N. The calculated value is fairly close to the experimentally obtained result: $\vartheta_0 = -1.7\times10^{-11}$ N. The value was obtained, according to eq. (X.4), from the dependence of contact angles on the film's radii [327]. Further development of the theory of line tension was performed using the same approach by Dobbs and Indekeu [331].

The line tension leads, according to eq. (X.4), to the relationship between the contact angle and the drop radius r. However, in many cases [332–334] the dependence of $\cos\theta_0$ on r gives values of line tension that are larger by several orders compared with known data [327, 335]. The dependences $\cos\theta_0$ vs. r are of another nature here and associated with the formation of a border near to the line of wetting (see Fig. 10.9 below) on a deformable surface [332]

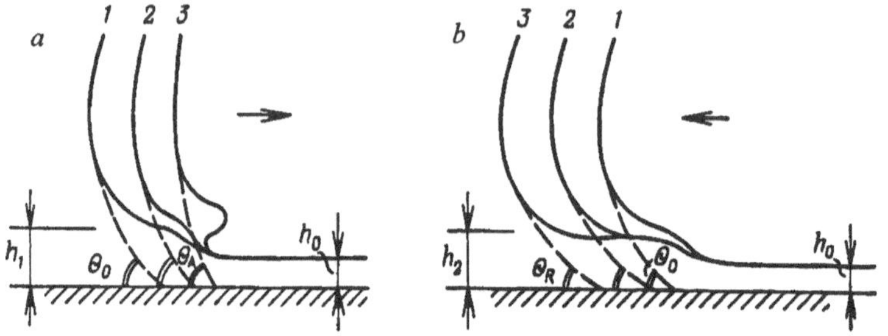

Figure 10.8 States of mechanical meniscus equilibrium in the range of contact angles $\theta_R < \theta < \theta_A$ in the cases of advancing (*a*) and receding (*b*) motion of a meniscus.

or with the effect of surface roughness producing an irregular wetting line. When the amplitude *A* of deviation from the round form is constant, then at *A* close or smaller than *r* the shape of the drop and its contact angle will depend on its size [333, 334].

§2 Hysteresis of Contact Angles

We have considered the equilibrium values of contact angle θ_0 corresponding to the thermodynamic equilibrium state of the drop or liquid meniscus on a solid undeformable substrate. However, on attempting to move the drop or meniscus, it has been found experimentally that (when $\theta_0 < 90°$) the contact angle should be increased to $\theta_A > \theta_0$, when the liquid advances onto a free part of the surface. Recession of the liquid takes place when the contact angle $\theta_R \leq \theta_0$. This phenomenon is called the static hysteresis of the contact angle.

Strictly speaking, only the values of the advancing θ_A or the receding θ_R contact angles can be measured experimentally, because the establishment of the thermodynamically equilibrium state controlled by the mass transfer processes between the bulk liquid and the wetting or adsorption films on the surface usually requires a long time. In the range of contact angles $\theta_R < \theta < \theta_A$ many states of mechanical equilibrium may be realized (Fig. 10.8). Here the curves *1* correspond to the meniscus profile in a state of thermodynamic equilibrium, when $\theta = \theta_0$. When $\theta > \theta_0$ (Fig. 10.8, *a*, curve 2), an equilibrium is quickly established in the bulk (beneath the meniscus and in the transition zone) where the viscous resistance is weak. Here there is no equilibrium with a thin film h_0. When $\theta > \theta_0$, the liquid would flow into the film, however due to the high viscous resistance to flow in the thin film, this process goes slowly. As a result, for each moment of time the profile of the meniscus or transition zone takes the shape corresponding to the state of their mechanical equilibrium in the absence of equilibrium with the thin film. The thicker the film, the quicker it reaches in equilibrium with the meniscus, but thick films form only in the case of complete wetting. Thus, no hysteresis was observed at small contact angles. The thinner the film, the worse is the wetting, and the more pronounced is the hysteresis of the contact angle.

In [336] the profiles corresponding to the state of mechanical equilibrium of the meniscus and transition zone were calculated by eq. (II.12). This enabled the limiting profile corresponding to $\theta = \theta_A$ to be found when mechanical equilibrium becomes impossible and the meniscus comes into motion (Fig. 10.8, curve *3*). This motion takes place as a result of the liquid overflow onto a practically immobile film. This mechanism was first discussed by Frenkel [337] in application to a drop sliding down an inclined surface.

Values of θ_A may be determined from the isotherm $\Pi(h)$ using the equation [336]:

$$\cos\theta_0 - \cos\theta_A = 1 + (1/\sigma)\int_{h_0}^{h_1} \Pi(h)dh\,, \qquad \text{(X.6)}$$

where h_1 is the film thickness in the region of the bending of the transition zone profile (Fig. 10.8, *a*).

During the meniscus recession, the limiting state $\theta = \theta_R$ corresponds to the formation of a horizontal region in the transition zone (Fig. 10.8, *b*, curve *3*), when a metastable β-film remains behind the receding meniscus. As this film is completely wetted by the bulk liquid, the receding angle (at not too high θ_0) is close to zero.

The mechanism of hysteresis discussed is associated with impeded mass transfer between the film and the meniscus. It may take place even on molecularly smooth and energetically uniform surfaces.

A strong hysteresis of a different nature takes place on rough or strongly heterogeneous surfaces. To overcome steps or hydrophobic regions, higher values of contact angle θ_1 are necessary. On the other hand, the recession is facilitated and depends on the equilibrium contact angle θ_2 of hydrophilic regions or the slope of inclines. Thus the motion back and forth occurs at different values of the contact angle and $\theta_1 > \theta_2$. Both are considered as equilibrium angles for different regions of the surface. Generally, it is assumed that different local equilibrium values θ_i arise on heterogeneous surfaces. The existence of regions of heterogeneity (geometrical or chemical) on a macroscopic scale leads to stepped movement of the line of wetting.

It should be noted that roughness may arise spontaneously on an initially smooth but deformable surface under the action of an uncompensated normal component of the force F (Fig. 10.9). The value of this force acting on the perimeter of wetting is given by the Young equation:

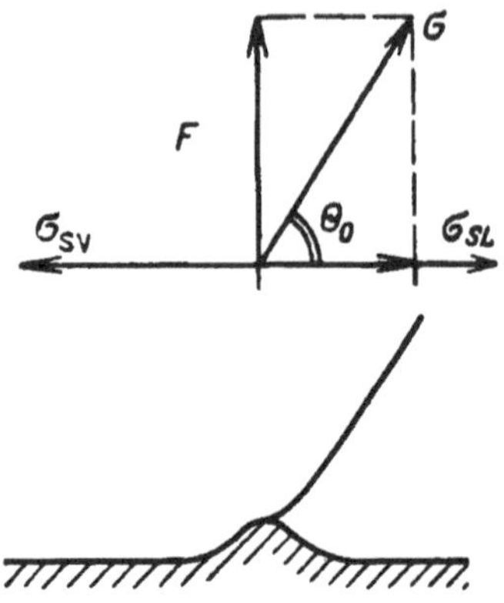

Figure 10.9 Deformation of a non-rigid surface in the region of the wetting line.

$$F = \sigma \sin \theta_0. \tag{X.7}$$

Its action is compensated by the stresses generated in the solid substrate. The height of the protrusion formed is determined by the value of the modulus of elasticity of the substrate. The deformation is easily revealed visually, for example at the surface of gelatine gel [338]. Estimations give the following values of the protrusion height: of about 0.1 μm for rubber and of atomic size for mica [339].

The extension of the perimeter of contact is accompanied by performance of the work of deformation. According to Rusanov [340], this effect may be reduced to line tension of a special type $\kappa_d > 0$, being of about 10^{-5} N that is many orders higher than that of the true line tension. Meniscus movement is accompanied by a moving wave of surface deformation. The existence of the transition zone between the meniscus and the film leads to a redistribution of the applied force F over the surface region several microns wide near to the contact line [341].

The surface roughness not only influences the onset of the motion of drops or menisci, but causes a change in the equilibrium contact angle θ_{0R} compared to θ_0 on a similar smooth substrate. The contact angle changes because the actual area of contact between the liquid and a solid substrate changes. Here the equilibrium conditions lead to the Wenzel–Derjaguin equation [342]:

$$\cos \theta_{0R} = K_R \cos \theta_0, \tag{X.8}$$

where K_R is the roughness coefficient equal to the ratio of the actual contact area and its projection on a horizontal plane. For contact angles $\theta_0 < 90°$,

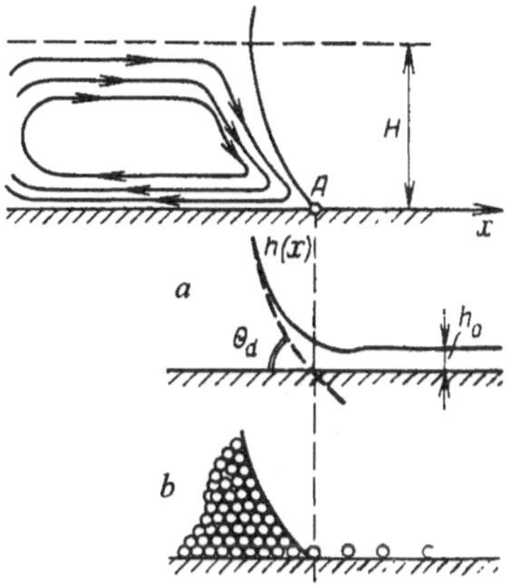

Figure 10.10 Scheme of streamlines near to an advancing meniscus in a capillary and the shape of contact zone near to the point *A* in the case when wetting films (*a*) or adsorption films (*b*) are formed.

increasing values of K_R cause the contact angle to be reduced compared to a smooth surface. At high values of K_R, $\theta_{0R} \rightarrow 0$ which leads to complete wetting. For practical purposes, rough surfaces are often used to provide complete wetting. The absence of hysteresis is related here to the filling of cavities with liquid under the action of capillary forces that facilitates the mass transfer between the films and the bulk liquid and eliminates the hysteresis.

For surfaces consisting of *m* micro-regions with different equilibrium contact angles θ_i, the mean value of the angle is determined from the Cassie equation:

$$\cos\theta_0 = \sum_{i=1}^{m} \varphi_i \cos\theta_i, \qquad \text{(X.9)}$$

where φ_i is the fraction of different regions.

The peculiarities of contact angle hysteresis in melts, when chemical reactions take place with a solid, in particular with metal substrates, are discussed in the monograph [343].

§3 Dynamic Contact Angles

When the static value θ_A or θ_R has been overcome, the meniscus begins to move relative to the surface. As this takes place, the distribution of pressure changes due to viscous losses, changing the shape of the meniscus. Extending now the undisturbed part of the profile of the moving meniscus up to the intersection with the substrate, we obtain a value of the contact angle that depends on the rate v of displacement of the perimeter of wetting. Advancing dynamic contact angles θ_d, as shown in experiments, begin to differ from static values exceeding these when $v \geq 10^{-3}$ cm/s.

Hydrodynamic calculations and direct observations with small suspended particles give the picture of the velocity field close to the moving meniscus shown in Fig. 10.10. The main viscous resistance exhibit the thinnest liquid layers in the region A near to the line of wetting. On a larger scale this region is shown in Fig. 10.10, below: for wetting (a) and adsorption (b) films. For hydrophilic surfaces, when the isotherms $\Pi(h)$ give an equilibrium thicknesses h_0 corresponding to wetting films (Chapter II, §2) a hydrodynamic analysis is possible assuming the film h_0 is a Newtonian viscous liquid.

An equation for the pressure gradient in the flowing liquid accounting for both capillary and surface forces may be obtained from eq. (II.12). For the one-dimensional flow in the direction of the x-axis we get:

$$dP/dx = -\sigma(d^3h/dx^3) - d\Pi(h)/dx = -[(\sigma d^3h/dx^3) + (d\Pi(h)/dh)(dh/dx)], \quad \text{(X.10)}$$

where the approximation of slightly sloping layer profile was used for surface curvature, $K \approx d^2h/dx^2$.

If the equation of viscous flow for a plane film (V.1) and the conditions of discontinuity at a stationary meniscus movement at a rate v = const in a slit capillary are applied locally, we obtain the following equation [344, 345]:

$$-(h^3/3\eta)(dP/dx) = v[h(x) - h_0]. \quad \text{(X.11)}$$

After substituting of eq. (X.10), this takes the form [346]:

$$h^3\{(d^3h/dx^3) + (1/\sigma)[d\Pi(h)/dh](dh/dx)\} = 3\eta v(h - h_0)/\sigma, \quad \text{(X.12)}$$

where σ is the surface tension and η is the liquid viscosity.

In the case of very thick films $h_0 \geq 0.1$ μm that arise through perfect wetting, we may restrict our consideration to the capillary effects only. Here on the left side of eq. (X.12) only the first term is retained. The solution of eq. (X.12) both in this case and especially when the surface forces are taken into account requires numerical methods to be applied [346, 347]. The only exception is the case of small values of the criterion $N = \eta v/\sigma$, where the analytical solution can be obtained [348]:

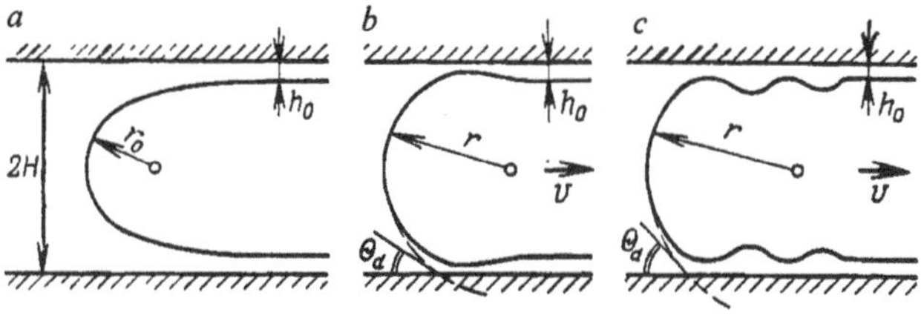

Figure 10.11 Meniscus and transition zone profiles in equilibrium state (*a*) and in the case of the increasing rate of motion (*b* and *c*).

$$\tan \theta_d = 3.4(N)^{1/3}. \quad \text{(X.13)}$$

As seen from this equation, the dynamic contact angle grows with the rate of displacement of the advancing meniscus.

Another analytical form of the hydrodynamic solution for the case of small Reynolds numbers has been obtained by Voinov [349]:

$$\theta_d^3 = \theta_0^3 + 9N \ln (L/L_s),$$

where L and L_s are the characteristic lengths for the macroscopic and microscopic regions of flow. This equation also follows from a more general analysis of the hydrodynamic problem performed by Cox [350].

In the general case the calculations based on eq. (X.12) require identification of an isotherm of disjoining pressure of wetting films $\Pi(h)$. In [346] a numerical solution has been obtained for the isotherm in the form $\Pi = A/h^n > 0$ corresponding to complete wetting of surfaces of a plane slit of width $2H$. The profile shape of a flowing transition zone depends here on the parameter α equal to:

$$\alpha = nA^{1/n}\Pi_0^{n(n-1)}/\sigma(3N)^{2/3}, \quad \text{(X.14)}$$

where $\Pi_0 = P_c = \sigma/r_0$ is the capillary pressure of the immobile meniscus (Fig. 10.11, *a*) also determining the film thickness when $x \to \infty$:

$$h_0 = (A/\Pi_0)^{1/n}.$$

When $\alpha > 1.89$, the thickness of the flowing layer has a single minimum, as shown in Fig. 10.11, *b*. For $\alpha < 1.89$, corresponding to increasing values of the flow rate or to thin films (i.e. small values of the surface forces constant A) wavy films form in front of the advancing meniscus (Fig. 10.11, *c*). When

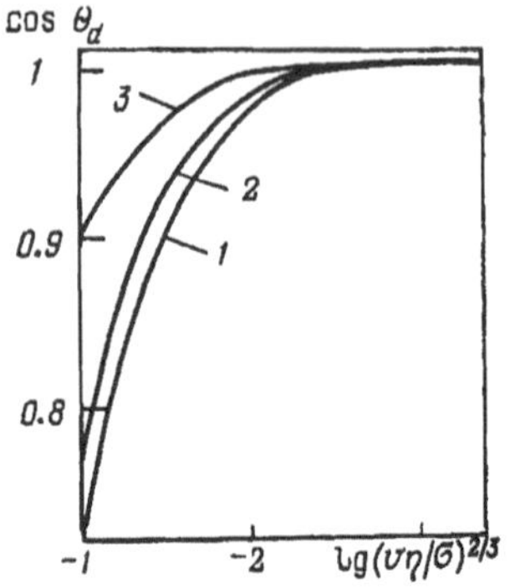

Fig. 10.12

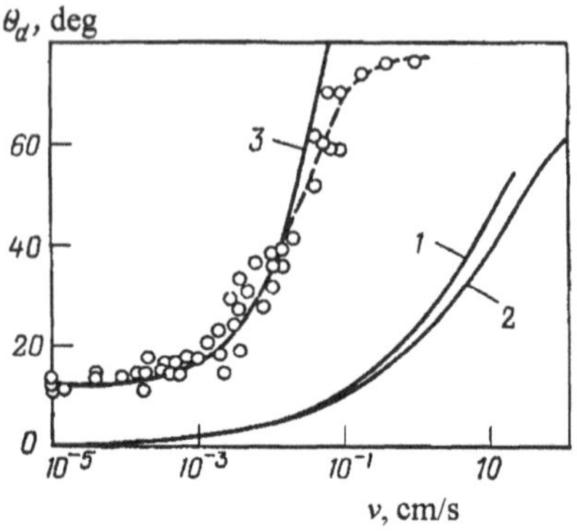

Fig. 10.13

Figure 10.12 Calculated dependences of a dynamic contact angle, θ_d, on the parameter $N = \eta v/\sigma$ for different slit width H and equilibrium wetting film thickness h_0. 1) $H = 10^{-2}$ cm, $h_0 = 15$ nm; 2) $H = 1.25 \cdot 10^{-3}$ cm, $h_0 = 7.4$ nm; *3)* $H = 1.25 \cdot 10^{-5}$ cm, $h_0 = 1.6$ nm.

Figure 10.13 Experimental dependence $\theta_d(v)$ (points) in comparisons with values calculated using eq. (X.12) (curve *1*) and eq. (X.13) (curve *2*) for the case of complete wetting and with eq. (X. 19) (curve *3*).

$A = 0$, and capillary forces only act, the wavy films emerge, i.e. the profile shown in Fig. 10.11, *c*, persists at all flow rates. The action of surface forces quenches the capillary waves due to an increase in the positive disjoining pressure tending to level off the film thickness.

Figure 10.12 shows the results of numerical calculations for a dynamic contact angle θ_d as a function of the criterion $N = \eta v/\sigma$. The calculations were made for $n = 3$, $\sigma = 30$ mN/m, $\eta = 10^{-3}$ N s/m^2, and $A = 10^{-21}$ J. The results are applicable not only for a flat slit, but also for cylindrical capillaries, when the condition $H >> h_0$ is fulfilled. To determine θ_d, the segment of the profile $h(x)$ of constant curvature $d^2h/dx^2 = \sigma/r =$ const adjacent to the transition zone was used and, therefore, located in the region of smooth profile ($dh/dx << 1$). At a low rate of flow ($N < 10^{-3}$), the values of $H/r = \cos\theta_d$ are close to 1 that corresponds to the equilibrium contact angle $\theta_0 = 0$. When $N > 10^{-3}$, the values of H/r become less than 1 and the dynamic contact angle grows. The larger the slit width H or the capillary radius determining the thickness of equilibrium wetting film h_0, the earlier the difference between θ_d and θ_0 appears and θ_d grows more rapidly with N. Thus the curve *1* in Fig. 10.12 corresponds to the film thickness $h_0 = 15$ nm; the curve 2 to $h_0 = 7.4$ nm; the curve *3* to $h_0 = 1.6$ nm. At the greatest film thickness $h_0 = 15$ nm, the results of calculations agree with those performed without taking account of the disjoining pressure [347], and at small N with those calculated by eq. (X.13).

When $N \leq 0.2$ and $\cos\theta_d > 0.6$, good agreement with the theory was achieved for glycerol ($\sigma = 62$ mN/m, $\eta = 14$ N s/m^2) in quartz capillaries of radii from 53 to 412 μm [347], at the thicknesses h_* of preliminary deposited wetting films ranging from 0.1 to 8 μm. These thicknesses exceeded by one order of magnitude that of thermodynamically equilibrium films h_0 of glycerol. When h_* approached h_0 deviation from the theory was observed due to the action of surface forces ignored in [347]. Satisfactory agreement with hydrodynamic theory was also achieved for viscous paraffin oil [351].

The area of application of hydrodynamic solutions is restricted to the cases of perfect wetting, when the meniscus has the shape presented in Fig. 10.11, *a*.

For water and other liquids of low viscosity the area of application of the hydrodynamic theory is determined by the thickness of the films emerging in front of an advancing meniscus. The measurements of θ_d for water were carried out in plane slits and capillaries [352, 353]. The values of $\cos\theta_d$ were determined from the known Washburn equation for capillary rise [354]. This represents a modification of the Poiseuille equation in which the difference between the capillary ($P_c = 2\sigma\cos\theta/r$) and hydrostatic pressures ($P_h = \rho g H$), divided by the height of the liquid column H in the capillary, is taken as the pressure gradient. The current value of $\cos\theta_d$ is determined from the experimentally measured rate of liquid rise $v = dH/dt$ at a given value of $H(t)$:

$$\cos\theta_d = (H/\sigma)[(4v\eta/r) + (\rho g r/2)], \tag{X.15}$$

where r is the capillary radius, g is the acceleration due to gravity and ρ is the density of the liquid.

The experimental results obtained for water in vertically fixed quartz capillaries of radii from 15 to 25 μm and up to 1 m long are shown by points in Fig. 10.13 [353]. An α-film of thickness close to $h_0 = 4–5$ nm, taken from the curve *4* in Fig. 10.3, was first formed in the course of vapor adsorption at $p/p_s \approx 1$ on the capillary surface. Static advancing contact angles formed between the bulk water and this film measured by the height of capillary rise H_0 (at $t \to \infty$) ranged from $\theta_A = 8°$ to 14° for different capillaries.

Hydrodynamic theory gives the dependence θ_d vs. v represented by the curve *1* through numerical calculations by eq. (X.13). The curve *2* shows the results of calculations by eq. (X.13). As seen from the comparison of experimental data and hydrodynamic theory, there is only qualitative agreement. The same conclusion follows from the analysis of other experimental data for water [352].

Quantitative deviations primarily arise because the calculations by eq. (X.12) were made only for the case of complete wetting and smooth isotherms $\Pi(h)$ of type $\Pi = A/h^n$. For water films on quartz (when $\theta_0 \approx 10°$) S-type isotherms are characteristic (Fig. 2.6) which substantially changes the profile of the transition zone and, therefore, the dependence of θ_d on v.

To explain the experimental rate dependences of the dynamic contact angle $\theta_d(v)$ in the case of partial or poor wetting, the molecular-kinetic model of Blake and Haynes [355] is used. In doing this, it is assumed that the main resistance of the meniscus arises in the contact region (Fig. 10.10, *b*). Let us consider further modification of this model developed by Hoffman [356] in which the Eyring theory of reaction rates is used. At equilibrium, at $\theta = \theta_0$, the exchange between the molecules of liquid under the meniscus and in the adsorption layer is regulated by the symmetrical potential barrier. Here the number of molecules transferring from the adsorption layer into the bulk is equal to the number of molecules moving in the opposite direction. In the absence of equilibrium, at $\theta > \theta_0$ or $\theta < \theta_0$, the potential barrier is no longer symmetric, and the line of wetting displaces at the rate:

$$v = a_c(j_F - j_B), \qquad \text{(X.16)}$$

where j_F and j_B are the frequencies of molecule jumps in the forward and the reverse directions, and a_c is the length of the jumps. For $\theta > \theta_0$, $j_F > j_B$, and the line of wetting moves along the x-axis at the rate v equal, according to the Eyring theory of viscosity [357], to:

$$v = 2a_c[(kT/\hbar)\exp(-G/kT)]\sinh[\sigma(\cos\theta_0 - \cos\theta_d)/nkT], \qquad \text{(X.17)}$$

where $\hbar$ is the Planck constant, k is the Boltzmann constant; G is the height of the potential barrier and n is the number of molecules per unit surface area in the vicinity of the line of wetting. The following expression is then used for the viscosity of the bulk liquid η according to the Eyring theory:

$$\eta = (\hbar/a_c^2\delta)\exp(G_0/kT), \qquad \text{(X.18)}$$

where δ is the molecular diameter, G_0 is the activation energy of viscous flow.

Applying the expression (X.18), we obtain the following equation relating θ_d to the rate of displacement v:

$$v = (2kT/a_c\delta\eta)\exp(-\Delta G/kT)\sinh[(\sigma/nkT)(\cos\theta_0 - \cos\theta_d)], \qquad \text{(X.19)}$$

where $\Delta G = G - G_0$.

At a low rate of meniscus displacement, when the values of $\cos\theta_d$ are close to those of $\cos\theta_0$, eq. (X.19) simplifies:

$$v = \sigma[(\cos\theta_0 - \cos\theta_d)/(a_c n\delta\eta)]\exp(-\Delta G/kT). \qquad \text{(X.20)}$$

As seen from this expression, in the case of adsorption films the difference of the cosines of the equilibrium and dynamic contact angles depends linearly on the rate of displacement v.

The results of calculations of θ_d by eq. (X.19) are shown in Fig. 10.13 by curve *3*. The island model of the α-film on quartz [358] which is thicker over the hydrophilic regions of quartz surface covered with hydroxyl groups and thinner over the less hydrophilic siloxane surface was used for this purpose. The thin adsorption layers, presumably 1–2 molecules thick, determine the kinetics of molecular transfer between the meniscus and the surface layers. We can then assume $n = 2\times10^{15}$ cm^{-2}, $\delta = 0.3$ nm and $a_c \approx 2\delta = 0.6$ nm, bearing in mind that according to Eyring the transfer proceeds through the vacancy defects of a regular lattice. Then the bulk values of $\sigma = 72$ mN/m and $\eta = 10^{-3}$ N s/m^2 for water are used, because only these enter into eq. (X.19). The values of ΔG are used as a fitting parameter to reconcile the theory with the experiment results. These characterize the difference in mobility of molecules in the bulk and in the surface layers. Satisfactory agreement with experiment has been achieved, as seen from Fig. 10.13, assuming $\Delta G = 11kT$ that corresponds to a sharply reduced molecular mobility in the adsorption phase compared to the bulk liquid.

It is difficult to know how close this model is to reality in the case of water α-films with mean thickness of 4–5 nm. It may be more successfully applied to surfaces less hydrophilic than quartz, when adsorption films with mean thickness smaller than 1–2 molecular layers are formed.

Consider now the results first obtained on the basis of the Eyring model by Blake and Haynes [355]. They assumed that meniscus displacement is controlled by surface diffusion of liquid molecules that gives the following equation:

$$v = 2a_c D \sinh[\pm\sigma(\cos\theta_0 - \cos\theta_d)/nkT)]. \qquad \text{(X.21)}$$

Here the plus sign refers to an advancing meniscus and the minus sign to the receding one; D is the number of molecular displacements in unit of time per unit length of the line of wetting; n is the surface concentration of adsorption centres; and a_c is the mean distance between them. The product $a_c D$ is the reciprocal of the surface viscosity η_s being the force impeding the meniscus displacement.

In [352] good agreement between the calculations of $\theta_d(v)$ by eq. (X.21) and the experimental data for water on glass surface has been achieved with the assumption that the number of OH-groups (centres of adsorption for water molecules) per unit surface is equal to $n = 1.4\times10^{14}$ cm^{-2}. This value of n is close to the known surface concentration of hydroxyls, 4.2×10^{14} cm^{-2}.

For the water–benzene system in a hydrophobized capillary ($\theta_0 = 119°$), the distance between the surface centres was obtained as equal to $a_c = 1.4$ nm and $D = 500\ s^{-1}$ [355]. For water on a surface of polyethylene terephtalate ($\theta_0 = 60°$) the values were $a_c = 1.2$ nm and $D = 1.1\times10^6\ s^{-1}$, and for the water–glycerol system: $a_c = 1.7$ nm and $D = 4\times10^4\ s^{-1}$. Close values of a_c and D were also obtained for receding menisci [359]. It was shown here that the changes in the form of the dependence $\theta_d(v)$ with increasing rate of meniscus displacement may be explained rather by the transition from one set of parameters in eq. (X.21) to another than by the transition to the hydrodynamic regime of flow.

In the case of meniscus motion along a rough lyophilic surface, it is necessary to take into account the possibility of capillary imbibition into channels formed by irregularities on the surface in front of the advancing meniscus [360].

Therefore, the theory of dynamic contact angles may be considered satisfactorily developed for practical use in the case of complete wetting, when the hydrodynamic theory is valid, and for the cases when there are thin adsorption layers on the surface in front of the advancing meniscus. In the first case the dependences $\theta_d(v)$ are determined by the isotherms of disjoining pressure $\Pi(h)$ and the liquid viscosity, and in the second by the isotherms of adsorption and the surface mobility of molecules. The intermediate case is more complicated and needs further examination.

The problem of a receding dynamic contact angle $\theta_R(v)$ has been solved in a similar way. In the framework of the molecular-kinetic approach the results are symmetrical and differ only in the sign in eq. (X.21). For perfect wetting and a small rate of displacement, v, an equilibrium film h_0 remains behind the receding meniscus, and $\theta_R = \theta_0$. In the case of S-shaped isotherms $\Pi(h)$ (Fig. 10.2, curve 2), metastable β-films may form that creates the conditions for complete wetting: $\theta_0 > \theta_R = 0$. However, the β-films have a limited life-time and they may rupture after a time transforming into the thermodynamically stable α-films. The capillary surface behind the receding meniscus becomes covered with drops produced in the transition $\beta \rightarrow \alpha$.

As the rate of meniscus recession increases, a non-equilibrium dynamic liquid film forms, having a thickness $h > h_0$ that depends on the capillary radius r and on the rate of recession v [345, 361]:

$$h = Cr(v\eta/\sigma)^{2/3}. \qquad \text{(X.22)}$$

For isotherms $\Pi(h)$ of type $\Pi = A/h^n$ the coefficient C depends on the parameter:

$$\alpha = nA/\sigma h_0^{n-1}(3v\eta/\sigma)^{2/3}. \qquad \text{(X.23)}$$

For $A = 0$, when the action of surface forces may be neglected, $\alpha = 0$ and $C =$

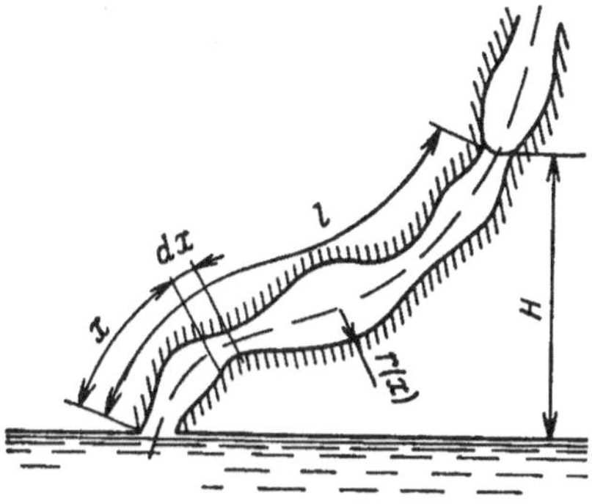

Figure 10.14 For use in the derivation of equations for capillary rise and impregnation.

1.32 [345]. When $\alpha = 0.1$, $C = 1.37$, for $\alpha = 1$, $C = 1.76$ and for $\alpha = 10$, $C = 7.4$ [361]. Non-equilibrium films of thickness $h > h_0$ are capillary-unstable. The system comes into an equilibrium state in two ways: evaporation of excess liquid $(h - h_0)$ or by local undulation and coalescence in thin slits and capillaries. The process of development of capillary bridges and formation of columns of capillary liquid has been studied in [252].

Equation (X.22) has been experimentally checked and is widely applied in calculations of the thicknesses of liquid films formed on the walls of vessels after drainage, on photo films, on rods and threads pulled out of the liquid [362].

§4 Capillary Rise and Imbibition

Calculations of capillary rise and imbibition are based on the Poiseuille equation for viscous flow and the expression for the meniscus capillary pressure [354]. The application of the first equation is possible at the length of impregnation $l >> r$, when the lines of flow become parallel (Fig. 10.10). Generally, the dynamic value of the contact angle θ_d should be used in the Laplace equation, that depends on the rate of meniscus displacement $v = Q/\pi r^2(l)$. Here Q is the liquid flux in the capillary, $r(l)$ is the local value of radius that may depend on the length of impregnation $l(t)$ and on the time, t. For capillaries of variable cross-section and arbitrary orientation (Fig. 10.14), the Poiseuille equation is applied locally on an elemental segment dx. Then for the loss of pressure across the length of impregnation $P(l)$ we obtain:

$$P(l) = (8\eta Q/\pi)\int_0^l dx/r(x)^4 = (2\sigma \cos \theta_d/r(l)] + P_0, \qquad \text{(X.24)}$$

where $r(x)$ expresses the radius variations along the length of the irregular cylindrical capillary and P_0 is an additional external pressure. In the case of the capillary rise $P_0 = -\rho g H$, where H is the height of the meniscus above the liquid level. In the course of imbibition into a horizontal capillary, $H = 0$, and the rate of meniscus displacement is equal to:

$$v(l) = [2\sigma\cos\theta_d/8\eta r^3(l)]/\int_0^l dx/r^4(x)\,. \tag{X.25}$$

For a capillary of constant radius, this expression is transformed into the well-known equation of capillary impregnation:

$$v = r\sigma\cos\theta_d/4\eta l. \tag{X.26}$$

At a low rate of imbibition ($v = dl/dt \leq 10^{-3}$ cm/s), when $\theta_d = \theta_0 = \text{const}$, the expression for the length l of the imbibed liquid in a horizontal capillary over the time t, counted from the moment when the capillary has touched the bulk liquid, can be obtained from eq. (X.26):

$$l(t) = [rt\sigma\cos\theta_0/2\eta]^{1/2} = K\sqrt{t}\,. \tag{X.27}$$

This equation is also applicable for imbibition in an arbitrary oriented capillary for the condition $H \ll 2\sigma\cos\theta_0/\rho g r$, applicable to capillaries of small radius or for the initial stage of capillary rise, at small H.

For straight cylindrical capillaries oriented at the angle α to the horizon, for the rate of capillary rise at $\theta = \theta_0$ we obtain from eq. (X.24):

$$v = (r^2\sin\alpha/8\eta H)[(2\sigma\cos\theta_0/r) - \rho g H]. \tag{X.28}$$

For the vertical orientation of a capillary, $\sin\alpha = 1$, the rate of rise depends only on H. Since $v = dH/dt$, the expression for the time of capillary rise to the height H can be obtained from eq. (X.29):

$$t = (8\eta/r^2\rho g)\{H_0\ln[H_0/(H_0 - H)] - H\}. \tag{X.29}$$

As $t \to \infty$, the ultimate height of the capillary rise is achieved $H_0 = H(\infty)$ equal to:

$$H_0 = 2\sigma\cos\theta_0/r\rho g. \tag{X.30}$$

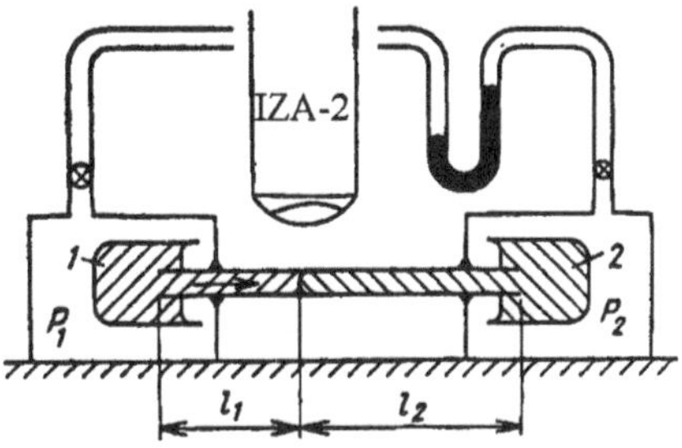

Figure 10.15 Apparatus for studying the kinetics of capillary displacement of immiscible fluids.

To accelerate the capillary imbibition, the application of an additional constant external pressure, P_0, may be used. Then, instead of (X.27), from eq. (X.24) for forced imbibition into a horizontal capillary, we will get:

$$l(t) = [r^2 t(P_c + P_0)/4\eta]^{1/2} = K\sqrt{t}, \qquad \text{(X.31)}$$

where $P_c = 2\sigma\cos\theta_0/r$ is the capillary pressure of the meniscus. For $P_0 = 0$ this equation transforms into eq. (X.27). When $P_0 >> P_c$, eq. (X.31) determines the length, l, of the liquid column entered into the capillary under the action of external constant pressure P_0. The larger the capillary radius r, the greater the length of imbibition, l, (when other factors remain the same) because the viscous resistance decreases with r in the case of Poiseuille flow.

The condition $P_0 \geq -2\sigma\cos\theta_0/r$, when $\theta_0 > 90°$ and $\cos\theta_0 < 0$, determines the possibility of liquid entering an unwettable capillary under the imposed external pressure.

The equations for capillary rise and capillary imbibition were first obtained by Washburn [354] who also suggested taking into account the influence of the viscous resistance of the fluid being displaced from the capillary. The account of this resistance is of great importance in the case of capillary displacement of one liquid by another. In a general case, the equation of capillary imbibition takes the following form:

$$v = r\sigma_{12}\cos\theta_d / 4(\eta_1 l_1 + \eta_2 l_2), \qquad \text{(X.32)}$$

where l_1 and l_2 are the lengths of the capillary occupied with liquid *1* and liquid *2* (Fig. 10.15) and η_1 and η_2 are their viscosities, respectively; σ_{12} is the interfacial tension. When $l_1 = 0$ or $l_2 = 0$ or in the case of low viscosity of one of the fluids (for instance a gas), eq. (X.32) is transformed into eq. (X.26).

When, besides the capillary pressure, there is a drop of external pressure, $\Delta P = P_1 - P_2$, the equation transforms in the following way:

$$v = r^2[\Delta P + (2\sigma_{12}\cos\theta_d/r)]/8(\eta_1 l_1 + \eta_2 l_2), \tag{X.33}$$

where $2\sigma_{12}\cos\theta_d/r = P_c$ is the capillary pressure of the meniscus. When $\Delta P >> P_c$, this equation describes the Poiseuille flow of immiscible liquid in a capillary:

$$v = r^2\Delta P/8(\eta_1 l_1 + \eta_2 l_2).$$

Expressing l_2 as $(L - l_1)$, where L is the capillary length, we obtain the dependence of flow rate v only on l_1, that is on the position of the meniscus (Fig. 10.15).

In a series of cases the influence of the retained gas should be taken into account, for example, when the pores or capillaries are blind or when the impregnation advances from the two sides of the porous body simultaneously. If the gas is insoluble in the liquid, the imbibition stops when the gas is compressed to the pressure balancing the capillary pressure of the meniscus. The length of the liquid penetration, l, depends here on the radius, r, and the length of the capillary L. The pressure of the compressed gas is determined by the Boyle–Mariotte law: $PV = P_0V_0$, where P_0 and V_0 are pressure and volume of the gas before the compression. Because the gas volume is proportional to the length of the free part of the capillary, then for the change in gas pressure relative to that in the surrounding medium P_0 we obtain: $P/P_0 = L(L - l)$. Consequently, for the relative length of imbibition (in the case of non-soluble gases) we have:

$$l/L = 1/[1 + (P_0 r/2\sigma\cos\theta_0)]. \tag{X.34}$$

The length of imbibition is larger, the smaller the capillary radius and the lower the initial gas pressure. Therefore, to accelerate capillary imbibition of porous bodies, a preliminary gas evacuation is often recommended. In the case of complete evacuation ($P_0 = 0$), the imbibition naturally goes to completion, and $l/L = 1$.

In practice gases are soluble in liquids. Therefore, a gas compressed in a blind capillary diffuses through the liquid column, l, under the pressure drop, $P - P_0$, and, consequently, as a result of the difference in the concentration of the dissolved gas $C - C_0$. The rate of gas diffusion depends on the coefficient of gas diffusion in the liquid, $D \sim 10^{-5}$ cm^2/s. As a result, the length of the gas column $(L - l)$ gradually decreases, and when $t \to \infty$ all the capillary is filled with the liquid. However, the rates of imbibition controlled by gas diffusion

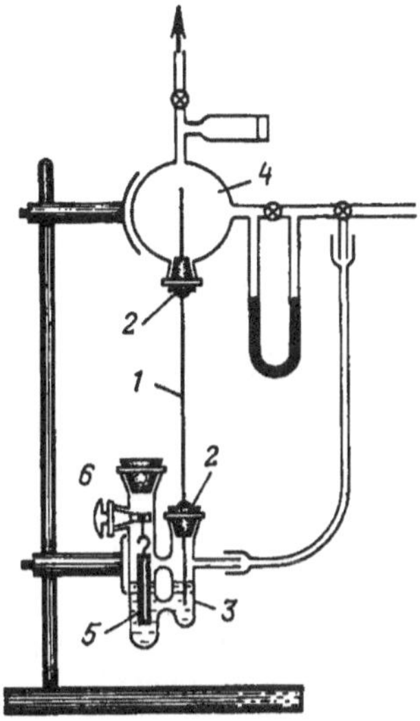

Figure 10.16 Apparatus for studying capillary rise of liquids in narrow capillaries.

are usually low. These processes, as well as those of diffusion extraction of soluble inclusions from pore surface in the course of capillary imbibition, are discussed in detail in the monograph [363].

The theory of capillary rise and impregnation has been experimentally checked many times. The most reliable results were obtained using straight cylindrical quartz capillaries with a molecularly smooth surface produced by the method of rapid stretching of heated silica tubes (99.99% SiO_2) [353]. The apparatus used is shown in Fig. 10.16. Capillary (*1*) of radius 15–20 μm and 1 m long was glued by epoxy resin into the glass plugs (*2*) held in joints in the vessel (*3*) filled with water and in the upper chamber (*4*). The level of liquid in the vessel (*3*) may be changed by the dipping of the cylinder (*5*), when rotating the plug (*6*) in the joint. The apparatus was pumped out to avoid air viscous resistance and to form quickly an equilibrium water film on the capillary surface. The start point was the moment when the capillary end touched the liquid rising in the vessel (*3*). The meniscus position H in respect to the liquid level was observed by the cathetometer KM-6.

The results of these experiments are presented in Fig. 10.17. To avoid damage, the capillaries were slightly bent when mounted. Due to this and also because of a small dipping of the capillary into the liquid, the length of the filled section of the capillary l entered into the Poiseuille equation slightly differed from the height of rise H. Experimental data are shown by points in Fig. 10.17. Corresponding theoretical curves 2, 5 and 7 are calculated by

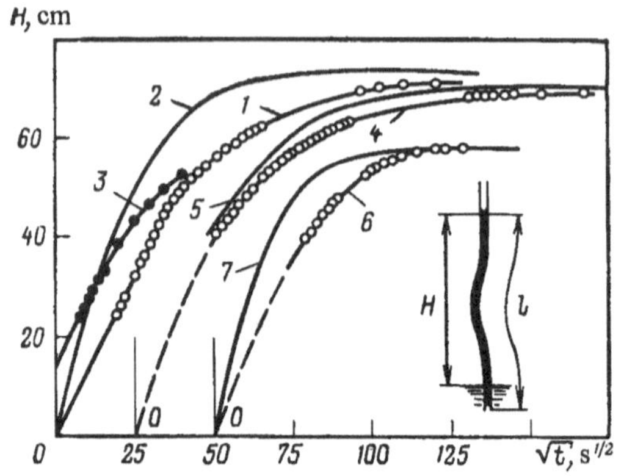

Figure 10.17 Height of water rise, H, vs. time in quartz capillaries $r = 20$ μm (*1* and *3*); $r = 18$ μm (*4*) and $r = 25$ μm (*6*). Solid curves 2, 5 and 7 are calculated using eq. (X.29) for the case of complete wetting.

eq. (X.29) at θ_0 = const with a correction for the small difference between l and H. Equilibrium contact angles, θ_0, were determined from eq. (X.30) by the maximal capillary rise H_0 established approximately within two days. The curve *1* corresponds to the values: $H_0 = 71.27$ cm and $\theta_0 = 12°20'$, the curve *4*: $H_0 = 70.48$ cm and $\theta_0 = 7°40'$, the curve *6*: $H_0 = 57.77$ cm and $\theta_0 = 10°30'$. The discrepancy between the theory and experiment is caused by the difference between the dynamic θ_d and equilibrium θ_0 contact angle (Fig. 10.12). Due to high values of θ_d in the initial stage of rising, when the rates of imbibition are high, the experimental curves $H(t)$ pass below the theoretical relationship calculated for θ_0 = const without corrections for dynamic effects.

The role of film thickness on the capillary walls was studied using repeat experiments on the rising of the water (curve *3*), for which purpose the water was depressed by the air pressure drawn into the chamber *4* (Fig. 10.16), after the curve *1* has been obtained. In this case a non-equilibrium wetting film $h > h_0$ remained on the capillary surface. The thickness of the film at the bottom was higher than at the top due to the increase in the rate of displacement with decreasing column length in the capillary. The rate of rising was in this case higher due to the complete wetting (curve *3*). Therewith the height of capillary rise H_0 remained the same, for in the upper section of the capillary there was an equilibrium film of thickness $h_0 \approx 4{-}5$ nm presented.

The differences in the height of capillary rise H_0 corresponding to the advancing contact angle θ_A and to receding θ_R are small. The latter values were measured after the meniscus was raised to the height $H > H_0$ equal to 75–80 cm and then allowed to fall. Static hysteresis of contact angles does not exceed several degrees.

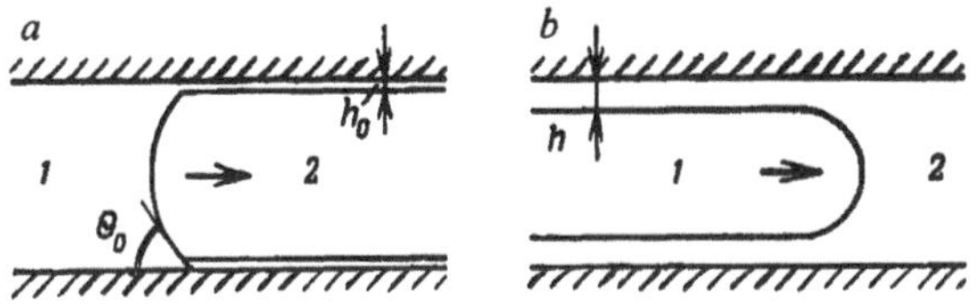

Figure 10.18 Capillary effects during displacement of immiscible liquids: *a*) spontaneous displacement; *b*) forced displacement under additional external pressure.

For highly viscous liquids (oil, glycerol) completely wetting the capillaries, the dynamic effects are insignificant. Their imbibition into the capillaries of radius from 1 to 10 μm follows eq. (X.27) at $\theta_0 = 0$ and using the bulk values of viscosity and surface tension.

The dynamic effects were observed during a capillary displacement of immiscible liquids, like oil/water. Experimental investigations carried out using the apparatus shown in Fig. 10.15 were focused on a study of the mechanism of displacement of oil by water [364]. Model systems and hydrophilic quartz capillaries of radius 10–20 μm were used in the experiments. Nonpolar liquids of various viscosity, from dibutylphthalate ($\eta = 2.5\times10^{-2}$ N s/m^2) to toluene ($\eta = 6\times10^{-4}$ N s/m^2), were used as "oils". The interfacial oil–water tension σ_{12} ranged here from 53 mN/m (water–tetradecane) to 19.2 mN/m (water–dibutylphthalate).

Both spontaneous displacement of oils by water and mutual displacement activated by an additional pressure drop $\Delta P = P_1 - P_2$ (Fig. 10.15) were studied. The meniscus displacement in the capillary was registered by the comparator IZA-2 with an accuracy of 1 μm.

During a slow movement of the meniscus ($v = 10^{-4}$–10^{-3} cm/s) of the better wetting liquid *1*, advancing on the equilibrium wetting film of thickness h_0 (Fig. 10.18, *a*) the displacement of liquid *2* proceeds at a constant contact angle θ_A and a constant drop of the capillary pressure $P_c = \sigma_{12} \cos\theta_A / r$, which is the driving force of the process. In the course of the spontaneous displacement of tetradecane, dibutylphthalate or toluene by water, the values of θ_A ranged from 83° to 86°. The kinetics of displacement are well described by eq. (X.32) using the bulk values of viscosity and interfacial tension σ_{12}.

For a forced displacement ($v = 10^{-2}$–10^{-1} cm/s) with an additional drop of external pressure ΔP (Fig. 10.18, *b*), the capillary pressure of the meniscus remained constant but equal to $P_c = 2\sigma_{12}/r$ which corresponds to the condition of complete wetting by an oil. This pressure hinders displacement especially in narrow capillaries. In this case the condition of complete wetting is dynamic and connected with formation of thick non-equilibrium films of liquid *2* behind the quickly receding meniscus. The presence of thick films of bulk liquid provides their complete wetting by the displaced liquid *2*. When the flow stops, non-equilibrium films rupture forming first local thickenings and then even drops of liquid 2 in liquid *1*. At a higher rate of displacement ($v \geq 1$ cm/s) the rupture of the non-equilibrium films occurs during the course of displacement. As a result, the disintegration of the displacing liquid in drops with formation of an emulsion occurs.

An estimation of the thickness of non-equilibrium films was performed by eq. (X.22) at $C = 1.32$. The values of h from 70 (at $v = 10^{-2}$ cm/s) to 300 nm (at $v = 10^{-1}$ cm/s) were obtained for the displacement of viscous dibutyl-phthalate by water. In the case of forced displacement of water by dibutyl-phthalate the film thicknesses were lower: from 10 to 40 nm, respectively. The dynamic effects shown up at a mutual displacement are stronger, the higher the viscosity of the displaced liquid.

Similar effects were also observed in hydrophobized quartz capillaries [365]. A comprehensive review of publications on displacement of immiscible liquids is presented in [366].

The addition of surfactants adsorbing on the oil–water interfacial surface and on the capillary surface markedly influences the kinetics of displacement [367–369]. Surfactants adsorb on the walls of the capillaries in the course of imbibition. At a high rate of imbibition, this causes a decrease to zero of the concentration of surfactant at the moving meniscus. In this case the kinetics of impregnation is actually controlled only by interfacial oil–water tension. At a slow imbibition the surfactant concentration on the meniscus is controlled by diffusion of surfactant from the bulk solution. This may lead to a step-wise movement of the wetting line [370] or its slow advance next to the front of the surface diffusion of surfactant ahead of the meniscus [371]. This effect will be discussed in detail in the next section.

The influence of surfactants on the kinetics of mutual displacement of liquids such as oil–water in the presence of surfactants was studied once the surfactant solution had been balanced with the capillary surface by the preliminary adsorption [367–369]. For this purpose the solution of surfactant at the required concentration was pumped through the capillary for a long time before the experiments. After such a pretreatment only the processes associated with changes in capillary pressure may influence the kinetics of displacement. The transition of surfactant molecules from the surface of the receding meniscus onto that of forming wetting film was observed. As a result, the surfactant concentration at the meniscus decreases and the receding dynamic angle tends to zero in the range of rates of displacement $v < 10^{-4}$ cm/s. The capillary pressure of the receding meniscus, calculated by eq. (X.33), rises with v. When $v > 10^{-4}$ cm/s, values of the dynamic capillary pressure stabilize and remain constant in the range of flow rates up to 2×10^{-2} cm/s.

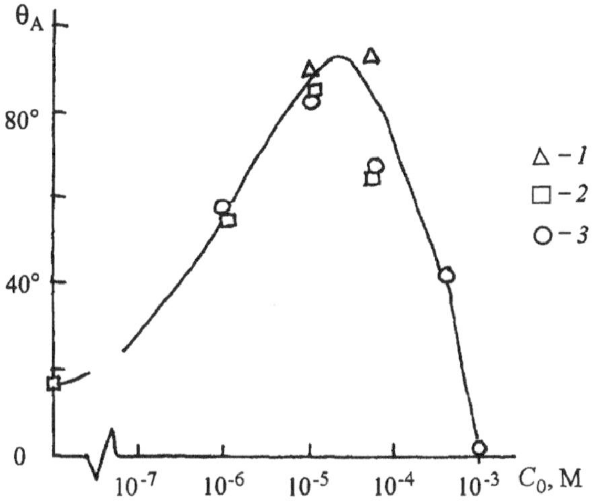

Figure 10.19 Dependence of advancing contact angle, θ_A, on CTAB concentration, C_0, obtained during displacement of tetradecane (points *1*) and silicone oil, M = 2200, (points *2*) by aqueous CTAB solution from narrow quartz capillaries. Points *3* were obtained by the method of sessile drop of variable volume for CTAB solutions on a quartz plate in a medium of silicone oil.

Analysis of the processes of convective and diffusive transfer of surfactant molecules between the film and the receding meniscus leads to the following dependence of surfactant concentration at the meniscus C_m on the rate of its movement v [368, 369]:

$$C_m/C_0 = 1/\{1 + [kPe/r(1 + Pe_s)]\}. \tag{X.35}$$

Here C_0 is the concentration of the bulk solution; r is the capillary radius; $k = \Gamma_m/C_m$ is the Henry constant; $Pe = v\Delta/D$ and $Pe_s = v\Delta_s/D_s$ are the bulk and surface Peclet criteria, where Δ and Δ_s, D and D_s are characteristic lengths and coefficients of surfactant diffusion, respectively.

During a rising rate of recession, the ratio C_m/C_0 tends to the constant value equal to:

$$C_m/C_0 = 1/[1 + (k\Delta D_s/r\Delta_s D)]. \tag{X.36}$$

The surfactant concentration near to the meniscus decreases more sharply in the narrow capillaries and with a high adsorbability of surfactant characterized by the k value. Equations (X.35) and (X.36) agree well with experimental data [368, 369].

The dependence of the dynamic advancing contact angle, θ_A, on the surfactant concentration may be obtained from the values of capillary pressure of the advancing meniscus determined by eq. (X.33). Data obtained for the displacement of tetradecane (points *1*) and silicone oil (points *2*) by aqueous CTAB solution are shown as an example in Fig. 10.19. These data agree well with the results of measurements of θ_A by the method of sessile drop of variable volume (points *3*) [368]. The observed dependence of θ_A on the concentration of cationic surfactant is similar to that presented in Fig. 10.6, *a*. As seen from Fig. 10.19, complete wetting of the capillary by aqueous CTAB solution is provided at a surfactant concentration close to CMC. Maximal values of the contact angle θ_A are achieved at $C_0 \approx 10^{-5}$ M, when the quartz surface has already become hydrophobic and the oil–water interface still remains hydrophobic. In this case the hydrophobic attraction of surfaces is more pronounced, which, together with forces of electrostatic attraction ($\psi_1\psi_2 < 0$) provides the highest values of the contact angle (see Chapter 10, § 1).

In real porous bodies, where pores are of various sizes, the displacement is more complex due to the effects of hysteresis associated both with pore branching and variable radius and heterogeneity of their surface. As a result the displacement front appears to be irregular and so-called "viscous fingers" form [372]. When the displaced liquid more effectively wets the porous body, the width of the viscous fingers is commensurate with pore radius. This is explained by the formation of stable wetting films damping the impact of heterogeneity. When a less effective wetting liquid is being displaced, the width of viscous fingers may be much greater than the pore size due to the rupture of non-equilibrium films and the formation of plugs of one liquid in another.

Various methods may be applied to intensify the capillary rise and imbibition. One of them is the improvement of wetting by the pretreatment of the capillary surface to make it lyophilic. In the case of hydrophobic surfaces, surfactants are usually used for this purpose, and this will be discussed in detail in the next section.

An ultrasonic field considerably influences the capillary rise [373]. This is ascribed to a variety of reasons, one being the sonic pressure which, however, damps with the distance from the capillary entrance, and is confined to this zone of about several diameters [374]. The effect of asymmetry of viscous resistance occurring when the liquid flows in or out of the capillary may also influence the capillary rise [373, 375]. The height of a capillary rise may exceed or be beyond the equilibrium height, H_0, depending on the shape of the capillary bore near to its entrance. In experiments with a conical perforation in a diaphragm, when the effect of asymmetry was considerably intensified, it has been shown that the diaphragm possesses a pumping ability [373]. When the diaphragm vibrates, the liquid flows towards the narrow end of the conical channel. However the effect takes place only at a rather high frequency of vibration, when the laminar flow in the perforation is disturbed. In capillaries with a flat end, the asymmetry is created by the abrupt transition from the bulk liquid to the narrow capillary bore. Usually this leads to an additional capillary rise, which is larger, the higher the frequency of vibrations of the capillary in respect to the bulk liquid. Cavitation bubbles formed at the capillary end may further enhance the effect of asymmetry.

In liquid dielectrics (with high values of dielectric constant ε_0) the rate of imbibition may be changed under the action of an external non-uniform electric field (NEF) [376].

Let us consider the motion of liquid in a capillary of radius r_0 under the simultaneous action of two gradients: pressure $\partial P/\partial z$ and external mass force $f(z)$. In the case of stationary laminar flow of viscous uncompressible liquid along the z-axis of a cylindrical capillary, the Navier–Stokes equation takes the form:

$$\partial^2 v/\partial r^2 + (1/r)\partial v/\partial r = (1/\eta)\partial P/\partial z - (\rho/\eta)f(z), \quad \text{(X.37)}$$

where $v(r)$ is the local velocity, r is the radial coordinate, ρ and η are density and viscosity of the liquid.

The right side of the equation does not depend on r and may be denoted as:

$$(1/\eta)\partial P/\partial z - (\rho/\eta)f(z) = \lambda. \quad \text{(X.38)}$$

The value of λ may be found from the boundary conditions for capillary imbibition (Fig. 10.20). Here $z = 0$ corresponds to the liquid level and $z = l$ to the meniscus position in the capillary. All further calculations are performed in a quasi-stationary approximation assuming that for each l value the stationary equation (X.37) may be used.

Figure 10.20 exemplifies a possible distribution of the mass force $f(z)$ along the capillary, caused by the NEF. The latter is created by two electrodes (needle-plate) separated by a distance L. The curve $E(z)$ shows schematically the distribution of the electric field along the capillary axis. The disposition of the capillary and the electrodes may be different. For further consideration, it is important that the NEF may act on some part of dielectric liquid filling the capillary.

For application to the scheme presented in Fig. 10.20, the following boundary conditions may be written:

$$P = P_0,\ z = 0, \quad \text{(X.39)}$$

$$P = P(l),\ z = l. \quad \text{(X.40)}$$

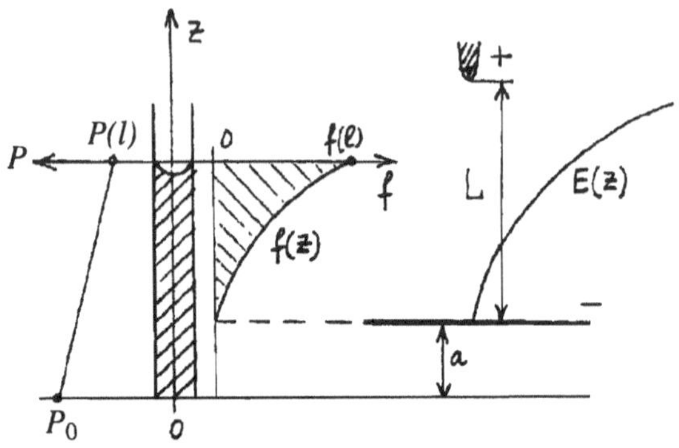

Figure 10.20 For the calculation of capillary imbibition under the action of a non-uniform electric field.

Here P_0 = const corresponds to the pressure acting on the bulk liquid surface, and $P(l)$ represents the sum of hydrostatic and capillary pressure for the general case. Integrating both parts of eq. (X.38) from 0 to z with account of the boundary condition (X.39), we obtain the following expression for the distribution of pressure:

$$P(z) = P_0 + \lambda\eta z + \rho\int_0^z f(z)dz\,.$$

Using the second boundary condition (X.40) we obtain:

$$P(l) = P_0 + \lambda\eta l + \rho\int_0^l f(z)dz\,.$$

These equations allow us to determine the value of λ:

$$\lambda = \left[P(l) - P_0 - \rho\int_0^l f(z)dz\right]/\eta l\,. \qquad \text{(X.41)}$$

As seen from the expression obtained, λ depends on the difference of pressures $P(l) - P_0$, on the length of the liquid column l and on the value of an integral, numerically equal to the area shaded in Fig. 10.20. A particular value of $\lambda(l)$ corresponds to each meniscus position $l(t)$, where t is the time.

The velocity distribution, $v(r, l)$, can be found from eq. (X.37) which may be reduced to an ordinary differential equation:

$$\partial^2 v/\partial r^2 + (1/r)\partial v/\partial r = \lambda . \qquad (X.42)$$

After double integration of eq. (X.42) with account of the ordinary boundary conditions ($v = 0$ at $r = r_0$, $dv/dr = 0$ at $r = 0$) we obtain, replacing λ by its value from eq. (X.41):

$$v(r, l) = r_0^2[1 - r^2/r_0^2]\left[\Delta P + \rho\int_0^l f(z)dz\right]/4\eta l ,$$

where $\Delta P = P(l) - P_0$ and r_0 is the capillary radius. The flow rate of liquid in the capillary, averaged over the cross-section, is equal to:

$$V = (1/\pi r_0^2)\int_0^{r_0} v(r)2\pi r dr = r_0^2\left[\Delta P + \rho\int_0^l f(z)dz\right]/8\eta l . \qquad (X.43)$$

In the absence of the external mass force ($f(z) \equiv 0$) this equation simplifies to the Poiseuille formula.

As seen from eq. (X.43), the value of the integral of $f(z)$ will change in relation to the length of the liquid column in the capillary (Fig. 10.20). However, this will be true only in the case, when the meniscus is in the region of action of the NEF. When the meniscus is out of this region, and the field is localized in the region between $l_1 - l_2 = \Delta l < l$, the value of the integral will remain constant and independent of l. In this case eq. (X.43) takes the form:

$$V = dl/dt = r_0^2\left[\Delta P + \rho\int_{l_1}^{l_2} f(z)dz\right]/8\eta l . \qquad (X.44)$$

Because both terms in the brackets are independent of l, it follows that, after integration, we obtain:

$$l^2 = r_0^2 t\left[\Delta P + \rho\int_{l_1}^{l_2} f(z)dz\right]/4\eta , \qquad (X.45)$$

where the coordinates l_1 and l_2 define the region of action of the NEF.

The value of the additional driving force $f(z)$ acting on the liquid dielectric is equal to:

$$f(z) = [\varepsilon_b(\varepsilon_0 - 1)/\rho]E(z)dE(z)/dz, \qquad (X.46)$$

where ε_b is the dielectric constant of the vacuum. The distribution of $E(z)$ for a system of electrodes needle-plate is described by the formula:

$$E(z) = 2Ub/\{(b^2 - z^2)\ln[(b+L)/(b-L)]\}. \qquad (X.47)$$

Here b is the focal length of the hyperboloid describing the shape of the needle end (Fig. 10.20), L is the distance between the electrodes, U is the voltage on the electrodes. As shown previously [208], results of calculations of the field strength by eq. (X.47) for the system of electrodes used agree well with direct measurements of $E(z)$.

To check the theory, the rates of imbibition of various liquids into thin capillaries have been measured. Ethylene glycol with a dipole moment $\mu_e = 7.4 \times 10^{-30}$ C·m and nonpolar transformer oil ($\mu_e = 0$) were used as working liquids. These liquids were used because of their high viscosity ($\eta = 0.019$ N s/m^2 for ethylene glycol and $\eta = 0.023$ N s/m^2 for oil at 20°) that impeded the imbibition enabling reliable measurements to be performed for the meniscus positions l as a function of time t. On the other hand, it is worthwhile to make comparisons of the imbibition of liquids different in ($\varepsilon_0 - 1$) values entered into eq. (X.46) and determining the contribution of the NEF into the capillary imbibition. For ethylene glycol $\varepsilon_0 = 38.7$, and for oil $\varepsilon_0 = 2.2$ which gave a ratio of ($\varepsilon_0 - 1$) values equal to 31.4.

The experiments were performed in the thermostatted chamber with a transparent front wall designed for observations. The displacement of menisci in vertically placed capillaries was measured by a cathetometer KM-6.

The impeding action of the NEF is shown in Fig. 10.21. Measurements were carried out in the same capillary with and without the imposed NEF. First the imbibition was allowed to proceed without the field, then the field was switched on. After a time, the NEF was switched off and then switched on again. Because the region of action of the field was within the length l of the filled part of the capillary (Fig. 10.21), the value of the integral from $f(z)$ might be considered constant during these experiments.

As seen from the graphs, when $\nabla E^2 < 0$ the field considerably impedes the rate of imbibition. The functions $l^2(t)$ for imbibition with the NEF, as in the case of imbibition without NEF, are close to linear. This agrees with eq. (X.44) and demonstrates that the integral force $f(z)$, applied to a definite capillary region with imbibed liquid, is constant.

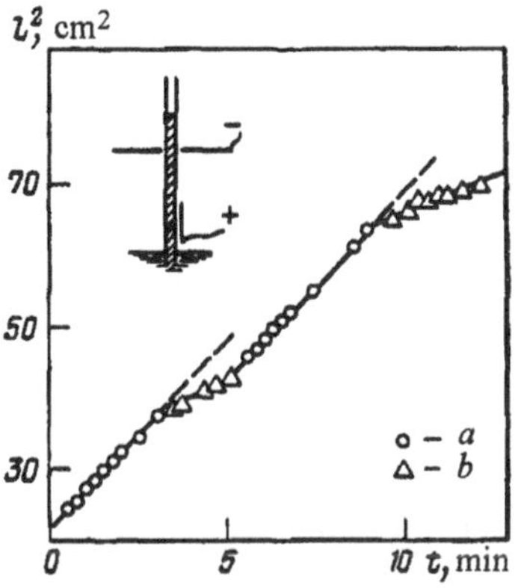

Figure 10.21 Dependence $l^2(t)$ for ethylene glycol imbibition into capillary $r = 14.6$ μm without the field (points *a*) and under the applied NEF (points *b*). $U = 7.5\times10^3$ V; $\int_{l_1}^{l_2} f(z)dz = -1.45\times10^3$ N/m·kg.

At the same values of ∇E^2, the influence of the field on imbibition is more pronounced in capillaries of greater radius. This may be explained by the fact that the values of ΔP in eq. (X.45) decrease with growing r_0, whereas the term in brackets associated with the action of NEF does not depend on r. Thus it follows that a considerable acceleration of imbibition can be achieved just for those impregnating liquids when the values of ΔP are small (poor wetting, low interfacial tension, large values of r_0). The field can hold and displace a non-wetting liquid ($\theta > 90°$) under the condition that the negative values of ΔP are compensated by positive values of the second term in eq. (X.44). On the other hand, on changing the sign of the field, the inflow of the wetting liquid into the capillary may be impeded.

Figure 10.22 shows the results of observations of the imbibition of ethylene glycol (curves *1–3*) and transformer oil (curve *4*) for another disposition of the external NEF in respect to the capillaries. The needle electrode was placed over the open end of the capillary, and the plane electrode (fixed above the liquid level in the cell) had a hole into which the capillary was placed.

First, when the field was switched off, the normal imbibition occurred (points *a*). Linear graphs $l^2(t)$ agree well the theoretical eq. (X.45). When the NEF is switched on (the meniscus is in the region of field action), the dependence $l^2(t)$ becomes non-linear and the rate of imbibition rises due to the growth of the second term in eq. (X.43). In experiments with the NEF, the form of the curves $l^2(t)$ in the first stage of imbibition (when the meniscus had not yet entered into the region of the field action) does not differ from those in the absence of the field. Then, when the liquid column enters into the

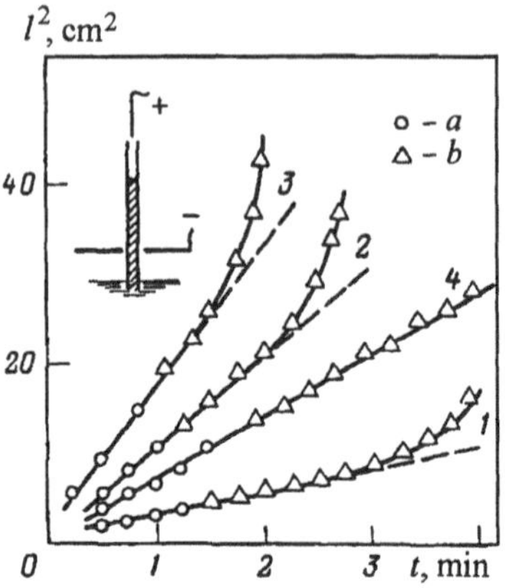

Figure 10.22 Imbibition of ethylene glycol (curves *1–3*) and transformer oil (curve *4*) into the capillaries without (points *a*) and with NEF (points *b*). r = 7.5 μm (*1*); r = 23.1 μm (2); r = 37.2 μm (*3*); r = 25.1 μm (*4*).

region of the NEF action, the experimental points are above those obtained in the absence of the field. The increase in the rate of imbibition under the action of the field is the stronger, the larger the value of l, i.e. the closer is the meniscus to the needle-electrode. This agrees with eq. (X.43) and is mainly associated with growing values of E and ∇E near to the needle-electrode (Fig. 10.20) and also with the growth of integral value of mass force $f(z)$.

When the field is switched off the form of the curves $l^2(t)$ returns to the theoretical relation described by eq. (X.27). Experiments have shown that the field begins to affect markedly the rate of imbibition at the values of $E \geq 3.2\times10^5$ V/m, $\nabla E \geq 2.8\times10^7$ V/m^2 in capillaries of radius r_0 = 7.5 μm; at $E \geq 2.7\times10^5$ V/m, $\nabla E \geq 2.3\times10^7$ V/m^2 in capillaries of radius r_0 = 23.1 μm and at $E \geq 1.2\times10^5$ V/m, $\nabla E \geq 1.3\times10^7$ V/m^2 in capillaries of radius r_0 = 37.2 μm. Furthermore, the effect of the NEF ($E \leq 2.2\times10^6$ V/m, $\nabla E \leq 3.5\times10^8$ V/m^2) on the rate of imbibition of transformer oil in capillaries of radii ranging from 3 to 45 μm (curve *4*) was not found due to a much lower value of the dielectric constant $\varepsilon_0 = 2.2$.

Good qualitative agreement between experiments and theory has been achieved. Especially marked is the action of the field in the final stage of imbibition, when the ordinary process advances slowly. Some quantitative discrepancies are related to the influence of the NEF on the contact angle of the meniscus and also to some deviations of actual E and ∇E values inside the quartz capillary from those in air calculated by eq. (X.47).

Therefore, the NEF may substantially influence the kinetics of capillary imbibition of liquids into fine capillaries modelling the porous structure. Application of the NEF is more effective in the case of slow spontaneous

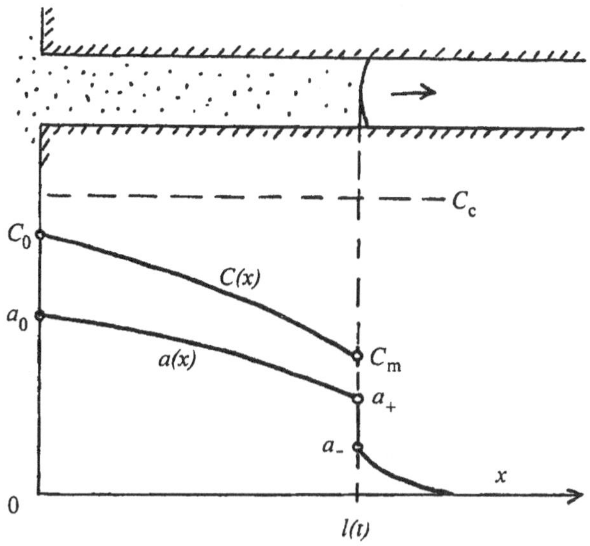

Figure 10.23 Distribution of surfactant concentration inside a capillary $C(x)$ and in the adsorbed layer $a(x)$ during imbibition at $C_0 < C_c$.

capillary imbibition. The results obtained demonstrate that the NEF may be used to control capillary impregnation or capillary emptying for liquids with high dielectric constant. For example in experiments on capillary impregnation of a firebrick with water, a 30% surplus in the rate of imbibition was obtained under the action of the NEF, $\nabla E = 7.6\times10^8$ V/m^2 [376].

§5 Acceleration of Capillary Impregnation of Hydrophobic Porous Systems by the Addition of Surfactants

As shown in the previous section, a lyophilization of pore surface that decreases the contact angle, θ_0, is required to accelerate spontaneous capillary imbibition. The problem of impregnation of porous bodies with badly wetting or non-wetting ($\theta_0 > 90°$) liquids can be solved by addition of surfactants into the impregnating liquid, because they adsorb on the pore surface and make the contact angle less than 90°. In aqueous solutions, adsorbed surfactant molecules are directed with the polar heads towards the liquid phase. Thus any increase in surfactant adsorption leads to a gradual decrease of the contact angle.

Investigations of the rates and the mechanism of impregnation of hydrophobized quartz capillaries with solutions of non-ionic surfactant, syntamide-5, have been performed in [377–380].

Capillary imbibition of pure liquids in the absence of surfactants is described by eq. (X.27). Addition of a surfactant to accelerate the imbibition makes sense, only if the resulting decrease in the contact angle prevails over the counteraction of the decreased surface tension. This is the case for the imbibition of aqueous solutions in hydrophobic capillaries ($\theta \geq 90°$). Our research has shown that not only the bulk, but also surface diffusion of surfactant molecules, including that on the unwetted portion of the capillary, can play an important role [379–381].

Theory

Let us consider the spontaneous imbibition of a surfactant solution into a cylindrical capillary taking into account the possibility of the eventual passing of some surfactant molecules through the meniscus surface to the unwetted portion of the capillary (Fig. 10.23). Let $C(x, t)$ and $a(x, t)$ be the local surfactant concentrations (in g/cm^3) in the bulk solution and on a capillary surface (adsorbed state), respectively. A constant surfactant concentration, $C(0, t) = C_0 = \text{const}$, is maintained at the capillary inlet. We shall consider first the case with $C_0 < C_c$, where C_c is the CMC. The surfactant transport inside the solution-filled capillary will be given by the following equation:

$$\partial(a + C)/\partial t = D_1\partial^2 C/\partial x^2 + D_{1s}\partial^2 a/\partial x^2 - v\partial C/\partial x, \qquad \text{(X.48)}$$

where D_1 and D_{1s} are diffusion coefficients of surfactant molecules in the bulk and over the wetted surface; $v = dl/dt$ is the rate of meniscus displacement.

Since the imbibition of surfactant solutions into a hydrophobized capillary is rather slow, the kinetics of adsorption may be ignored and it is possible to include $a = f(C)$, an equilibrium adsorption isotherm. For simplicity, we will use here a linear isotherm:

$$a(x, t) = (2/r)G_1 C(x, t) = F_1 C(x, t), \qquad \text{(X.49)}$$

where G_1 is the Henry constant contained in the adsorption equation $\Gamma = G_1 C$ (g/cm^2). In the case of a cylindrical capillary $a = 2\Gamma/r$.

Substituting eq. (X.49) into eq. (X.48) and introducing the effective diffusion coefficient $D = (D_1 + F_1 D_{s1})/(1 + F_1)$ for the region $0 \leq x \leq l(t)$, we obtain:

$$\partial C/\partial t = D\partial^2 C/\partial x^2 - [(dl/dt)/(1 + F_1)]\partial C/\partial x. \qquad \text{(X.50)}$$

For the empty part of the capillary $l(t) \le x \le \infty$, the corresponding equation takes the form:

$$\partial a/\partial t = D_{s2}\partial^2 a/\partial x^2, \tag{X.51}$$

where D_{s2} is the coefficient of surface diffusion over the unwetted capillary surface.

Equations (X.50) and (X.51) may be solved for the following boundary and initial conditions:

$$C(0, t) = C_0 = \text{const}; \; a(\infty, t) = 0. \tag{X.52}$$

$$l(0) = 0; \; a(x, 0) = 0. \tag{X.53}$$

When the rates of penetration are controlled by the viscosity of the liquid and eq. (X.27) is valid, the purpose of the theory is to determine the solution concentration at the meniscus $C_m < C_0$, where C_0 is the surfactant concentration at the capillary inlet. The respective values of $\sigma(C_m)$ and $\theta(C_m)$ may be determined from experimental dependences $\sigma(C)$ and $\theta(C)$ for a given system. In this case the value of advancing contact angle, $\theta = \theta_A$, should be used in eq. (X.27).

Equation (X.48) under the conditions (X.52) and (X.53) has a solution, only if C_m= const, that corresponds to $l = K\sqrt{t}$ law of spontaneous imbibition. The following mass-balance condition on the moving meniscus surface is valid:

$$[D_{s2}\partial a/\partial x]_{l_+} - [D_l\partial C/\partial x + D_{s1}\partial a/\partial x]_{l_-} = (a_- - a_+)\partial l/\partial t. \tag{X.54}$$

Here, l_+ and l_- represent positions on the opposite sides of the meniscus: for the liquid phase side (l_-) and on the unwetted one (l_+). The corresponding values of a_- and a_+ characterize the discontinuous change in adsorption (Fig. 10.23).

When C_m= const, the value of a_- is also constant because these values are related by the adsorption isotherm (X.49): $a_- = F_l C_m$. Assuming that the imbibition goes rather slowly to preserve the equilibrium between C_m and a_+, we can use an adsorption equation similar to eq. (X.49):

$$a_+ = (2/r)G_2 C_m = F_2 C_m, \tag{X.55}$$

where G_2 is the corresponding Henry constant.

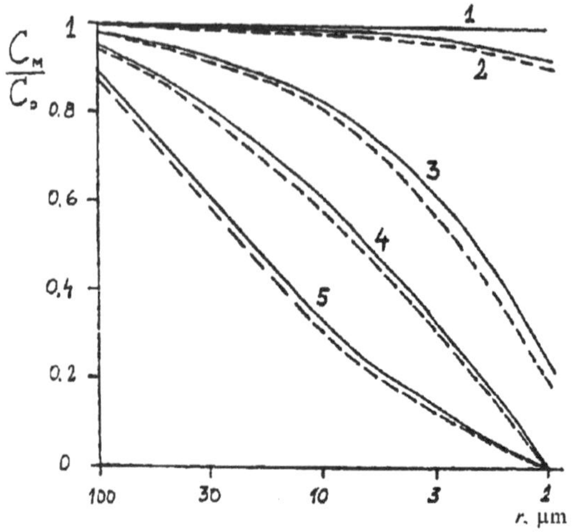

Figure 10.24 Relative decrease of surfactant concentration near to meniscus C_m/C_0 in dependence on penetration rates (K) and capillary radii (r). The adopted values of $K = 10^{-4}$ (*1*); 10^{-3} (2); $5\cdot10^{-3}$ (*3*); 10^{-2} (*4*), and $2\cdot10^{-2}$ (*5*) cm $s^{-1/2}$.

When C_m= const, the resolution of eqs. (X.50) and (X.51) can be found in a parametric form [378]: $C = C(\xi)$ and $a = a(\xi)$, where $\xi = x/\sqrt{t}$. After transformation we get:

$$(-\xi/2)\partial C/\partial\xi = D\partial^2 C/\partial\xi^2 - [K/2(1+F_1)]\partial C/\partial\xi\,; \qquad \text{(X.56)}$$

$$(-\xi/2)\partial a/\partial\xi = D_{s2}\partial^2 a/\partial\xi^2\,. \qquad \text{(X.57)}$$

Solving eqs. (X.55) to (X.57), we get a transcendental equation that contains two unknown variables: K and C_m. To determine K and C_m, eq. (X.27) should be added, which can be written as follows:

$$K = \sqrt{(r\varphi_m/2\eta)}\,, \qquad \text{(X.58)}$$

where $\varphi_m = \sigma(C_m)\cos\theta_A(C_m)$ is the wetting stress depending on the surfactant concentration, C_m, at the meniscus.

Figure 10.24 shows the results of calculations of C_m/C_0 vs. capillary radii r for different values of K characterizing the rate of penetration. The following values of parameters entered into eqs. (X.54)–(X.57) were used:

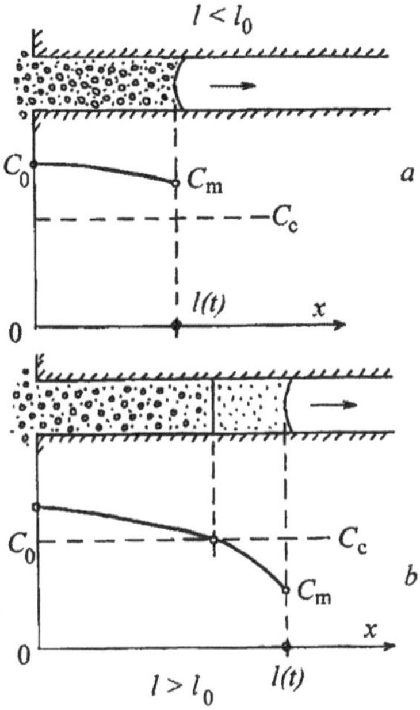

Figure 10.25 Distribution of surfactant concentration inside a capillary $C(x)$ at $C_0 > C_c$ in the cases of $l < l_0$ (a) and $l > l_0$ (b).

$G_1 = 10^{-4}$ cm, $G_2 = 10^{-5}$ cm, $D_1 = 10^{-5}$ cm^2/s, $D_{s2} = 10^{-6}$ cm^2/s (solid curves) and 10^{-5} cm^2/s (dashed curves).

At a very low rate of penetration (curves *1* and *2*), the concentration C_m does not differ appreciably from C_0 due to the small Peclet number ($Pe = vl/D$) and the predominance of diffusion. At very high values of K, the concentration close to the meniscus tends to zero. This also takes place when the penetration is rapid under the action of external pressure [382], at $Pe \gg 1$. In this case surfactant molecules cannot reach the meniscus and influence the contact angle. The capillary radius behaves similarly: the smaller the values of r, the higher is the influence of surfactant adsorption. An increase in D_{s2} values also leads to some decrease in C_m due to the enhanced diffusion along the unwetted capillary surface.

It should be noted that, because the values of K and r are interrelated

$$K = (r\sigma_m \cos\theta_m / 2\eta)^{1/2}, \tag{X.59}$$

the results shown in Fig. 10.24 should be considered at r = const. In this case K values depend only on σ_m and $\cos\theta_m$.

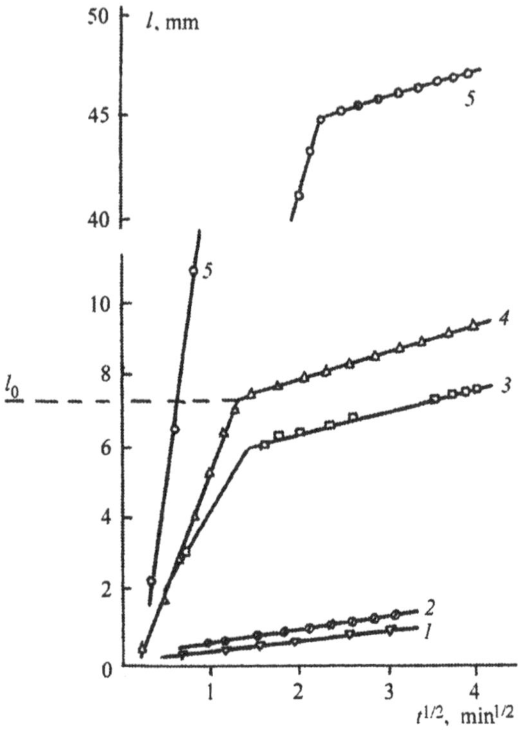

Figure 10.26 Dependences of the penetration length l on the time t for syntamide-5 solutions of various concentration: $C_0 = 0.05$ (*1*); 0.1 (*2*); 0.4 (*3*); 0.5 (*4*); and 1% (*5*). Hydrophobized quartz capillary, $r = 16$ μm, $\theta_A = 101°$, $\theta_R = 97°$.

The case of high concentration $C_0 > C_c$, when the adsorption is accompanied by micelle destruction, should be considered separately (Fig. 10.25). Here the wetting stress $\varphi_0 = \sigma_m \cos\theta_A$ is constant and the rate of penetration is controlled by a K value equal to:

$$K_0 = \sqrt{r\varphi_0/2\eta} = \text{const.} \tag{X.60}$$

Unlike the situation discussed above, the equation of the meniscus displacement is known: $l = K_0\sqrt{t}$. In this case the surfactant adsorption on the capillary surface is accompanied by the continuously decreasing concentration, C_m (near to the meniscus): from C_0 at the start of imbibition ($t = 0$) to $C_m = C_c$. The solid curve in Fig. 10.25, *a* represents the distribution of concentration $C(x)$ at the moment of time $t > 0$, when $C_0 > C_m > C_c$.

When C_m becomes lower than C_c (Fig. 10.25, *b*, solid curve), the condition K_0 = const is no longer valid. The values of K fall below K_0, and the rate of penetration decreases. As soon as the condition $C_m = C_c$ is satisfied at the meniscus, a further decrease in concentration, C_m, will cause separation of the micelle front from the meniscus surface (Fig. 10.25, *b*). The next stage of imbibition, occurring when $C_m < C_c$ has been described above.

Experimental Results

The results of measurements of meniscus displacement l with time t in a hydrophobed quartz capillary, r = 16 μm, are shown in Fig. 10.26. The changes in the penetration length were measured using the comparator IZA-2 with an accuracy of ±1 μm. All the experiments were performed with aqueous solutions of non-ionic surfactant syntamide-5 (CMC = 0.1%).

Let us consider first the case of high concentration of the surfactant solution, $C_0 > C_c$ (curves *3–5*). A sharp decrease in the penetration rate at $l = l_0$ (see Fig. 10.26) corresponds to the transition from $C_m > C_c$ to $C_m < C_c$, where C_m is the surfactant concentration at the meniscus. The length l_0 may be estimated from the mass-balance condition at the meniscus surface [378, 379]:

$$a_k(dl/dt) = -(dU_M/dt)r/2\,. \tag{X.61}$$

The left-hand side of this equation characterizes the rate of surfactant adsorption on the wetted surface (g/cm^{2}s), and the right-hand side characterizes the rate of micelle disintegration per unit surface area. It is assumed that at $C = C_c$ the limiting adsorption $a_k = F_1 C_c$ = const is achieved. The difference $U_M = C_M - C_c$ is the micelle concentration, and C_M is the concentration of micellar solution. The value (r/2) represents the ratio of the elementary volume of the solution close to the meniscus, $\pi r^2 dl$, to its surface area, $2\pi r dl$.

From eq. (X.61) solved for the initial condition $t = 0$, $U_M(0) = C_0 - C_c$ follows:

$$U_M = (C_0 - C_c) - (4C_c G_1 K_0 \sqrt{t})/r^2\,. \tag{X.62}$$

Values of U_M decrease as all micelles near the meniscus disintegrate, i.e. as the concentration C_M reduces to C_c and $U_M = 0$. This enables the time, $t = t_0$, corresponding to $U_M = 0$ to be found from eq. (X.61) and then it is used to calculate the length l_0 of the rapid imbibition:

$$l_0 = K_0\sqrt{t} = [(C_0/C_c) - 1]r^2/4G_1\,. \tag{X.63}$$

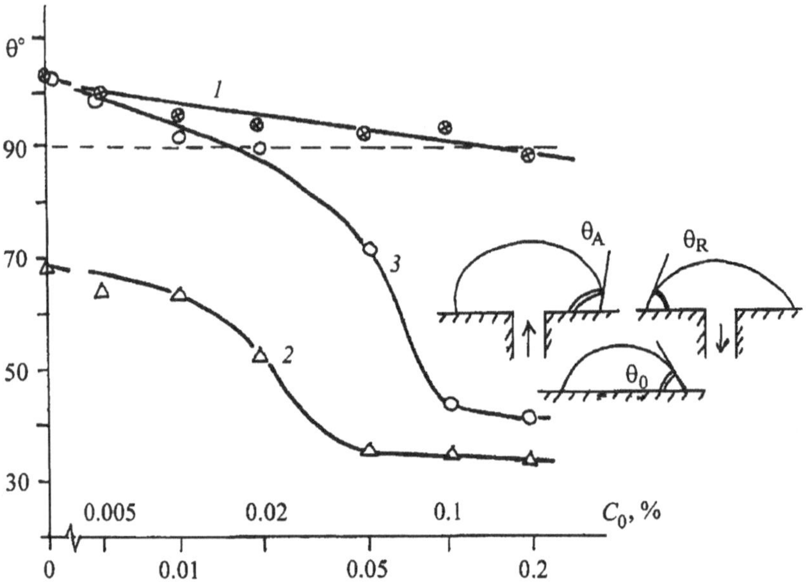

Figure 10.27 Advancing θ_A (curve *1*), receding θ_R (curve *2*) and static θ_0 (curve *3*) contact angles on a flat methylated quartz surface vs. concentration of non-ionic surfactant, syntamide-5.

According to this equation, values of l_0 decrease with C_0 in agreement with experimental results (Fig. 10.26). We can calculate $\varphi_0 = \sigma_m \cos\theta_A$ from eq. (X.59) and, using the known values of $\sigma = \sigma(\text{CMC}) = 30$ mN/m, determine the advancing contact angle. The slope of the linear part of the graphs for the first stage of imbibition ($l < l_0$) gives values of $\cos\theta_A$ equal to 0.02–0.04 that corresponds to $\theta_A \approx 88°$. The values of K_0 for $C_0 \geq 1\%$ are of the order of 10^{-1} cm/s$^{1/2}$.

The values of K for the second stage of imbibition (when $l > l_0$) increase with C_0 due to an increase in the absolute value of C_m. The increase in K values (from 3×10^{-3} to 1.7×10^{-2} cm/s$^{1/2}$) is limited because the rate of flow influences C_m in the opposite direction (Fig. 10.24).

Using eq. (X.58), we have calculated values of the wetting stress $\varphi = \sigma_m \cos\theta_A$ from the experimental data. At low values of $K \approx 10^{-3}$ cm/s$^{1/2}$ (Fig. 10.26, curves *1* and *2*) and $r \approx 10$ μm, surfactant concentrations at the meniscus C_m do not much differ from C_0 values (Fig. 10.24). In this case it was possible to calculate the value of $\cos\theta_A$ from φ using the known dependence $\sigma = \sigma(C_0)$ for the bulk solution. The values of $\cos\theta_A$ obtained are of the order of 10^{-7} that corresponds to θ_A values near to 90°.

Values of the advancing θ_A, receding θ_R and static θ_0 contact angles measured for sessile droplets on a methylated quartz plate [383] are shown in

Fig. 10.27. To determine θ_A and θ_R, the drops were disposed on the polished flat cut surface of a thick-walled quartz capillary over its entrance (Fig. 10.27). For solution concentration $C_0 < 0.1\%$, the advancing angles are greater than 90° and, therefore, capillary suction is impossible. However, as shown in Fig. 10.26, penetration by solutions of low concentration takes place. Similar results were obtained in other experiments with hydrophobized capillaries [384, 386] and also hydrophobized quartz powders [385]. It may be supposed that here the rate of imbibition is controlled not by the capillary forces but by a diffusion mechanism.

Diffusion Mechanism of Penetration of Surfactant Solutions

Let us first discuss the results obtained in the experiments on capillary rise. Figure 10.28 shows the experimental time-dependences of the meniscus position, l, for surfactant solution in vertically oriented capillaries. The linear dependences $l(\sqrt{t})$ correspond to the constant value of $K = l/(\sqrt{t}) = 2.36\times10^{-3}$ cm/s$^{1/2}$, (curve *1*) and $K = 1.3\times10^{-3}$ cm/s$^{1/2}$ (curve 2). These values are close to those calculated from curve 2 (Fig. 10.26) for the same bulk solution concentration, $C_0 = 0.1\%$. This gives grounds for supposing that the same mechanism of penetration was operating in both cases.

If the known mechanism of imbibition, governed by the capillary forces, is assumed here, the values of K obtained may be used to calculate the wetting stress $\varphi = \sigma\cos\theta_A = 2\eta K^2/r$. Calculations give the values of $\varphi = 10^{-4}$ mN/m and $\cos\theta_A = 3.4\times10^{-6}$ (curve *1*), 2.2×10^{-5} mN/m and 7.3×10^{-7} (curve 2), respectively. The calculated values of contact angles are also only slightly lower than 90°, as in experiments with horizontally oriented capillaries.

Assuming the mechanism of capillary forces, the rise of the liquid cannot exceed a limiting value (see eq. (X.30)): $l_{max} = 2\sigma\cos\theta_A/\rho g r$, where ρ is the density of the liquid and g is the acceleration due to gravity. Substituting the values of wetting stress calculated above, φ, into this formula, we obtain $l_{max} = 2\times10^{-3}$ mm (for curve *1*) and 3×10^{-4} mm (for curve 2). However, the meniscus rises up to 2–3 cm and continues to move (Fig. 10.28).

The only reasonable explanation of this effect is that the meniscus moves following the front of surface diffusion of surfactant molecules hydrophilizing the capillary surface. The meniscus is held at a given height, l, by the receding contact angle formed to equilibrate the hydrostatic pressure, $\rho g l$.

The coefficient of surface diffusion, D_s, may be estimated from the equation of surface diffusion:

$$d\Gamma/dt = D_s(d^2\Gamma/dx^2), \tag{X.64}$$

where Γ is the adsorption and x is the distance from the capillary entrance. Under the boundary conditions:

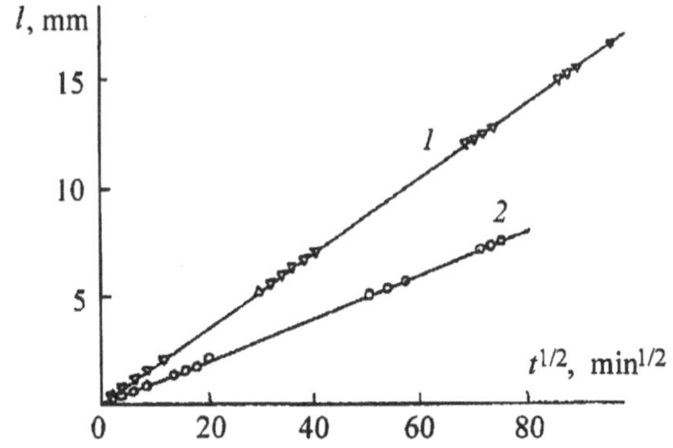

Figure 10.28 Dependences of the rise of height, l, of syntamide-5 solution ($C_0 = 0.1\%$) on the time, t, in vertically oriented hydrophobized capillaries: $r = 11$ μm, $\theta_A = 122°$, $\theta_R = 80°$ (curve *1*); r = 15 μm, $\theta_A = 120°$, $\theta_R = 86°$ (curve 2).

$$\Gamma(0) = \Gamma_0 = \text{const}, \qquad \Gamma(\infty) = 0, \tag{X.65}$$

the solution of eq. (X.64) takes the form:

$$\Gamma_* = \Gamma_0[1 - \text{erf}(K/2D_s^{1/2})]. \tag{X.66}$$

Here the adsorption Γ_* corresponds to the position of the moving meniscus (Fig. 10.29). When $\Gamma > \Gamma_*$ the advancing contact angle approaches 90°, which makes it possible to set the meniscus in motion.

We may suppose that in the case under consideration $\Gamma_* << \Gamma_0$. This leads to the well-known Einstein equation $l \approx \sqrt{(2D_s t)}$, or $K \approx \sqrt{(2D_s)}$. The experimental values of K for curves *1* and *2* (Fig. 10.26) give reasonable values of D_s ranging from 10^{-6} to 3×10^{-6} cm^2/s.

Direct evidence of surface diffusion of syntamide-5 over the unwetted part of the hydrophobized quartz capillary was obtained in experiments [386] when the meniscus was stopped and held for a long time at a fixed position. After 25 days the contact angles of pure water were measured along the capillary surface close to the meniscus. Advancing contact angles θ_A measured in the 7 mm zone are lower than those far from the meniscus. The estimated value of the coefficient of surface diffusion is of order of 10^{-7} cm^2/s.

Surfactant diffusion affecting the front of the wetting line may be a reason for the unstable movement of the latter [370, 387]. Surface diffusion of surfactants also influences the time it takes for droplets to form an equilibrium contact angle on a mica surface [388].

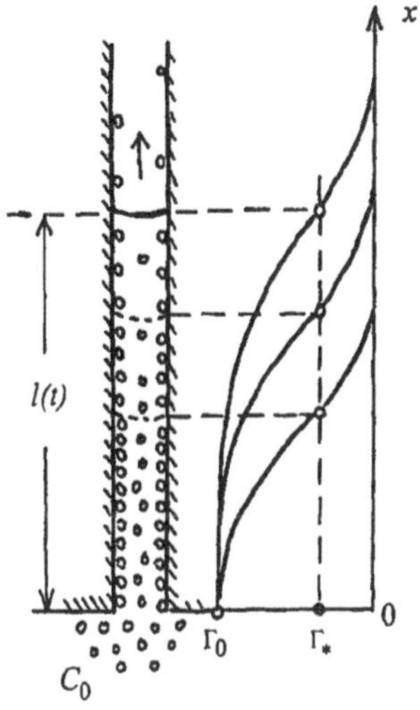

Figure 10.29 Schematic representation of the diffusion mechanism of penetration of a surfactant solution into a hydrophobized capillary, having an advancing contact angle of water more than 90°.

Not only surfactant molecules may diffuse in advance of the wetting front. The spreading of pure liquids by surface diffusion of molecules from a microdroplet over a solid surface was comprehensively studied using ellipsometric measurements [389–394]. It has been observed how, on top of the first monolayer, a second and subsequent layers form, and the corresponding coefficients of surface diffusion were calculated. For liquid PDMS on a hydrophobized silicone wafer, the coefficients of surface diffusion in the first monolayer grow with decreasing molecular mass M of the PDMS from $D_s = 4\times10^{-8}$ cm^2/s for M = 28,400 to 7×10^{-7} cm^2/s for M = 6700. Correlation between the D_s values and bulk viscosity of PDMS have been established.

Forced Imbibition

The influence of external pressure on imbibition of surfactant solution has been studied using a horizontally oriented hydrophobed capillary (r = 6 μm, θ_A= 96°). The capillary was first washed three times with a C_0 = 0.1% solution of syntamide-5 up to the length l = 3.7 cm (Fig. 10.30, point c). So, we may suppose this part of the capillary to be in equilibrium with the surfactant solution.

We have then examined, at the same concentration C_0, the spontaneous imbibition of surfactant solution into the capillary. The linear a–b part of the

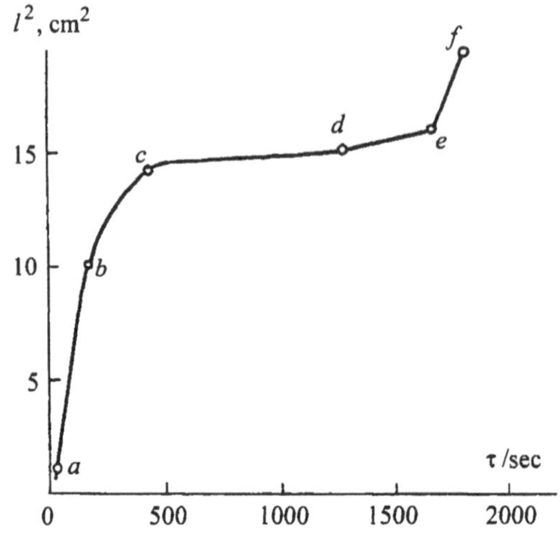

Figure 10.30 Imbibition of syntamide-5 solution (C_0 = 0.1%) into the hydrophobized quartz capillary $r = 6$ μm, $\theta_A = 96°$. The part *a–c* of the capillary was previously equilibrated with the solution. The part *c–d* corresponds to spontaneous imbibition of the solution into untreated part of the capillary. The parts *d–e* and *e–f* correspond to forced imbibition into the untreated part of the capillary under the applied external pressures 1.3 N/m^2 and 2.6 N/m^2, respectively.

graph in Fig. 10.30 corresponds to imbibition at a constant value of advancing contact angle, $\theta_A = 86°$, considerably lower than $\theta_A = 96°$ for pure water. Near to the point *c*, the spontaneous imbibition is retarded, because the complete adsorption equilibrium in *b–c* section of the capillary has not yet been reached. The next part of the graph, *c–d*, corresponds to extremely slow movement of the meniscus. The kinetics, as discussed above, is determined here by the surface diffusion of surfactant molecules in front of the meniscus. The slope of this region of the graph gives the value of K equal to 6.7×10^{-4} cm/s$^{1/2}$ and $\cos\theta_A = 5\times10^{-7}$, as in the cases considered above.

Subsequently, an external gas pressure was applied to the capillary inlet equal to 1.3 N/m^2 (Fig. 10.30, part of the graph *d–e*) and 2.6 N/m^2 (part *e–f*). The calculated values of θ_A for these regions equal 94° and 96°, respectively, i.e. are practically the same as for water.

The latter results show that in the case of forced imbibition the addition of surfactant has little or no effect. At a high Peclet number, the surfactant cannot approach the meniscus zone, because surfactant molecules adsorb on capillary walls. This explains, for instance, why the addition of surfactant solutions alone cannot enhance the recovery of oils, especially when the solution is forced by a pressure gradient.

Spontaneous imbibition of microemulsions into hydrophobized capillaries occurs at much higher rates ($K \approx 0.1$ to 1 cm/s$^{1/2}$), because under these conditions the complete wetting of the capillary surface is realized [384].

The results obtained lead to the conclusion that three mechanisms of penetration of surfactant solutions into hydrophobized capillaries are possible. The first takes place at a high concentration of surfactant, $C_0 > C_c$. Spontaneous imbibition advances at a high rate but is limited to some length l_0, that depends on C_0, r and G_1. The second mechanism is realized when $C_0 < C_c$. The rate of penetration in this case is much lower being controlled by the reduced concentration, C_m = const, near to the meniscus. At a still lower concentration C_0 in thin capillaries, the diffusion mechanism of penetration takes place. The rate of penetration is determined here by the surface diffusion of surfactant molecules in advance of the meniscus.

Penetration of surfactant solutions into the hydrophobized capillaries under external pressure is not effective, because the surfactant molecules, being adsorbed, cannot approach the surface of the advancing meniscus (at high Peclet numbers).

Although these conclusions were made on the basis of experiments with non-ionic syntamide-5 and hydrophobic surfaces ($\theta_0 > 90°$), they are likely to be more general. The analogous acceleration of imbibition was achieved by addition of various surfactants in experiments with coal powders and dried turf [395, 396].

CONCLUSION

This monograph considers liquid and vapor equilibrium and transport in porous bodies influenced by surface phenomena and taking into account conditions of restricted geometry. Theoretical treatment is based on disjoining pressure isotherms $\Pi(h)$ that include the action of molecular, electrostatic, structural and steric components of long range surface forces in thin liquid films and interlayers, depending on their thickness h.

Some comments need to be made about the modern state of the theory of surface forces. Although the theory of molecular forces is fairly well developed, the Hamaker constant A may only be estimated rather than calculated. Further progress in more accurate prediction of the A values depends on receiving more extended and detailed information about the spectral data of interacting condensed phases across a wide range of frequencies.

With regard to electrostatic forces, their further definition requires more information about dielectric properties of the boundary layers of electrolyte solutions within the EDL. This is especially important in the case of interlayers and films thinner than 10 nm. The theory of structural forces still remains insufficiently advanced. All known attempts are based on highly simplified assumptions and cannot be considered as a solution of the problem. The difficulties arising here are caused by the absence of a general theory that may predict local changes in water structure in thin interlayers depending on their thickness and boundary conditions, namely surface hydrophilicity and charge. As a result, at present we have to limit ourself to experimentally obtained parameters that contain the exponential approximation of the $\Pi_s(h)$ isotherm.

Steric forces, which are only briefly mentioned in the monograph, deserve much more attention. These forces are of special importance in thin interayers of polymer and surfactant solutions. In this case it is also necessary to take into account the so called structural-mechanical barrier, first introduced by P. Rehbinder and discussed recently in the review paper [397].

Usually the complete form of the $\Pi(h)$ isotherm is calculated as an algebraic sum of independent components. This simple approach is not always correct. For instance, molecular forces are influenced by the distribution of ions within the EDL and by structural changes in the water within thin interlayers. The structural changes of water, that give rise to structural forces, also influence the electrostatic forces. Electrostatic and molecular interactions in turn influence the structural forces. As was shown recently [398], the superposition approximation in application to the action of electrostatic and structural forces is limited to the cases of low surface charges and high electrolyte concentration.

Many difficulties arise when considering the interactions in very thin, molecular size interlayers and films. In these cases macroscopic theories of the surface forces become unacceptable and we need to look for a more adequate description. For instance, adsorption films, as distinct from wetting films, cannot be considered as a part of a thinned out bulk liquid phase. There is no sharp boundary between these two kinds of surface films, as in the general case of transition from the three-dimensional to two-dimensional state of matter.

The data about physical properties of thin liquid layers collected in this monograph remain incomplete, and rather outline the tendency for the changes occurring in density, viscosity and solvent power. This is connected with the high sensitivity of the properties to the detailed state of the adjacent solid surface, its chemical and geometrical heterogeneity, and the adsorption of admixtures. The theory of the liquid state, especially in application to polar liquids with hydrogen bonds, is also far from complete. Some progress may be expected by using methods of numerical simulation, molecular dynamics and Monte Carlo, in applications to thin liquid layers.

The effect of slippage of water over hydrophobic surfaces needs further understanding and theoretical treatment. The manifestation of this effect is of great importance when measuring hydrophobic attraction forces [399], rate of coalescence of bubbles and oil droplets in emulsions.

Non-freezing water in porous bodies still attracts much attention both in theoretical and practical aspects. The first attempts to investigate thin non-freezing interlayers, using thin molecularly smooth quartz capillaries as model bodies, are briefly described in the monograph. However, it would be interesting to extend the investigations on the non-freezing behavior of electrolyte solutions. In this case the effects of ion redistribution between aqueous phase and ice become important. The momentary state of a frozen system becomes dependent on the rate of movement of a freezing front that may influence the amount of ions trapped by ice.

In porous bodies incompletely saturated by liquids, the flow of a wetting film is described in this monograph on the basis of hydrodynamic equations that include the gradient of disjoining pressure as a driving force. The theory was verified using model wetting films formed on the inner surface of quartz capillaries. Prediction of the contribution of the film flow to the total mass transfer in real porous bodies needs further refinement.

A similar situation takes place in the case of such non-stationary processes as evaporation and drying of porous bodies, capillary imbibition and displacement of nonmiscible liquids. Comparing the predictions formulated for model systems with the results obtained for real porous bodies will make it possible to determine the limits of application of theoretical approaches. As a result, further refinement of the theory may be achieved, taking into account the peculiarities of porous networks of real objects. More complex models of porous structures may be used in this case.

The effect of shrinkage of deformable bodies during drying also needs further understanding. In this monograph only some simple model systems were considered and compared with experimental data. Further progress may be made on the basis of a micro-rheological approach that takes into account forces of interaction between particles in concentrated dispersions. As a result, correlation between the structural state of a dispersion and its macroscopic rheological characteristics may be established.

The phenomenon of wetting also requires a much more detailed description. The data mentioned in this monograph are limited to some simple model cases. The problem of wetting is worthy of a separate monograph due to the great body of published results concerning contact angles and surface free energies of substrates. The data need to be systematized and discussed from a unified point of view. The first attempts have been made in recently published monographs [400, 401].

Modern colloid chemistry is subject to intense progress, transforming more and more towards physics of interfaces and colloids [402]. This is one of the reasons why in this monograph the properties of porous bodies and processes of mass transfer are considered rather from the physical position. In particular great attention is being paid now to so called mesoscopic systems, which occupy an intermediate position between colloidal particles and molecules. Nanoparticles and nanolayers with characteristic sizes lower than 10 nm have received ever increasing attention by material scientists, physicists, chemists and biologists [403]. This demands the development of new approaches for describing correctly the interparticle interactions in such systems.

The account of heterogeneity of porous bodies is also necessary in calculation of heat transfer, especially in the case when the skeleton of a porous body differs markedly in thermal conductivity from that of the fluid filling the pores. Therefore, a general problem arises in consideration of the kinetics of mass and heat transfer on the basis of a unified approach, which takes into account the peculiarities of porous structure and surface forces.

The goal of this monograph may be considered to have been achieved when it attracts the attention of the specialists working in the field of theory and practice of mass transfer processes in porous bodies to the effects of surface phenomena. The latter influence markedly the state of liquids in porous bodies and kinetics of liquid and vapor flow in thin pores. This monograph represents a first attempt to include the basic ideas of surface phenomena and colloid science into the theory of mass transport in porous bodies.

REFERENCES

1. Derjaguin B. V., Churaev N. V., Muller V. M., *Surface Forces*, Cons. Bureau/Plenum, N. Y., 1987.
2. Lifshits E. M., *Zh. Eksp.Teor. Fiz.*, 1955, vol. 29, no. 1, p. 94.
3. Dzyaloshinskii I. E., Lifshits E. M., Pitaevskii L. P., *Zh. Eksp.Teor. Fiz.*, 1959, vol. 37, no. 2, p. 229.
4. *Colloid Science*,V. 1, Ed. by H. R. Kruyt, Elsevier, 1952, p. 207.
5. Churaev N. V., *Colloid and Polym. Sci.*, 1975, vol. 253, no. 2, p. 120.
6. Derjaguin B. V., Landau L. D., *Zh. Eksp.Teor. Fiz.*, 1941, vol. 11, no. 12, p. 802; Verwey E. J. W., Overbeek J. Th. G., *Theory of the Stability of Lyophobic Colloids*, Amsterdam, Elsevier, 1948.
7. Derjaguin B. V., *Kolloid. Zh.*, 1954, vol. 16, no. 6, p. 425.
8. Derjaguin B. V., Churaev N. V., *J. Colloid Interface Sci.*, 1974, vol. 49, no. 2, p. 249.
9. Churaev N. V. and Derjaguin B.V., *J. Colloid Interface Sci.*, 1985, vol. 103, no. 2, p. 542.
10. Israelachvili J. N., Pashley R. M., *Nature*, 1982, vol. 300, p. 341; Pashley R. M., Mc Guiggan P. M., Horn R. G., Ninham B. W., *J. Colloid Interface Sci.*, 1988, vol. 125, no. 2, p. 569; Christenson H. K., Claesson P. M., Berg J., Herder P. C., *J. Phys. Chem.*, 1989, vol. 93, no. 4, p. 1482; Israelachvili J. N., *Surf. Sci. Reports,* 1992, vol. 14, no. 3, p. 109.
11. Derjaguin B.V., *Kolloid-Ztschr.*,1934, Bd. 69, no 2, S. 155.
12. Rehbinder P. A., Shchukin E. D., Margolis L. D., *Dokl. Akad. Nauk SSSR*, 1964, vol. 154, no. 3, p. 695.
13. Kheifets L. I., Neimark A. V., *Mnogofaznye protsessy v poristykh telakh* (Multiphase Processes in Porous Bodies), Moscow: Khimiya, 1982.
14. *Physical and Chemical Aspects of Adsorbents and Catalysts*, Ed. by B. G. Linsen, Acad. Press, Lond., N. Y., 1970.
15. Dubinin M. M., *Osnovnye problemy teorii fizicheskoi adsorbtsii* (Major Problems of the Theory of Physical Adsorption), Moscow: Nauka, 1970, p. 251–269; *The Modern Theory of Capillarity*, Ed. by A. I. Rusanov and F. C. Goodrich, Leningrad: Khimiya, 1980, p. 100; Haynes J. M., In: *Colloid Science*, vol. 2, London, 1975, p. 101.
16. Vol'fkovich Y. M., *Kolloid. Zh.*, 1979, vol. 41, no. 4, p. 640.
17. Agrest M. M., Aleksandrov M. M., Reifman L. S., *Zh. Fiz. Khimii*, 1976, vol. 50, no. 9, p. 2425.

18. Ermakova T. P., Sobolev V. D., Khadakhane N. E., Churaev N. V., Ananich N. I., *Kolloid. Zh.*, 1977, vol. 39, no. 4, p. 755.
19. Avgul' N. N., Kiselev A. V., Poshkus D. P., *Adsorbtsiya gazov i parov na odnorodnykh poverkhnostyakh* (Adsorption of Gases and Vapors on Uniform Surfaces), Moscow: Khimiya, 1975.
20. Derjaguin B. V., Zakhavaeva N. N., Talaev M. V., Fillipovskii V. V., *Opredelenie udel'noi poverkhnosti poroshkoobraznykh tel po soprotivleniyu filtratsii razrezhennogo vozdukha* (Determination of Specific Surface of Porous Bodies from the Resistance of Filtration of Rarefied Air), Moscow: Izd. Akad. Nauk SSSR, 1957.
21. Derjaguin B. V., Fedoseev D. V.,Vnukov S. P., *Kolloid. Zh.*, 1986, vol. 48, no. 3, p. 410; *Powder Technol.*, 1976, vol. 14, no. l, p. 169.
22. *The Solid–Gas Interface*, Ed. by E. A. Flood, N. Y., Marcel Dekker, 1967.
23. Churaev N. V., Il'in N. I., *Radioindikatomye metody issledovaniya dvizheniya podzemnykh vod* (Radiotracer Methods of Investigation of Motion of Ground Water), Moscow: Atomizdat, 1973.
24. Mosolov A. B., Dinariev O. Y., *Zh. Tekhn. Fiz.*, 1987, vol. 57, no. 9, p. 1679.
25. Feder J., *Fractals*, Plenum Press, N. Y., 1988.
26. Derjaguin B. V., Churaev N. V., *Smachivayushchie plenki* (Wetting Films), Moscow: Nauka, 1984.
27. Derjaguin B. V., Churaev N. V., *Kolloid. Zh.*, 1976, vol. 38, no. 6, p. 1082; *J. Colloid Interface Sci.*, 1982, vol. 54, no. 2, p. 157.
28. Fisher L. R., Israelachvili J. N., *Colloids and Surfaces*, 1981, vol. 3, no. 4, p. 303; *Adv. Colloid Interface Sci.*, 1982, vol. 16, p. 117.
29. Thompson S., Gubbins K. E., *et al.*, *J. Chem. Phys.*, 1984, vol. 81, no. 1, p. 530.
30. Radushkevich L. V., *Izv. Akad. Nauk SSSR, Ser. Khim.*, 1952, no. 6, p. 1006.
31. Melrose J. C, Wallick G. C., *J. Phys. Chem.*, 1967, vol. 71, no. 11, p. 367.
32. Neimark A. V., *Kolloid. Zh.*, 1985, vol. 47, no. 3, p. 531.
33. Antonchenko V. Y., Davydov A. S., Il'in V. V., *Osnovy fiziki vody* (General Physics of Water), Kiev: Naukova Dumka, 1991.
34. Christou H. I., Whitehouse J. S., Nicholson D., Parsonage N. G., *Faraday Symp. Chem. Soc.*, 1981, no. 16, p. 139.
35. Anastasion N., Fincham D., Singer K., *JCS Faraday Trans.*, p. II, 1983, vol. 79, no. 11, p. 1639.
36. Marchesi M., *Chem. Phys. Lett.*, 1983, vol. 97, no. 2, p. 224.
37. Sonnenschein P., Heinzinger K., *Chem. Phys. Lett.*, 1983, vol. 102, no. 6, p. 550.
38. Lee C. Y., McCammon J. A., Rossky P. J., *J. Chem. Phys.*, 1984, vol. 80, no. 9, p. 4448.
39. Il'in V. V., Makovskii N. N., Khryapa V. M., Churaev N. V., *Dokl. Akad. Nauk SSSR*, 1986, vol. 283, no. 3, p. 667.
40. Horn R. G., Israelachvili J. N., *J. Chem. Phys.*, 1981, vol. 75, no. 3, p. 1400.
41. Derjaguin B. V., Karasev V. V., in: *Surface and Colloid Sci.*, vol. 15, Ed. E. Matijevic, Plenum Press, N. Y., 1993, p. 221.
42. Churaev N. V., Sobolev V. D., Zorin Z. M., in: *Thin Liquid Films and Boundary Layers*, N. Y., London, Acad. Press, 1971, p. 213.
43. Warnock J., Awschalom D. D., Shafer M. W., *Phys. Rev. B.: Condens. Matter*, 1986, vol. 34, no. 1, p. 475.
44. Martini G., *Colloids Surf.*, 1984, vol. 11, no. 3–4, p. 409.
45. Apel' P. Y., Kolikov V. M., Kuznetsov B. I., *Kolloid. Zh.*, 1985, vol. 47, no. 1, p. 3; no. 4, p. 772.
46. Bitsanis I., Somers S. A., Davis H. T., Tirrell M., *J. Chem. Phys.*, 1990, vol. 93, no. 5, p. 3427.

47. Churaev N. V., Sobolev V. D., Somov A. N., *J. Colloid Interface Sci.*, 1984, vol. 97, no. 2, p. 574.
48. Dukhin S. S., *Electroprovodnost' i electrokineticheskie svoistva dispersnykh sistem* (Electroconductivity and Electro-kinetic Properties of Disperse Systems), Kiev: Naukova Dumka, 1975.
49. Voznyi P. A., Churaev N. V., *Kolloid. Zh.*, 1977, vol. 39, no. 2, p. 264; no. 3, p. 438.
50. Khadakhane N. E., Sobolev V. D., Churaev N. V., *Kolloid. Zh.*, 1980, vol. 42, no. 5, p. 911.
51. Somov A. N., Churaev N. V., *Kolloid. Zh.*, 1982, vol. 44, no. 3, p. 614.
52. Derjaguin B. V., Popovskii Y. M., *Kolloid. Zh.*, 1982, vol. 44, no. 5, p. 863.
53. Dumanskii A. V., *Liofil'nost' dispersnykh sistem* (Lyophilicity of Disperse Systems), Kiev: Izd. Akad. Nauk USSR, 1960.
54. Martynov G. A., Starov V. M., Churaev N. V., *Kolloid. Zh.*, 1980, vol. 42, no. 3, p. 489; no. 4, p. 657.
55. Aitkuliev K., Sobolev V. D., Churaev N. V., *Kolloid. Zh.*, 1985, vol. 47, no. 4, p. 647; 1984, vol. 46, no. 2, p. 211.
56. Churaev N. V., *Kolloid. Zh.*, 1985, vol. 47, no. 1, p. 112.
57. Dukhin S. S., Derjaguin B. V., *Elektroforez* (Electrophoresis), Moscow: Nauka, 1976.
58. Dytnerskii Y. I., *Baromembrannye protsessy* (Baro-membrane Processes), Moscow: Khimiya, 1986.
59. Samoilov O. Y., *Struktura vodnykh rastvorov electrolitov i gidratatsiya ionov* (Structure of Aqueous Electrolyte Solutions and Ionic Hydration), Moscow: Izd. Akad. Nauk SSSR, 1957 ; Yashkichev V. I., Samoilov O. Y., *Zh. Strukt. Khim.*, 1962, vol. 3, no. 2, p. 211; Buslaeva M. N., Samoilov O. Y., *Zh. Strukt. Khim.*, 1963, vol. 4, no. 4, p. 502; no. 5, p. 682.
60. Lashnev V. I., Sobolev V. D., Churaev N. V., *Teor. Osn. Khim. Tekhn.*, 1976, vol. 10, no. 6, p. 926.
61. Sobolev V. D., Sergeeva I. P., Churaev N. V., *Zh. Fiz. Khim.*, 1986, vol. 10, no. 7, p. 1828.
62. Kawaguchi M., Suzuki C., Takahashi A., *J. Colloid Interface Sci.*, 1988, vol. 121, no. 2, p. 585.
63. Happel J., Brenner H., *Low Reynolds Number Hydrodynamics*, Prentice-Hall, 1965.
64. Karasev V. V., Derjaguin B. V., Efremova E. N., *Kolloid. Zh.*, 1962, vol. 24, no. 4, p. 471.
65. Derjaguin B. V., Karasev V. V., Khromova E. N., *Kolloid. Zh.*, 1986, vol. 48, no. 4, p. 671.
66. Rakhmukov B. V., Seliverstova I. I., Serpinskii V. V., Fomkin A. A., *Izv. Akad. Nauk SSSR, Ser. Khim.*, 1979, no. 11, p. 2419.
67. Brodskaya E. N., Rusanov A. I., Kuni F. M., *Kolloid. Zh.*, 1970, vol. 32, no. 6, p. 836.
68. Nabutovskii V. M., Belosludov V. R., Korotkikh A. M., *Kolloid. Zh.*, 1979, vol. 41, no. 5, p. 876.
69. Anderson D. M., Low P. F., *Soil Sci. Soc. Am. Proc.*, 1958, vol. 22, no. 2, p. 99.
70. Low P. F., *Soil Sci. Soc. Am. Proc.*, 1976, vol. 40, no. 4, p. 500; 1979, vol. 43, no. 5, p. 651.
71. Oliphant J. L., Low P. F., *J. Colloid Interface Sci.*, 1983, vol. 95, no. 1, p. 45.
72. Belouschek P., Zuppa M., Maier S., *Kolloid. Zh.*, 1986, vol. 48, no. 6, p.1067.
73. De Groot S., Mazur P., *Nonstationary thermodynamics* (translated from English), Moscow: Mir, 1964.

74. Haase R., *Thermodynamik der irreversiblen Prozesse*, Moscow: Mir, 1967.
75. Kiseleva O. A., Sobolev V. D., Starov V. M., Churaev N. V., *Kolloid. Zh.*, 1979, vol. 41, no. 2, p. 245.
76. Bogomolov V. N., *Usp. Fiz. Nauk*, 1978, vol. 124, p. 171.
77. Hocking L. M., *J. Fluid Mech.*, 1976, vol. 76, no. 10, p. 801.
78. Churaev N. V., Sergeeva I. P., Sobolev V. D., Zorin Z. M., *Colloids Surf. A*, 1993, vol. 76, p. 23; *Izv. Akad. Nauk, Ser. Khim.*, 1994, no. 1, p. 31.
79. Churaev N. V., Sergeeva I. P., Sobolev V. D., Derjaguin B. V., *J. Colloid Interface Sci.*, 1981, vol. 84, no. 2, p. 451.
80. Bondarenko N. F., *Fizika dvizheniya podzemnykh vod* (Physics of Ground Water Motion), Leningrad: Gidrometeoizdat, 1973.
81. Muratov I. B., Sobolev V. D., Churaev N. V., *Kolloid. Zh.*, 1991, vol. 53, no. 1, p. 131.
82. Skawinski R., Lasowska A., *Bull. Acad. Polonaise Sci., Ser. sci. techniq.*, 1974, vol. 22, no. 5, p. 235.
83. Golubev V. S., *Dokl. Akad. Nauk SSSR*, 1978, vol. 238, no. 6, p. 1318.
84. Churaev N.V., *Inzh.-Fiz. Zh.*, 1983, vol. 45, no. 1, p. 154.
85. Moskvin L. N., Gorshkov A. I., Gumerov M. F., *Dokl. Akad. Nauk SSSR*, 1982, vol. 265, no. 2, p. 378.
86. Burgreen D., Nakache F. R., *Trans. ASME*, 1965, vol. E32, no. 3, p. 675.
87. Lidorenko N. S., Il'in B. I., Pet'kin N. V., Kuz'menko B. B., *Elektrokhimiya*, 1980, vol. 16, no. 11, p. 1680.
88. Voronkov G. Y., Gurevich M. A., Fedorin V. A., *Khemotronnye ustroistva* (Chemotronic Devices), Moscow: VNII Elektromekhaniki, 1965.
89. Borovinskaya L. B., *Pochvovedenie*, 1964, no. 5, p. 69.
90. Tikhomolova K. P., *Elektroosmos* (Electroosmosis), Leningrad: Khimiya LO, 1989.
91. Starov V. M., Churaev N. V., *Kolloid. Zh.*, 1979, vol. 41, no. 2, p. 297.
92. Muller V. M., Sergeeva I. P., Sobolev V. D., Churaev N. V., *Kolloid. Zh.*, 1986, vol. 48, no. 4, p. 718.
93. Churaev N. V., Derjaguin B. V., *Dokl. Akad. Nauk SSSR*, 1966, vol. 169, no. 2, p. 396; in: *Research in Surface Forces*, Ed. by B. V. Derjaguin, vol. 3, Cons. Bureau, N. Y., London, 1971, p. 261.
94. Rice C. L., Whitehead R., *J. Phys. Chem.*, 1965, vol. 69, no. 11, p. 4017.
95. Derjaguin B. V., Sidorenkov G. P., Zubashchenko E. A., Kiseleva E. V., *Kolloid. Zh.*, 1947, vol. 9, no. 5, p. 335.
96. Derjaguin B. V., Dukhin S. S., Korotkova A. A., *Kolloid. Zh.*, 1961, vol. 23, no. 1, p. 53.
97. Dukhin S. S., Derjaguin B. V., *Dokl. Akad. Nauk SSSR*, 1964, vol. 159, no. 2, p. 401; no. 3, p. 636.
98. Muller V. M., *Kolloid. Zh.*, 1978, vol. 40, no. 2, p. 259.
99. Derjaguin B. V., Dzyaloshinsky I. E., Koptelova M. M., Pitajevsky L. P., *Disc. Faraday Soc.*, 1965, no. 40, p. 246.
100. Derjaguin B. V., Churaev N. V., *J. Colloid Interface Sci.*, 1977, vol. 62, no. 3, p. 369.
101. Pitaevskii L. P., *Zh. Eksp. Teor. Fiz.*, 1959, vol. 37, no. 2, p. 577.
102. Derjaguin B. V., Koptelova M. M., *Kolloid. Zh.*, 1969, vol. 31, no. 5., p. 692; in: *Research in Surface Forces*, Ed. by B. V. Derjaguin, vol. 4, Cons. Bureau, N. Y., London, 1975, p. 182.
103. Lechnick W. J., Shaewitz J. A., *J. Colloid Interface Sci.*, 1984, vol. 102, no. 1, p. 71.
104. Korotkova A. A., Churaev N. V., *Kolloid. Zh.*, 1987, vol. 49, no. 5, p. 1004.

105. Churaev N. V., *Kolloid. Zh.*, 1980, vol. 42, no. 2, p. 295.
106. Sasidhar V., Ruckenstein E., *J. Colloid Interface Sci.*, 1981, vol. 82, no. 2, p. 439.
107. Derjaguin B. V., Churaev N. V., Martynov G. A., *J. Colloid Interface Sci.*, 1980, vol. 75, no. 2, p. 419.
108. Sabbatovskii K. G., Sobolev V. D, Churaev N. V., *Kolloid. Zh.*, 1993, vol. 55, no. 5, p. 142.
109. Dorokhov V. M., Martynov G. A., Starov V. M., Churaev N. V., *Kolloid. Zh.*, 1984, vol. 46, no. 2, p. 238; no. 4, p. 651; no. 6, p. 1088.
110. Starov V. M., Churaev N. V., *Adv. Colloid Interface Sci.*, 1993, vol. 43, p. 145.
111. Starov V. M., Churaev N. V., *Teor. Osn. Khim. Tekhn.*, 1980, vol. 14, no. 4, p. 509.
112. Dukhin S. S., Churaev N. V., Yaroshchuk A. E., *Khim. Tekhn. Vody*, 1984, vol. 6, no. 4, p. 291.
113. Dukhin S. S., Churaev N. V., Shilov V. N., Starov V. M., *Usp. Khim.*, 1988, vol. 57, no. 6, p. 1013.
114. Dukhin S. S., Sidorova M. P., Yaroshchuk A. E., *Elektrokhimiya membran i obratnyi osmos* (Electro-Chemistry of the Membrane and Reverse Osmosis), Leningrad: Khimiya, 1991.
115. Derjaguin B. V., Sidorenkov G. P., *Dokl. Akad. Nauk SSSR*, 1941, vol. 32, no. 9, p. 622.
116. Kjellander R., Marcelja S., *Chem. Phys. Lett.*, 1985, vol. 120, no. 4–5, p. 393.
117. Il'in V. V., Khryapa V. M., Churaev N. V., *Fiz. Mnogochastichnykh Sistem*, 1991, no. 18, p. 50.
118. Nieuwoudt J. C., Kirkpatrick T. R., *J. Statist. Phys.*, 1984, vol. 34, no. 1–2, p. 203.
119. Haase R., Steiner C., *Z. Phys. Chem.*, 1959, vol. 21, no. 2, p. 270; 1965, vol. 44, no. 3, p. 301.
120. Dirksen C., *Soil Sci. Soc. Am. Proc.*, 1969, vol. 33, no. 6, p. 821.
121. Srivastava R. S., Abrol I. P., *Indian J. Chem.*, 1969, vol. 7, no. 11, p. 1121.
122. Dariel M. S., Kedem O., *Isr. Atom. Energy Com.*, 1971, no. 1262, p. 105; *J. Phys. Chem.*, 1975, vol. 79, no. 3, p. 336.
123. Sanyal S. K., Mukherjee A. K., *Indian J. Chem.*, 1985, vol. A24, no. 4, p. 338.
124. Churaev N. V., Derjaguin B. V., Zolotarev P. P., *Dokl. Akad. Nauk SSSR*, 1968, vol. 183, no. 5, p. 1139.
125. Fletcher N. H., *Phil. Mag.*, 1962, vol. 7, no. 2, p. 255.
126. Kvlividze V. I., Kiselev V. F., Ushakova L. A., *Dokl. Akad. Nauk SSSR*, 1970, vol. 191, no. 5, p. 1088.
127. Valeri S., Mantovani S., *J. Chem. Phys.*, 1978, vol. 69, no. 11, p. 5207.
128. Ocampo J., Klinger J., *J. Colloid Interface Sci.*, 1982, vol. 86, no. 2, p. 377; *J. Phys. Chem.*, 1983, vol. 87, no. 21, p. 4325.
129. Beaglehole D., Nason D., *Surface Sci.*, 1980, vol. 96, no. 1–3, p. 357.
130. Furukawa Y., Yamamoto M., Kuroda T., *J. Cryst. Growth*, 1987, vol. 82, no. 4, p. 665.
131. Elbaum M., Lipson S. G., Dash J. G., *J. Cryst. Growth*, 1993, vol. 129, p. 491.
132. Vershinin P. V., Derjaguin B. V., Kirilenko N. V., *Izv. Akad. Nauk SSSR, Ser. Geogr. Geofiz.*, 1949, vol. 13, no. 1, p. 108.
133. Jellineck H. H. G., *J. Colloid Interface Sci.*, 1967, vol. 25, no. 2, p. 192.
134. Kvlividze V. I., Kurzaev A. B., in: *Poverkhnostnye sily v tonkikh plenkakh* (Surface Forces in Thin Films), Moscow: Nauka, 1979, p. 211.
135. Churaev N. V., Bardasov S. A., Sobolev V. D., *Colloids Surf. A*, 1994, vol. 79, p. 11; *Langmuir*, 1994, vol. 10, no. 11, p. 4203.

136. Barer S. S., Churaev N. V., Derjaguin B. V., Kiseleva O. A., Sobolev V. D., *J. Colloid Interface Sci.*, 1980, vol. 74, no. 1, p. 173.
137. Dedyulya I. V., Churaev N. V., *Kolloid. Zh.*, 1987, vol. 49, no. 3, p. 559.
138. Bardasov S. A., Sobolev V. D., Churaev N. V., *Kolloid. Zh.*, 1993, vol. 55, no. 4, p. 15.
139. Chernov A. A., Yakovlev V. A., *Poverkhnost'*, 1986, no. 2, p. 39.
140. Derjaguin B. V., Churaev N. V., *J. Cold Regions Sci. Technol.*, 1986, vol. 12, no. 1, p. 57.
141. Khaimov-Mal'kov V. Y., In: *Rost kristallov* (Growth of Crystals), Moscow: Izd. Akad. Nauk SSSR, 1959, p. 5.
142. Vignes M., Dijkema K. M., *J. Colloid Interface Sci.*, 1974, vol. 49, no. 2, p. 165.
143. Biermans M., Dijkema K. M., De Vries D. A., *J. Hydrology*, 1978, vol. 37, no. 1–2, p. 137.
144. Everett D. H., Haynes J. M., *Bull. RILEM*, 1965, no. 27, p. 31.
145. Loch P. G., Miller R. D., *Soil Sci. Soc. Am. Proc.*, 1975, vol. 39, no. 6, p. 1036.
146. Hoekstra P., Miller R. D., *J. Colloid Interface Sci.*, 1967, vol. 25, no. 2, p. 166.
147. Romkens M. J. M., Miller R. D., *J. Colloid Interface Sci.*, 1973, vol. 42, no. 1, p. 103.
148. Chernov A. A., Bronshtein V. L., *Dokl. Akad. Nauk SSSR*, 1976, vol. 231, no. 4, p. 849.
149. Gilpin R. R., *J. Colloid Interface Sci.*, 1980, vol. 74, no. 1, p. 44.
150. Kononenko V. G., Geguzin Y. E., *Kristallografiya*, 1981, vol. 26, no. 1, p. 157.
151. Chernov A. A., Yakovlev V. A., *Dokl. Akad. Nauk SSSR*, 1985, vol. 285, no. 2, p. 353; *Langmuir*, 1987, vol. 3, no. 5, p. 635.
152. Hiroi H., Mizusaki T., Tsuneto T., Hirai A., Eguchi K., *Phys. Rev. B*, 1989, vol. 40, no. 10, p. 6581.
153. Derjaguin B. V., Nerpin S. V., *Dokl. Akad. Nauk SSSR*, 1954, vol. 99, no. 6, p. 1029; 1955, vol. 100, no. 1, p. 17.
154. Ershov A. P., Churaev N. V., *Teor. Osn. Khim. Tekhn.*, 1968, vol. 2, no. 2, p. 317.
155. Zorin Z. M., Kolarov T., Platikanov D., *Kolloid. Zh.*, 1986, vol. 48, no. 4, p. 677; no. 5, p. 901.
156. Ingram B. T., *JCS Faraday Trans.* p. I, 1974, vol. 70, no. 5, p. 868.
157. Derjaguin B. V., Samygin M. M., In: *Vyazkost' zhidkostei i kolloidnykh rastvorov* (Viscosity of Liquids and Colloid Solutions), Moscow–Leningrad: Izd. Akad. Nauk SSSR, 1941, vol. 1, p. 59.
158. Karpovich I. N., Churaev N. V., Panchenko M. S., *Kolloid. Zh.*, 1984, vol. 46, no. 1, p. 114.
159. Karpovich I. N., Churaev N. V., Panchenko M. S., *Kolloid. Zh.*, 1980, vol. 42, no. 4, p. 634.
160. Derjaguin B. V., Mel'nikova M. K., in: *Voprosy agrofiziki* (Agrophysical Problems), Leningrad: Sel'khozgiz, 1957, p. 30.
161. Kisileva O. A., Rabinovich Y. I., Sobolev V. D., Churaev N. V., *Kolloid. Zh.*, 1979, vol. 41, no. 6, p. 1074.
162. Ershov A. P., Churaev N. V., *Teor. Osnovy Khim. Tekhn.*, 1969, vol. 3, no. 4, p. 583.
163. Churaev N. V., Ershov A. P., *Kolloid. Zh.*, 1972, vol. 34, no. 5, p. 744.
164. Derjaguin B. V., Churaev N. V., Sobolev V. D., Barer S. S., *J. Colloid Interface Sci.*, 1981, vol. 84, no. 1, p. 182.
165. Cazabat A. M., *Adv. Colloid Interface Sci.*, 1991, vol. 34, p. 73.
166. Carles P., Cazabat A. M., *J. Colloid Interface Sci.*, 1993, vol. 157, no. 1, p. 196.
167. Derjaguin B. V., Churaev N. V., Kisileva O. A., Sobolev V. D., *Langmuir*, 1987, vol. 3, no. 5, p. 631.

168. *Voprosy kriologii zemli* (Problems of Earth Cryology), Moscow: Nauka, 1976.
169. Chizmadzhev Y. A., Markin V. S., *Elektrokhimiya*, 1966, vol. 2, no. 11, p. 1360.
170. Chesnokov S. V., *Elektrokhimiya*, 1968, vol. 4, no. 12, p. 1498.
171. Radoev B. P., Dimitrov D. S., Ivanov I. B., *Colloid Polym. Sci.*, 1974, vol. 252, no. 1, p. 50.
172. Manev E. D., Vassilieff C. S., Ivanov I. B., *Colloid Polym. Sci.*, 1974, vol. 254, no. 1, p. 99.
173. Vassilieff C. S., Manev E. D., Ivanov I. B., *Abh. Akad. Wiss. DDR*, 1986(87), no. 1, S. 465.
174. Ivanov I. B., *Pure Appl. Chem.*, 1980, vol. 52, p. 1241.
175. Pesach D., Marmur A., *Langmuir*, 1987, vol. 3, no. 4, p. 519.
176. Princen H. M., Cazabat A. M., Cohen Stuart M. A., Heslot F., Nicolet S., *J. Colloid Interface Sci.*, 1988, vol. 125, no. 1, p. 84.
177. Marra J., Huethorst J. A. M., *Langmuir*, 1991, vol. 7, p. 2748.
178. Tshchapek M., Wasowski C., *Colloids Surf.*, 1982, vol. 5, p. 65; 1991, vol. 55, p. 1.
179. Tschapek M., *Kolloid. Zt. und Zt. Polymere*, 1964, vol. 194, no. 2, S. 151.
180. Astrakhantseva N. P., Us'yarov O. G., in: *Poverkhnostnye sily v tonkikh plenkakh i ustoichivost' kolloidov* (Surface Forces in Thin Films and Stability of Colloids), Moscow: Nauka, 1974, p. 155.
181. Grigorov O. N., Frenkel' I. I., *Zh. Prikl. Khim.*, 1956, vol. 29, no. 2, p. 201.
182. Gorokhov M. M., Lebedev D. M., Churaev N. V., *Kolloid. Zh.*, 1967, vol. 29, no. 1, p. 55.
183. Ekserova D., Zakharieva M., in: *Poverkhnostnye sily v tonkikh plenkakh i dispersnykh sistemakh* (Surface Forces in Thin Films and Disperse Systems), Moscow: Nauka, 1972, p. 234.
184. Churaev N. V., *Kolloid. Zh.*, 1988, vol. 50, no. 6, p. 1167.
185. Kuz'mina G. L., Sobolev V. D., Churaev N. V., *Kolloid. Zh.*, 1990, vol. 52, no. 3, p. 493.
186. Dukhin S. S., Mishchuk N. A., Tarovskii A. A., Baran A. A., *Kolloid. Zh.*, 1987, vol. 49, no. 3, p. 616.
187. Devien M., *Techeniya i teploobmen rasrezhennykh gazov* (Flow and Heat Transfer of Rarefied Gases), Moscow: Izd. Inostr. Lit., 1962.
188. Shidlovskii V. P., *Vvedenie v dinamiku razrezhennogo gaza* (Introduction to the Dynamics of Rarefied Gases), Moscow: Nauka, 1965.
189. Pavlyukevich N. V., *Fizicheskaya kinetika i protsessy perenosa pri fazovykh prevrashcheniyakh* (Physical Kinetics and Transfer Processes in Phase Transformations), Minsk: Nauka i Tekhn., 1980.
190. Derjaguin B. V., Fedyakin N. N., in: *Research in Surface Forces*, Ed. by B. V. Derjaguin, vol. 2, Cons. Bureau, N. Y., London, 1966, p. 220.
191. Burganos V. N., *J. Chem. Phys.*, 1993, vol. 98, no. 3, p. 2268.
192. Barrer R. M., Nicholson D., *Canad. J. Chem.*, 1965, vol. 43, no. 4, p. 896.
193. Lund L. M., Berman A. S., *J. Appl. Phys.*, 1966, vol. 37, no. 6, p. 2489.
194. Ershova I. G., Churaev N. V., *Teor. Osnovy Khim. Tekhn.*, 1968, vol. 2, no. 1, p. 97.
195. Derjaguin B. V., Bakanov S. P., *Dokl. Akad. Nauk SSSR*, 1957, vol. 115, no. 2, p. 267; *Zh. Tekhn. Fiz.*, 1957, vol. 27, no. 9, p. 2056.
196. Zotov S. N., Rabinovich Y. I., Churaev N. V., *Inzh.-Fiz. Zh.*, 1978, vol. 34, no. 6, p. 1035.
197. Grivtsova L A., Grivtsov A. G., Churaev N. V., Chuikova L. F., *Kolloid. Zh.*, 1982, vol. 44, no. 5, p. 969; *Inzh.-Fiz. Zh.*, 1984, vol. 4, no. 1, p. 82.
198. Grivtsova L. A., Ergunova R. M., Churaev N. V., Khazin L. G., Preprint no. 136

Inst. Prikl. Matem. Akad. Nauk SSSR, Moscow: 1984.
199. Nicholson D., Petropoulos J. H., *J. Colloid Interface Sci.*, 1973, vol. 45, no. 2, p. 459; *Berichte Bunsen-Gesellschaft,* 1975, B. 79, no. 9, S. 796.
200. Nicholson D., *JCS Faraday Trans.*, p. I, 1975, vol. 71, no. 2, p. 238.
201. Nicholson D., Petropoulos J. N., *JCS Faraday Trans.*, p. I, 1984, vol. 80, no. 5, p. 1069.
202. Kruger A. J., Ouens S. S. D., de Vries D. A., In: *Teplo i massoperenos* (Heat and Mass Transfer), Kiev: Naukova Dumka, 1968, p. 295.
203. Zolotarev P. P, Churaev N. V., *Zh. Fiz. Khim.*, 1972, vol. 46, no. 5, p. 1123.
204. Zorin Z. M., Sobolev V. D., Churaev N. V., *Zh. Fiz. Khim.*, 1972, vol. 46, no. 5, p. 1127.
205. Zolotarev P. P., Rabinovich Y. I., Churaev N. V., *Inzh.-Fiz. Zh.*, 1977, vol. 32, no. 4, p. 649.
206. Derjaguin B. V., Bakanov S. P., *Dokl. Akad. Nauk SSSR*, 1962, vol. 144, no. 3, p. 535.
207. Panasyuk A. L., Panchenko M. S., Starov V. M., Churaev N. V., *Inzh.-Fiz. Zh.*, 1978, vol. 35, no. 1, p. 93.
208. Karpovich I. N., Churaev N. V., Panchenko M. S., *Kolloid. Zh.*, 1980, vol. 42, no. 4, p. 634.
209. Levdanskii V. V., Pavlyukevich N. V., *Inzh.-Fiz. Zh.*, 1971, vol. 20, no. 6, p. 1093.
210. Gaidukov M. N., Churaev N. V., Yalamov Y. I., *Zh. Tekhn. Fiz.*, 1976, vol. 46, no. 10, p. 2142.
211. Vargaftik N. V., *Spravochnik po teplofizicheskim svoistvam gasov i zhidkostei* (Handbook of Thermophysical Properties of Gases and Liquids), Moscow: Fizmatgiz, 1963.
212. Malyshev V. L., Gaidukov M. N., Gamayunov N. I., in: *Fizika aerodispersnykh sistem i fizicheskaya kinetika* (Physics of Aerodisperse Systems and Physical Kinetics), Moscow: 1979, p. 197.
213. Malyshev V. L., *Issledovanie osobennostei ispareniya vlagi iz kapillyarov pri vysokikh temperaturakh* (Study of the Peculiarities of Water Evaporation from Capillaries at High Temperatures): Cand. Sci. (Phys.-Math.) Dissertation, Kalinin: 1980.
214. Gamayunov N. I., Malyshev V. L., Uvarova L. A., Fei'dblyum A. S., *Inzh.-Fiz. Zh.*, 1984, vol. 47, no. 4, p. 647.
215. Gamayunov N. I., Uvarova L. A., Malyshev V. L., Fei'dblyum A. S., *Inzh.-Fiz. Zh.*, 1984, vol. 46, no. 3, p. 442.
216. Fei'dblyum A. S., Malyshev V. L., Uvarova L. A., *Voprosy fiziki formoobrazovaniya i fazovykh prevrashchenii* (Problems of Physics of Forming and Phase Transformations), Kalinin: KGU, 1983, p. 58.
217. Fei'dblyum A. S., *Gidrogazodinamika, teplo- i massoobmen v energeticheskikh ustanovkakh* (Hydrogasodynamics and Heat Transfer in Energetic Apparatus), Minsk: ITMO AN BSSR, 1984, p. 157.
218. Gamayunov N. I., Lankov A. A., Malyshev V. L., *Izv. Vuzov, Fizika*, 1984, no. 5, p. 65.
219. Churaev N. V., *Dokl. Akad. Nauk SSSR*, 1963, vol. 148, no. 6, p. 1361.
220. Derjaguin B. V., Nerpin S. V., Churaev N. V., *Kolloid. Zh.*, 1964, vol. 26, no. 3, p. 301; *Bull. RILEM*, 1965, no. 29, p. 93.
221. Nerpin S. V., Chudnovskii A. F., *Fizika pochv* (Soil Physics), Moscow: Nauka, 1967.
222. Derjaguin B. V., Nerpin S. V., Churaev N. V., In: *Fizika, khimiya, biologiya i mineralogiya pochv SSSR* (Physics, Chemistry, Biology and Mineralogy of Soils

of USSR), Moscow: Nauka, 1964, p. 43.
223. Churaev N. V., *Pochvovedenie*, 1969, no. 8, p. 21.
224. Churaev N. V., Yashchenko N. E., *Pochvovedenie*, 1973, no. 9, p. 105.
225. Potash M., Wayner P. C., *Int. J. Heat Mass Transfer*, 1972, vol. 15, p. 1851.
226. Wayner P. C., Kao Y. K, La Croix L. V., *Int. J. Heat Mass Transfer*, 1976, vol. 19, p. 487.
227. Moosman S., Homsy G. M., *J. Colloid Interface Sci.*, 1980, vol. 73, no. 1, p. 212.
228. Derjaguin B. V., Churaev N. V., Ershova I. G., *Dokl. Akad. Nauk SSSR*, 1968, vol. 182, no. 2, p. 368.
229. Churaev N. V., Ershova I. G., *Kolloid. Zh.*, 1971, vol. 33, no. 6, p. 913.
230. Zorin Z. M., Novikova A. V., Petrov A. K., Churaev N. V., In: *Poverkhnostnye sily v tonkikh plenkakh i ustoichivost' kolloidov* (Surface Forces in Thin Films and Stability of Colloids), Moscow: Nauka, 1974, p. 94.
231. Derjaguin B. V., Zorin Z. M., *Zh. Fiz. Khim.*, 1955, vol. 29, no. 10, p. 1755.
232. Kurzaev A. B., Kozlov S. N., Kiselev V. F., *Dokl. Akad. Nauk SSSR*, 1976, vol. 228, no. 4, p. 877.
233. Churaev N. V., *Pure Appl. Chem.*, 1989, vol. 61, no. 11, p. 1959.
234. Viktorina M. M., Derjaguin B. V., Ershova I. G., Churaev N. V., *Dokl. Akad. Nauk SSSR*, 1971, vol. 200, no. 6, p. 1306.
235. Kiseleva O. A., Starov V. M., Churaev N. V., *Kolloid. Zh.*, 1977, vol. 39, no. 6, p. 1164.
236. Lykov A. V., *Teplomassoobmen* (Heat and Mass Transfer), Moscow: Energiya, 1972.
237. Churaev N. V., *Inzh.-Fiz. Zh.*, 1970, vol. 19, no. 2, p. 224.
238. Hall A. C., *J. Phys. Chem.*, 1970, vol. 74, no. 14, p. 2742.
239. Starov V. M., Churaev N. V., *Inzh.-Fiz. Zh.*, 1975, vol. 29, no. 6, p. 1013.
240. Barrer R. M., *Diffuziya v tverdykh telakh* (Diffusion in Solid Materials), Moscow: Izd. Inostr. Lit., 1948.
241. Rhim H., Hwang S. -T., *J. Colloid Interface Sci.*, 1975, vol. 52, no. 1, p. 174.
242. Nerpin S. V., Churaev N. V., *Inzh.-Fiz. Zh.*, 1965, vol. 8, no. 1, p. 20.
243. Prozorov E. N., Starov V. M., Churaev N. V., *Kolloid. Zh.*, 1980, vol. 42, no. 4, p. 680.
244. Afanas'ev A. E., Gamayunov N. I., Churaev N. V., *Kolloid. Zh.*, 1971, vol. 33, no. 1, p. 11.
245. Zolotarev P. P., *Dokl. Akad. Nauk SSSR*, 1966, vol. 168, no. 1, p. 83.
246. Tishkova R. A., Churaev N. V., Ershov A. P., *Inzh.-Fiz. Zh.*, 1979, vol. 37, no. 5, p. 849.
247. Solov'ev V. M., Sobolev V. D., Churaev N. V., *Kolloid. Zh.*, 1987, vol. 49, no. 3, p. 503.
248. Pashley R. M., Israelachvili J. N., *J. Colloid Interface Sci.*, 1984, vol. 97, no. 2, p. 446.
249. Yatsimirskii K. B., Budarin L. I., Verlan' A. F., *Dokl. Akad. Nauk USSR, Ser. Geol. Khim. Biol. Nauk*, 1982, no. 8, p. 56.
250. Starov V. M., Churaev N. V., *Inzh.-Fiz. Zh.*, 1988, vol. 54, no. 4, p. 635.
251. Duguev S. V., Kiseleva O. A., Prozorov E. N., Churaev N. V., *Kolloid. Zh.*, 1981, vol. 43, no. 4, p. 646.
252. Berdinskaya N. V, Kusakov M. M., *Kolloid. Zh.*, 1972, vol. 34, no. 5, p.755.
253. La Mer V. K., Healy T. W., *Science*, 1965, vol. 148, no. 3666, p. 36.
254. Kochurova N. N., Geineman V. R., Rusanov A. I., *Zh. Fiz. Khim.*, 1975, vol. 49, no. 1, p. 20; 1977, vol. 51, no. 7, p. 1970.
255. Derjaguin B. V., Fedoseev V. A., Rozentzweig L. A., *Dokl. Akad. Nauk SSSR*,

1966, vol. 167, no. 3, p. 617.
256. Ershova I. G., Zorin Z. M., Churaev N. V., *Kolloid. Zh.*, 1976, vol. 38, no. 5, p. 973.
257. Derjaguin B. V., Zakhavaeva N. N., Lopatina A. M., *Pochvovedenie*, 1966, no. 2, p. 28.
258. Nerpin S. V., Pakshina S. M., in: *Materialy mezhvedomstvennogo soveshchaniya po probleme izucheniya i regulirovaniya ispareniya s vodnoi poverkhnosti i pochvy* (Proc. of an Interdepartment Meeting on Regulation of Evaporation from Water Surface and Soils), Inst. of Hydrology, Valdai, 1964, p. 204.
259. Karpovich I. N., Churaev N. V., Panchenko M. S., *Inzh.-Fiz. Zh.*, 1981, vol. 41, no. 6, p. 1049.
260. Panchenko M. S., Panasyuk A. L., Karpovich I. N., Churaev N. V., *Inzh.-Fiz. Zh.*, 1977, vol. 33, no. 2, p. 356.
261. Panchenko M. S., Panasyuk A. L., Tymchina I. I., *Elektronnaya obrabotka materialov* (Electronic Processing of Materials), 1984, no. 4, p. 41.
262. Volarovich M. P., Churaev N. V., *Issledovaniya torfa pri pomoshchi radioaktivnykh izotopov* (Investigations of Peat by Radiotracers), Moscow: Izd. Akad. Nauk SSSR.
263. Churaev N. V., Afanas'ev A. E., Gamayunov N. I., in: *Issledovaniya v oblasti poverkhnostnykh sil* (Investigations in the Field of Surface Forces), Moscow: Nauka, 1967, p. 395.
264. Gamayunov N. I., Afanas'ev A. E., Churaev N. V., *Issledovaniya protsessov obmena energiei i veshchestvom v sisteme pochva–rastenie–vozdukh* (Investigations of Energy and Mass Transfer Processes in the System Soil–Plant–Air), Leningrad: Nauka, 1972, p. 45.
265. Prozorov E. N., Starov V. M., Churaev N. V., *Inzh.-Fiz. Zh.*, 1978, vol. 34, no. 3, p. 423–430; no. 4, p. 660.
266. Lykov A. V., *Teplo i massoobmen v protsessakh sushki* (Heat and Mass Transfer in Drying Processes), Moscow–Leningrad: Gosenergoizdat, 1956; *Teoriya sushki* (Theory of Drying), Moscow: Energiya, 1968.
267. Churaev N. V., *Kolloid. Zh.*, 1965, vol. 27, no. 6, p. 908.
268. Churaev N. V., *Pochvovedenie*, 1966, no. 3, p. 45.
269. Lykov A. V., Mikhailov Y. A., *Teoriya teplo- i massoperenosa* (Theory of Heat and Mass Transfer), Moscow: Gosenergoizdat, 1963, 535 p.
270. Gamayunov N. I., *Issledovaniya v oblasti poverkhnostnykh sil* (Investigations in the Field of Surface Forces), Moscow: Nauka, 1967, p. 325.
271. Churaev N. V., Gorokhov M. M., *Pochvovedenie*, 1970, no. 6, p. 54.
272. Churaev N. V., *Inzh.-Fiz. Zh.*, 1972, vol. 23, no. 5, p. 807.
273. Timofeev D. P., *Kinetika adsorbtsii* (Adsorption Kinetics), Moscow: Izd. Akad. Nauk SSSR, 1962.
274. Zolotarev P. P., Rabinovich Y. I., Churaev N. V., *Inzh.-Fiz. Zh.*, 1977, vol. 32, no. 5, p. 800.
275. Zolotarev P. P., *Izv. Akad. Nauk SSSR, Ser. Khim.*, 1968, no. 10, p. 2403; 1969, no. 5, p. 1174; 1971, no. 9, p. 1882; 1972, no. 3, p. 526.
276. Lykov A. V., *Teoriya teploprovodnosti* (Theory of Heat Conductivity), Moscow: Vysshaya Shkola, 1967.
277. Dushchenko V. P., Panchenko M. S., *Inzh.-Fiz. Zh.*, 1968, vol. 16, no. 4, p. 654; 1969, vol. 16, no. 1, p. 67; 1972, vol. 22, no. 5, p. 715.
278. Zolotarev P. P., Dubunin M. M., *Dokl. Akad. Nauk SSSR*, 1972, vol. 203, no. 4, p. 1347.
279. Prusakov M. A., Rabinovich Y. I., Sazhin B. S., Churaev N. V., Shchadrina N. E., *Inzh.-Fiz. Zh.*, 1977, vol. 33, no. 4, p. 627.

280. Volarovich M. P., Davidovskii P. N., Gamayunov N. I., *Kolloid. Zh.*, 1965, vol. 27, no. 2, p. 167.
281. Globus A. M., *Fizika neizotermicheskogo vnutripochvennogo vlagoobmena* (Physics of Non-isothermic Moisture Transfer in Soil), Leningrad: Gidrometeoizdat, 1983.
282. Abramets A. M., Lishtvan I. I., Churaev N. V., *Inzh.-Fiz. Zh.*, 1982, vol. 43, no. 6, p. 976; 1984, vol. 46, no. 2, p. 253; *Vesti AN BSSR, Ser. Khim. Nauk*, 1983, no. 5, p. 21.
283. Efremov I. F., *Periodicheskie kolloidnye struktury* (Regular Colloidal Structures), Leningrad: Khimiya, 1971.
284. Ur'ev N. B., *Vysokokontsentrirovannye dispersnye sistemy* (Highly Concentrated Disperse Systems), Moscow: Khimiya, 1980.
285. Ur'ev N. B., *Fiziko-khimicheskie osnovy tekghnologii dispersnykh sistem i materialov* (Physicochemical Basis of Technologies of Disperse Systems and Materials), Moscow: Khimiya, 1988.
286. Ito K., Nakamura H., Yoshida H., Ise N., *J. Amer. Chem. Soc.*, 1988, vol. 110, no. 21, p. 6595.
287. Tadros T. F., *Chem. and Ind.*, 1985, no. 7, p. 210.
288. Sirota E. B., Ou-Yang H. D., Sinha S. K., Chaikin P. M., Axe J. D., Fujii Y., *Phys. Rev. Lett.*, 1989, vol. 62, no. 13, p. 1524.
289. Murray C. A., Sprenger W. O., Wenk R. A., *J. Phys.: Condens. Matter*, 1990, vol. 2, p. 385.
290. Rigos A. A., Wilemski G., *J. Phys. Chem.*, 1992, vol. 96, no. 10, p. 3981.
291. Cook R., Wilemski G., *J. Phys. Chem.*, 1992, vol. 96, no. 10, p. 4023.
292. Potanin A. A., Ur'ev N. B., Muller V. M., *Kolloid. Zh.*, 1987, vol. 49, no. 5, p. 915.
293. Ur'ev N. B., Potanin A. A., Mevis J., Moldenaers P., *Kolloid. Zh.*, 1987, vol. 51, no. 3, p. 535.
294. Churaev N. V., *Kolloid. Zh.*, 1988, vol. 50, no. 1, p. 108.
295. Bibik E. E., *Reologiya dispersnykh sistem* (Rheology of Disperse Systems), Leningrad: Izd. LGU, 1981.
296. Kul'chitskii L. I., Us'yarov O. G., *Fiziko-khimicheskie osnovy formirovaniya svoistv glinistykh porod* (Physicochemical Basis of Development of Properties of Clays), Moscow: Nedra, 1987.
297. Ostrikov M. S., Rostovtseva I. V., Dibrov G. D., Danilova E. P., *Kolloid. Zh.*, 1960, vol. 22, no. 4, p. 443.
298. Belopol'skii M. S., *Inzh.-Fiz. Zh.*, 1961, vol. 4, no. 4, p. 49.
299. Kazanskii M. F., Verezomskaya A. L., *Inzh.-Fiz. Zh.*, 1961, vol. 4, no. 10, p. 104.
300. Churaev N. V., *Inzh.-Fiz. Zh.*, 1962, vol. 5, no. 12, p. 41; 1963, vol. 6, no. 2, p. 31.
301. Ostrikov M. S., Dukhina T. P., Vlodavets I. N., Sinitsina G. M., *Kolloid. Zh.*, 1964, vol. 26, no. 5, p. 600.
302. Gamayunov N. I., Amusin L. G., *Kolloid. Zh.*, 1975, vol. 37, no. 6, p. 1156; 1976, vol. 38, no. 3, p. 551.
303. Gamayunov N. I., *Pochvovedenie*, 1985, no. 5, p. 147.
304. Abramets A. M., Ivashkevich L.S., Lishtvan I. I., Churaev N. V., *Kolloid. Zh.*, 1985, vol. 47, no. 2, p. 376.
305. Pievskii I. M., Grechina V. V., Nazarenko G. D., Stepanova A. I., *Sushka keramicheskikh stroimaterialov plasticheskogo formirovaniya* (Drying of Ceramic Building Materials of Plastic Formation). Kiev: Naukova Dumka, 1985.

306. Abrametz A. M., Lishtvan I. I., Churaev N. V., *Massoperenos v prirodnych dispersnych sistemach* (Mass Transfer in Natural Disperse Systems), Minsk: Navuka i Tekhnika, 1992.
307. Serebrovskaya M. V., Us'yarov O. G., Muller V. M., *Kolloid. Zh.*, 1972, vol. 34, no. 5, p. 730.
308. Low P. F., *Kolloid. Zh.*, 1986, vol. 48, no. 6, p. 1081.
309. *Fizikokhemicheskaya mekhanika prirodnykh dispersnykh sistem* (Physicochemical Mechanics of Natural Disperse Systems), Ed. E. D. Shchukin, N. V. Pertsov, V. I. Osipov, R. I. Zlochevskaja, Moscow: Moscow Univ., 1985.
310. Panasevich A. A., Nemchinskaya N. A., Tarasevich Yu. I., *Ukr. Khim. Zhurn.*, 1984, vol. 50, no. 9, p. 936.
311. Collins-George N., Rozeman J. M., *Austral. J. Soil Sci.*, 1970, vol. 8, no. 3, p. 239.
312. Churaev N. V., *Kolloid. Zh.*, 1994, vol. 56, no. 5, p. 707; 1955, vol. 57, no. 2 , p. 252.
313. Churaev N. V., *Mendeleev Comm.*, 1994, no. 6, p. 218.
314. Gee M. L., Healy T. W., White L. R., *J. Colloid Interface Sci.*, 1990, vol. 140, no. 2, p. 450.
315. Churaev N.V., *J. Colloid Interface Sci.*, 1995, vol. 172, no. 2, p. 479; *Adv. Colloid Interface Sci.*, 1995, vol. 58, no. 2/3, p. 87.
316. Churaev N. V., Sobolev V. D., *Adv. Colloid Interface Sci.*, 1995, vol. 61, p. 1.
317. Derjaguin B. V., Churaev N. V., *Langmuir*, 1987, vol. 4, p. 607.
318. Kornil'ev I. N., Zorin Z. M., Churaev N. V., *Kolloid. Zh.*, 1984, vol. 46, no. 5, p. 892.
319. Churaev N. V., *Rev. Phys. Appl.*, 1988, vol. 23, p. 975.
320. Zorin Z. M., Romanov V. P., Churaev N. V., *Kolloid. Zh.*, 1979, vol. 41, no. 6, p. 1066.
321. Aronson M. P., Princen H. M., *J. Colloid Interface Sci.*, 1975, vol. 52, no. 2, p. 345.
322. Pashley R. M., Israelashvili J. N., *Colloids Surf.*, 1981, vol. 2, p. 169.
323. Kekicheff P., Christenson H. K., Ninham B. W., *Colloids Surf.*, 1989, vol. 40, p. 31.
324. Zorin Z. M., Kolarov T., Platikanov D., *Kolloid. Zh.*, 1986, vol. 48, no. 4, p. 677; no. 5, p. 901; *Colloids Surf.*, 1987, vol. 22, p. 147.
325. Kolarov T., Zorin Z. M., Platikanov D., *Colloids Surf.*, 1990, vol. 51, p. 37.
326. Scheludko A., Toshev B. V., Platikanov D., in: *Sovremennaya teoriya kapillyamosti* (Modern Theory of Capillary Phenomena), Lenigrad: Khimiya, 1980, p. 274.
327. Kolarov T., Zorin Z. M., *Colloid Polym. Sci.*, 1979, vol. 257, p. 1292.
328. Scheludko A., Chakarov V., Toshev B. V., *J. Colloid Interface Sci.*, 1981, vol. 82, no. 1, p. 83.
329. Starov V. M., Churaev N. V., *Kolloid. Zh.*, 1980, vol. 42, no. 4, p. 703.
330. Churaev N. V., Starov V. M., Derjaguin B. V., *J. Colloid Interface Sci.*, 1982, vol. 89, no. 1, p. 16.
331. Dobbs H. T., Indekeu J. D., *Physica A*, 1993, vol. 201, p. 457.
332. Rusanov A. I., Ovrutskaya N. A., Akopyan L. A., *Kolloid. Zh.*, 1981, vol. 43, no. 5, p. 685.
333. Li D., Lin F. Y. H., Neumann A. W., *J. Colloid Interface Sci.*, 1991, vol. 142, no.1, p. 224.
334. Drelich J., Miller J. D., Hupka J., *J. Colloid Interface Sci.*, 1993, vol. 155, no. 2, p. 379.
335. Exerova D., Kashchiev D., Platikanov D., Toshev B.V., *Adv. Colloid Interface*

Sci., 1994, vol. 49, p. 303.
336. Martynov G. A., Starov V. M., Churaev N. V., *Kolloid. Zh.*, 1977, vol. 39, no. 3, p. 472.
337. Frenkel' Ya. I., *Zh. Eksp.Teor. Fiz.*, 1948, vol. 18, no. 7, p. 659.
338. Summ B. D., Mashnina N. V., Goryunov Yu. V., Izmailova V. N., *Kolloid. Zh.*, 1986, vol. 48, no. 3, p. 580.
339. Shanahan M. E. R., de Gennes P. G., In: *Adhesion*, vol. 11, Pap. 24th Ann. Conf. Adhes. and Adhesives. London, 1986, N. Y., 1987, p. 71.
340. Rusanov A. I., *Kolloid. Zh.*, 1977, vol. 39, no. 4, p. 704.
341. Derjaguin B. V., Starov V. M., Churaev N. V., *Kolloid. Zh.*, 1982, vol. 44, no. 5, p. 871.
342. Summ B. D., Goryunov Yu. V., *Fizikokhimicheskie osnovy smachivaniya i rastekaniya* (Physicochemical Bases of Wetting and Spreading), Moscow: Khimiya, 1976.
343. Bykhovskii A. I., *Rastekanie* (Spreading), Kiev: Naukova Dumka, 1983.
344. Landau L., Levich V., *Acta phys.-chim. URSS*, 1942, vol. 17, p. 42.
345. Derjaguin B. V., *Zh. Eksp.Teor. Fiz.*, 1945, vol. 15, no. 9, p. 503.
346. Starov V. M., Churaev N. V., Khvoroctyanov A. G., *Kolloid. Zh.*, 1977, vol. 39, no. 1, p. 201.
347. Zheleznyi B. V., *Dokl. Akad. Nauk SSSR*, 1972, vol. 207, no. 3, p. 647.
348. Friz G., *Z. angew. Phys.*, 1965, Bd. 19, no. 4, p. 374.
349. Voinov O. V., *Mekhanika zhidkosti i gaza* (Mechanics of Liquids and Gases), 1976, vol. 5, p. 76; *Dokl. Akad Nauk SSSR*, 1978, vol. 243, p. 1422.
350. Cox R. G., *J. Fluid Mech.*, 1986, vol. 1687, p. 1659.
351. Hansen R. J., Toong T. Y., *J. Colloid Interface Sci.*, 1971, vol. 37, no. 1, p. 196.
352. Gribanova E. V., Molchanova L. I., *Kolloid. Zh.*, 1978, vol. 40, no. 1, p. 30–36; no. 2, p. 217.
353. Berezkin V. V., Churaev N. V., *Kolloid. Zh.*, 1982, vol. 44, no. 3, p. 417.
354. Washburn E. W., *Phys. Rev.*, 1921, vol. 17, no. 3, p. 273.
355. Blake T. D., Haynes J. M., *J. Colloid Interface Sci.*, 1969, vol. 30, no. 3, p. 421.
356. Hoffinan R. L., *J. Colloid Interface Sci.*, 1975, vol. 50, no. 2, p. 228; 1983, vol. 94, no. 2, p. 470.
357. Glasstone S., Eyring H., *Theory of Rate Processes*, N. Y., London, McGraw-Hill, 1941.
358. Kvlividze V. I., Kiselev V. F., *Zh. Strukt. Khim.*, 1967, vol. 8, no. 2, p. 221.
359. Petrov J. G., Petrov P. G., *Colloids Surf.*, 1992, vol. 64, no. 2, p. 143.
360. Cazabat A. M., Cohen Stuart M. A., *Physicochem. Hydrodyn.*, 1987, vol. 9, no. 1–2, p. 23.
361. Zheleznyi B. V., *Zh. Prikl. Matem. Tekhn. Fiz.*, 1976, no. 3, p. 72.
362. Derjaguin B. V., Levi S. M., *Fizikokhimiya naneseniya tonkikh sloev na dvizhushchuyusya podlozhku* (Physicochemistry of Deposition of Thin Layers on a Moving Substrate), Moscow: Izd. AN SSSR, 1959.
363. Aksel'rud G. A., Al'tshuller M. A., *Vvedenie v kapillamo-khimicheskuyu tekhnologiyu* (Introduction to Capillary-Chemical Technology), Moscow: Khimiya, 1983.
364. Abbasov M., Zorin Z. M., Iskandaryan G. A., Churaev N. V., *Kolloid. Zh.*, 1989, vol. 51, no. 4, p. 627; p. 634.
365. Ershhov A. P., Zorin Z. M., Churaev N. V., *Kolloid. Zh.*, 1990, vol. 52, no. 2, p. 250.
366. Zorin Z. M., Churaev N. V., *Adv. Colloid Interface Sci.*, 1992, vol. 40, p. 85.
367. Ershhov A. P., Zorin Z. M., Svitova T. F., Churaev N. V., *Kolloid. Zh.*, 1993, vol. 55, no. 3, p. 39; no. 4, p. 45.

368. Churaev N. V., Ershhov A. P., Esipova N. E., Iskandaryan G. A., Madjarova E. A., Sergeeva I. P., Sobolev V. D., Svitova T. F., Zakharova M. A., Zorin Z. M., Poirer J. E., *Colloids Surf.*, 1994, vol. 91, p. 97.
369. Churaev N. V., Ershhov A. P., Zorin Z. M., *J. Colloid Interface Sci.*, 1996, vol. 177, no. 2, p. 589.
370. Cohen Stuart M. A., Cazabat A. M., *Progr. Colloid Polym. Sci.*, 1987, vol. 74, p. 64.
371. Churaev N. V., Martynov G. A., Starov V. M., *J. Colloid Polym. Sci.*, 1981, vol. 259, p. 747.
372. Stokes J. P., Weitz D. A., Gollub J. P., Dougherty A., Robbins M. O., Chaikin P. M., Lindsay H. M., *Phys. Rev. Lett.*, 1986, vol. 57, no. 14, p. 1718.
373. Kardashev G. A., Sobolev V. D., Churaev N. V., Shatalov A. L., *Kolloid. Zh.*, 1976, vol. 38, no. 3, p. 461.
374. Rozin Yu. P., Rozina E. Yu., Tikhonova V. S., *Ukr. Fiz. Zh.*, 1984, vol. 29, no. 10, p. 1522; vol. 30, no. 2, p. 235.
375. Yatsenko A. V., *Kolloid. Zh.*, 1986, vol. 48, no. 2, p. 390.
376. Karpovich I. N., Churaev N. V., Panchenko M. S., Panasyuk A. L., *Kolloid. Zh.*, 1982, vol. 44, no. 6, p. 1044; *Elektron. Obrab. Mater.*, 1982, no. 4, p. 62.
377. Berezkin V. V., Zorin Z. M., Frolova N. V., Churaev N. V., *Kolloid. Zh.*, 1975, vol. 37, no. 6, p. 1040.
378. Zolotarev P. P., Starov V. M., Churaev N. V., *Kolloid. Zh.*, 1976, vol. 38, no. 5, p. 895.
379. Churaev N. V., Martynov G. A., Starov V. M., Zorin Z. M., *J. Colloid Polym. Sci.*, 1981, vol. 259, p. 747.
380. Churaev N. V., Zorin Z. M., *Colloids Surf. A*, 1995, vol. 100, p. 131.
381. Berezkin V. V., Derjaguin B. V., Zorin Z. M., Frolova N. V., Churaev N. V., *Dokl. Akad. Nauk SSSR*, 1975, vol. 225, no. 1, p. 109.
382. Berezkin V. V., Ershhov A. P., Esipova N. E., Zorin Z. M., Churaev N. V., *Kolloid. Zh.*, 1979, vol. 41, no. 5, p. 849.
383. Zorin Z. M., Iskandaryan G. A., Churaev N. V., *Kolloid. Zh.*, 1978, vol. 40, no. 4, p. 671.
384. Esipova N. E., Zorin Z. M., Trapeznikov A. A., Shchegolev G. G., *Kolloid. Zh.*, 1984, vol. 46, no. 6, p. 1251.
385. Maevskaya L. N., Zorin Z. M., Churaev N. V., *Kolloid. Zh.*, 1977, vol. 39, no. 6, p. 1081.
386. Berezkin V. V., Zorin Z. M., Iskandaryan G. A., Churaev N. V., in: *Fizikokhimiya poverkhnostno-aktivnykh veshchestv* (Physicochemistry of Surfactants), vol. 2 , part 2, Moscow: Vneshtorgizdat, 1978, p. 329.
387. Princen H. M., Cazabat A. M., Cohen Stuart M. A., Heslot F., Nicolet S., *J. Colloid Interface Sci.*, 1988, vol. 126, no. 1, p. 84.
388. Mc Guiggan P. M., Pashley R. M., *Colloids Surf.*, 1987, vol. 27, p. 277.
389. Cazabat A. M., Heslot F., Fraysse N., Carles P., Levinson P., *Progr. Colloid Polym. Sci.*, 1990, vol. 82, p. 82.
390. Cazabat A. M., Heslot F., Fraysse N., *Progr. Colloid Polym. Sci.*, 1990, vol. 83, p. 52.
391. Cazabat A. M., *Adv. Colloid Interface Sci.*, 1991, vol. 34, p. 73.
392. Cazabat A. M., Fraysse N., Heslot F., *Colloids Surf.*, 1991, vol. 53, p. 1.
393. Cazabat A. M., *Adv. Colloid Interface Sci.*, 1992, vol. 42, p. 65.
394. Fraysse N., *Doct. Thes. Universite de Paris VI.* 1991.
395. Maevskaya L. N., Zorin Z. M., Churaev N. V., *Zh. Prikl. Khim.*, 1984, no. 6, p. 1413.
396. Lishtvan I. I., Maevskaya L. N., Zorin Z. M., Churaev N. V., *Kolloid. Zh.*, 1984,

vol. 46, no. 1, p. 29.
397. Shchukin E. D., *Kolloid. Zh.*, 1997, vol. 59, no. 2, p. 270.
398. Frink L. J. H. D., van Swol F., *J. Chem. Phys.*, 1996, vol. 105, no. 7, p. 2886.
399. Vinogradova O. I., *J. Phys.: Condens. Matter*, 1996, vol. 8, no. 47, p. 9491.
400. *Applied Surface Thermodynamics*, Ed. by A. W. Neumann and J. K. Spelt, Marcel Dekker, N. Y., Basel, Hong Kong, 1996.
401. Rusanov A. I., Prokhorov V. A., *Interfacial Tensiometry*, Elsevier, Amsterdam, 1996.
402. Lyklema J., *Fundamentals of Interface and Colloid Science*, Academic Press, vol. 1, 1991; vol. 2, 1995.
403. *Nanoparticles in Solids and Solutions*, Ed. by J. H. Fendler and I. Dekany, Kluwer Acad. Publ., Dordrecht, Boston, London, 1996.

SUBJECT INDEX

Adsorption layers 29, 40
  energy of adsorption 29, 30
  Langmuir and BET equations 30
  polymolecular adsorption 30

Capillary imbibition 273
  capillary rise 271, 275
  effect of surfactants 278, 287
  immiscible displacement 277
  mechanism of creeping 295

Capillary pressure 37
  capillary forces 37
  capillary penetration 275
  condensation in narrow pores 47
  hysteresis 42
  thermocapillary flow 132

Colloidal systems
  Derjaguin's approximation 16
  interparticle forces 14
  potential barrier 14
  potential wells 15

Combined transfer 171
  film flow and vapor diffusion 172
  film, vapor and condensate flow 187

Contact angles 249
  dynamic angles 263, 280
  electrolyte and surfactant solutions 256
  equilibrium values 40
  Frumkin–Derjaguin equation 40
  hysteresis 260

Diffusion
  surface diffusion 155
  vapor in capillaries 182
  vapor in porous bodies 214

Disjoining pressure 4
  in colloids 14
  in thin interlayers 4
  in wetting films 35
  shape of isotherms 14
  thermodynamic determination 4

Electrical double layers
  diffuse electrical layers 9
  electrical potentials 9
  electrokinetical phenomena 80, 85
  Helmholtz–Smoluchowski equation 83, 86
  surface charge 9, 86

ζ-potential 84, 85

Evacuation of non-volatile liquids
from porous bodies 210
theory and experimental
data 211–213

Evaporation from porous bodies
coefficients of internal mass
transfer 220
deep drying 226
mechanisms of mass transfer 220
model porous bodies 188, 214, 217
radiotracer technique 206
rate of drying 214
thermodiffusion 230

Evaporation of liquids
effect of film flow 182
electrolyte solutions 192
from capillaries 182
Hertz–Knudsen regime 200
polymer solutions 196
surfactant solutions 200

Flow in liquid-filled pores
capillary osmosis 92, 95, 102
cross coefficients 70
electroosmosis 80
electroviscosity 83, 89
filtration 71
Onsager's symmetry principle 70
reverse osmosis 96
thermodynamics of irreversible
processes 69
thermomechanical pressure 108
thermoosmosis 104, 107

Immiscible displacement
influence of surfactants 278
oil–water systems 273, 275

Liquids in thin pores
density 66
microstructure 51, 52
slippage 73, 75
solvent power 60
viscosity 56

Membrane rejection
concentration polarization 100
distribution coefficients 99
ionic solutions 104
osmotic pressure 103
selectivity coefficient 101
theory 101

Non-freezing water layers 112
flow of interlayers and films 138
ice sliding 113
influence of temperature and
pressure 115, 116
mass transfer in frozen bodies 120
technique of thickness
measurement 113
thermocrystallization flow 119, 137
viscosity 116

Non-uniform electrical field
capillary suction 281
film flow 129, 131
vapor diffusion 161

Porous structure
methods of characterization 20
model structures of porous bodies 18
rigid and non-rigid structures 15, 17

Shrinkage of porous bodies 237
containing plate particles 243
containing spherical particles 238
experimental data 246
mechanisms of shrinkage 241, 245,
246

Surface forces
electrostatic 8
molecular 6
structural 12

Transition zone 127
between liquid and wetting film 128,
258
line tension 258, 259

Vapor transport in pores
  free-molecular 147, 149
  non-isothermal flow 179
  self-diffusion coefficients 153
  slippage in thin capillaries 150
  surface diffusion 155, 159
  thermal slippage 160
  vapor–gas mixtures 156, 164
  viscous 151

Wetting films 31
  disjoining pressure 252, 35, 253
  electroosmotic flow 142, 143
  equilibrium thickness 32, 37
  hydrodynamics 123
  Marangoni flow 141
  pressure driven flow 125
  thermocapillary flow 133